Stochastic Thermodynamics

Stochastic Thermodynamics

An Introduction

LUCA PELITI &
SIMONE PIGOLOTTI

PRINCETON UNIVERSITY PRESS
Princeton and Oxford

Published by Princeton University Press,
41 William Street, Princeton, New Jersey 08540
6 Oxford Street, Woodstock, Oxfordshire OX20 1TR

press.princeton.edu

Library of Congress Control Number 2021936619
ISBN 978-0-691-20177-1
ISBN (e-book) 978-0-691-21552-5

British Library Cataloging-in-Publication Data is available

Editorial: Ingrid Gnerlich and Arthur Werneck
Production Editorial: Jill Harris
Text Design: Carmina Alvarez
Jacket Design: Wanda España
Production: Jacqueline Poirier
Publicity: Matthew Taylor and Amy Stewart
Copyeditor: Jennifer McClain

Jacket art: Shutterstock

This book has been composed in Minion Pro and Universe LT Std

Printed on acid-free paper. ∞

Printed in the United States of America

10 9 8 7 6 5 4 3 2

Wenn sich die hier zu behandelnde Bewegung samt den für sie zu erwartenden Gesetzmäßigkeiten wirklich beobachten läßt, so ist die klassische Thermodynamik schon für mikroskopisch unterscheidbare Räume nicht mehr als genau gültig anzusehen und es ist dann eine exakte Bestimmung der wahren Atomgröße möglich.

If the motion discussed here can actually be observed (together with the laws relating to it that one would expect to find), then classical thermodynamics can no longer be looked upon as applicable with precision to bodies even of dimensions distinguishable in a microscope, and an exact determination of the actual atomic dimensions is then possible.

—Albert Einstein, *Ann. Physik* **322** (8): 549 (1905)

Contents

Foreword

If Sadi Carnot's *Reflections on the Motive Power of Fire* (1824) marks the starting point of thermodynamics as a field of scientific inquiry, then this field is now nearly two centuries old. Its key elements—heat and work, entropy, thermodynamic potentials, the first and second laws—were in place by the late nineteenth century. This historical fact often conveys the sense that, while thermodynamics provides an undeniably powerful framework for scientists and engineers, there is nothing new to discover within the field itself.

Another familiar conception about thermodynamics is that it is inherently a field about macroscopic systems. In this view, statistical mechanics seeks to explain how thermodynamic properties arise from the interactions of vastly many atoms and molecules, but the properties themselves, and notions such as heat, work, and entropy, are inherently macroscopic.

The past few decades have seen much exciting research that goes against the grain of these two conceptions of thermodynamics. A growing community of physicists and chemists, theorists and experimentalists, take seriously the idea of applying thermodynamic principles to individual microscopic systems. Some of the impetus for this development comes from biology, where experimental data has led to a distinctly mechanical picture of biomolecules such as kinesin, myosin, and ATP synthase. These and similar molecules and molecular complexes are now frequently discussed, not generically as chemicals interacting with other chemicals within a living cell, but more pointedly as devices that consume fuel while they proceed through operating cycles that are not far different from those of macroscopic machines. Indeed, these biomolecules are often referred to as molecular *machines* or *motors*.

A parallel development has been the discovery of far-from-equilibrium *fluctuation relations* that express the second law of thermodynamics in a manner that focuses on random fluctuations, revealing unexpected features and symmetries of these fluctuations. These relations, together with more recent developments, contradict the notion that everything important in thermodynamics was discovered long ago.

Stochastic thermodynamics has emerged as a framework for describing how thermodynamic laws apply to individual mesoscopic systems, particularly away from equilibrium. The very name of this framework acknowledges that randomness is a defining feature of dynamics at these scales, where characteristic energies are on the order of k_BT. A central aim of stochastic thermodynamics is to define basic thermodynamic concepts, such as heat, work, and entropy production, at the level of individual, randomly fluctuating trajectories.

The tools of stochastic thermodynamics are now widely applied in a variety of physical, chemical, and biological contexts, and there is a vigorous push to extend the field into the realm of quantum mechanics. While excellent review and pedagogical articles have appeared, the need exists for a textbook on stochastic thermodynamics, at a level appropriate for a graduate-level course or as a reference for researchers wishing to enter the field.

In *Stochastic Thermodynamics: An Introduction*, Luca Peliti and Simone Pigolotti—both active scientists who have made important contributions to the field—aim to "provide a pedagogical introduction to stochastic thermodynamics for graduate students in physics." They succeed in accomplishing this goal and more, as this book provides a comprehensive reference to the most important facets of current research related to stochastic thermodynamics.

After a review of background material, the book lays out the central features of stochastic thermodynamics in chapter 3 and then discusses fluctuation relations in chapter 4. These chapters represent the original core of the field, and they highlight a key concept: the connection between *thermodynamic irreversibility* and *statistical irreversibility*. The former is quantified in terms of entropy production or dissipated work; the latter by comparing the probability of observing a particular sequence of events with that of observing the same events in time-reversed order. It is this connection that allows one to assign entropy production to individual, fluctuating trajectories. The book's subsequent chapters are devoted to an array of topics, including the thermodynamic implications of mesoscale information processing, large deviation theory, experimental investigations of stochastic thermodynamics, and the recently discovered thermodynamic uncertainty relations. Each chapter ends by pointing interested readers to further references, and provides exercises that are indispensable for readers wishing to master the subject.

Stochastic Thermodynamics: An Introduction provides an excellent and welcome resource for the graduate student wishing to learn about this active new field, for the lecturer teaching a course on the subject, and for the researcher seeking a detailed reference for the current state of the field. As Peliti and Pigolotti acknowledge in the preface, stochastic thermodynamics has developed rapidly. I expect that as it continues to develop, this book will serve as a standard reference and introduction to the field.

Christopher Jarzynski
College Park, Maryland, March 2020

Preface

The mesoscopic world is attracting a growing scientific interest. Technological miniaturization permits the construction of smaller and smaller devices, whose performance is limited by thermal fluctuations. In biology, proteins are a remarkable example of machines performing complex tasks at the nanoscale. Stochastic thermodynamics has emerged as a powerful framework to describe the thermodynamics of systems at these scales.

We are both interested in statistical mechanics and theoretical biophysics, and we have quite naturally ended up gravitating toward stochastic thermodynamics. During this journey, we have witnessed that more and more junior researchers are becoming interested in stochastic thermodynamics and its applications. However, these researchers face a substantial entry barrier. One reason for this difficulty is that stochastic thermodynamics has developed really quickly, particularly following important breakthroughs such as the discovery and understanding of fluctuation theorems.

We have confronted this problem by teaching stochastic thermodynamics at the graduate level. This book gathers the fruits of our experience. We hope that our book eases approaching stochastic thermodynamics and facilitates cross-fertilization with other fields of physics and biology.

Luca Peliti and Simone Pigolotti
Onna, Okinawa, October 2019

Acknowledgments

This book would not have been possible without many discussions we had over the course of the years with our collaborators, friends, and colleagues: Erik Aurell, Andre Barato, John Bechoefer, Stefano Bo, Daniel Busiello, Antonio Celani, Massimo Cencini, Davide Chiuchiù, Massimiliano Esposito, Jean-Baptiste Fournier, Alberto Imparato, Jean-François Joanny, Frank Jülicher, Ryoichi Kawai, Tetsuya J. Kobayashi, Florent Krzakala, Jorge Kurchan, David Lacoste, Amos Maritan, Carlos Mejía-Monasterio, Paolo Muratore-Ginanneschi, Izaak Neri, Matteo Polettini, Jacques Prost, Lorenzo Pucci, Andrea Puglisi, Riccardo Rao, Édgar Roldán, Lamberto Rondoni, Pablo Sartori, Udo Seifert, Ken Sekimoto, Gatien Verley, Angelo Vulpiani, and Frédéric van Wijlandt. These discussions significantly deepened our understanding of stochastic thermodynamics and motivated us to write this book. We fondly remember Christian van den Broeck, whose untimely death saddened us deeply. LP is grateful to Stefano Ruffo and Lamberto Rondoni for inviting him to lecture on the subject, respectively at SISSA and at the Turin Politecnico, an opportunity that helped structure the pedagogical trail we followed in this book.

We are especially grateful to Massimiliano Esposito, Akira Kawano, Andrea Puglisi, Édgar Roldán, Pablo Sartori, José Vila Chã Losa, and in particular to John Bechoefer, Giuseppe Gaeta, Ken Sekimoto, and Shoichi Toyabe for a critical reading of our manuscript. Special thanks are due to two anonymous readers and to Todd Gingrich for their careful reading and precious suggestions. We thank Bill Stern for proofreading and especially Jennifer McClain for copyediting. We feel deeply honored by Chris Jarzynski, who wrote the foreword to our book.

We have benefited all along from the assistance and suggestions of Princeton University Press. We warmly thank Arthur Werneck, Jill Harris, and in particular Ingrid Gnerlich, who closely followed this project from the beginning.

SP wishes to dedicate this book to Viola, who was born halfway through the writing and has been a constant source of joy ever since. He thanks Kate for her constant support and his parents, Marcello and Cristina. LP is grateful to Raya for her support and patience and to Margherita and Mira for everything.

Notation

$\asymp$	Leading exponential order.
$u = (u_x)$	Entire vector with components u_x.
$(\text{expression})_{xx'}$	Matrix element xx' of expression.
$\langle \ldots \rangle$	Average over probability distributions.
$\langle \ldots \rangle_{\mathrm{F}}$ and $\langle \ldots \rangle_{\mathrm{B}}$	Average over forward and backward process.
$\left. \dfrac{\partial X}{\partial Y} \right)_Z$	Partial derivative of X with respect to Y at Z fixed.
$\left. \dfrac{\partial f}{\partial x} \right\|_{x=0}$	Partial derivative of f with respect to x, evaluated at $x=0$.
$\int f(x) \cdot \mathrm{d}x, \int f(x) \circ \mathrm{d}x$	Ito and Stratonovich integrals of $f(x)$.
$a(\boldsymbol{x})$	Static observable, expressed as a function of the trajectory $\boldsymbol{x}$.
A_α	Affinity of cycle α.
$\vec{B}$	Magnetic field.
$(\mathcal{C}_\alpha)$	Set of fundamental cycles.
$C_{\alpha\beta}(t)$	Correlation function.
D	Diffusion coefficient.
$d_{xx'}$	Coefficient of a jump observable. Cost of phenotype switching.
$D_{\mathrm{KL}}(p\|q)$	Kullback-Leibler divergence of distributions p and q.
E, S, W, Q, P	Macroscopic (deterministic) energy, entropy, work, heat, pressure.
F	Equilibrium free energy.
F^{neq}	Nonequilibrium free energy.
$\mathcal{F}(x, t)$	Force acting on a Brownian particle.
$g_{\alpha\beta}$	Generalized friction coefficient.
h	Planck constant ($h = 6.626\,07015 \cdot 10^{-34}\,\mathrm{m}^2\,\mathrm{kg}\,\mathrm{s}^{-1}$ (exact)).
$H(\mathcal{S})$ or $H(p)$	Shannon entropy of system $\mathcal{S}$ with distribution $p = (p_x)$.
$H(\mathcal{S}_1, \mathcal{S}_2)$	Joint Shannon entropy of $\mathcal{S}_1$ and $\mathcal{S}_2$.
$H(\mathcal{S}_1 \| \mathcal{S}_2)$	Conditional Shannon entropy of $\mathcal{S}_1$ given $\mathcal{S}_2$.
$\mathcal{H}(p_r, r; \lambda)$	Hamiltonian.
i_{xy}	Stochastic mutual information.
$I(x)$	Rate (Cramér) function.

$I(\mathcal{S}_1 : \mathcal{S}_2)$	Mutual information of $\mathcal{S}_1$ and $\mathcal{S}_2$.
$J_{xx'}(t), J(x,t)$	Probability current (discrete, continuous).
J_{thr}	Threshold current.
$\mathcal{J}(\boldsymbol{x}), j(\boldsymbol{x})$	Integrated empirical current, empirical current.
j_s	Empirical entropy production rate.
$k_{xx'}$	Rate of jump from state x' to state x.
k_x^{out}	Escape rate $k_x^{\text{out}} = \sum_{x'} k_{x'x}$.
$\bar{k}_{xx'}, \widehat{k}_{xx'}$	Information-mediated jump rate and its reverse.
k_{B}	Boltzmann constant
	$\quad (k_{\text{B}} = 1.380\,649 \cdot 10^{-23}\ \text{m}^2\,\text{kg}\,\text{s}^{-2}\text{K}^{-1}$ (exact)).
$\mathcal{K}_{\alpha\beta}(t,t'), K_{\alpha\beta}(t)$	Linear response function.
$L_{xx'}, \mathcal{L}$	Generator of dynamics (discrete, continuous).
$L_{xx'}(q)$	Generator of tilted dynamics.
N_{A}	Avogadro number ($N_{\text{A}} = 6.022\,14076 \cdot 10^{23}\ \text{mol}^{-1}$
	(exact)).
$n_{xx'}(\boldsymbol{x})$	Number of jumps from x' to x in the trajectory $\boldsymbol{x}$.
$\mathcal{N}_{\boldsymbol{x}}$	Weight of the phenotypic trajectory $\boldsymbol{x}$.
$p_x, p(x)$	Probability of x, probability density of x.
$p_x(t), p(x;t)$	Probability of x at time t (discrete, continuous).
$p_{x\|y}, p(x\|y)$	Conditional probability of x given y (discrete,
	continuous).
$p_{x;t\|x';t'}, p(x;t\|x';t')$	Conditional probability of x at time t, given the state
	x' at time t'.
$p_x^{\text{eq}}, p^{\text{eq}}(x)$	Equilibrium probability (discrete, continuous).
$p_x^{\text{st}}, p^{\text{st}}(x)$	Stationary probability (discrete, continuous).
p_r, r	Particle momentum and location (1D).
$\vec{p}_r, \vec{r}$	Particle momentum and coordinates (vectors in
	physical space).
$\mathcal{P}_{\boldsymbol{x}}, \mathcal{P}(\boldsymbol{x})$	Trajectory probability density (discrete, continuous
	states).
$\mathcal{P}_{\boldsymbol{x}}\, \mathcal{D}\boldsymbol{x}$	Infinitesimal trajectory probability.
q, Q	Stochastic and average heat. Positive if released to
	the reservoir.
$q_{xx'}$	Mesoscopic heat released during a jump.
r	Probability of a binary variable.
r_x	Division rate of phenotype x.
$\mathcal{R}_{xx'}(\boldsymbol{x}), R_{xx'}$	Number and average rate of information-mediated
	transitions.
$\bar{s}$	Infimum of the entropy production.
$\mathcal{S}(\boldsymbol{x}; \boldsymbol{\lambda})$	Action.
$s^{\text{a}}, s^{\text{na}}$	Adiabatic and nonadiabatic entropy production.
$s^{\text{res}}, S^{\text{res}}$	Stochastic and average entropy released in the
	reservoir.
S^{stat}	Statistical entropy.
$s_x^{\text{sys}}, S^{\text{sys}}$	Stochastic and average system entropy.
$s^{\text{tot}}, S^{\text{tot}}$	Stochastic and average total entropy production.
$\dot{s}^{\text{tot}}, \dot{S}^{\text{tot}}$	Stochastic (instantaneous) and average entropy
	production rate.

T	Absolute temperature.
$\widehat{t}, \widehat{\boldsymbol{x}}, \widehat{\boldsymbol{\lambda}}$	Reverse time, backward trajectory, backward protocol.
t_0, t_f	Initial and final time.
$\mathcal{T}$	Duration of time interval $\mathcal{T} = t_f - t_0$.
$\mathcal{T}_{fp}, t_{fp}$	First-passage time: extensive ($\mathcal{T}_{fp}$) and intensive ($t_{fp} = \mathcal{T}_{fp}/|J_{thr}|$).
$\mathbb{T}$	Time-ordered product.
t_{rnd}	Random time.
t_{stop}	Stopping time.
$t_{xx'}$	Traffic between x' and x.
$U(x, \lambda)$	Potential energy.
$v(x, t), D(x, t)$	Drift and diffusion coefficients, respectively, of a Brownian particle.
w	Mesoscopic (stochastic) work. Positive if done on the system.
w^{diss}	Dissipated work ($w^{diss} = w - \Delta F$).
$W(t)$	Wiener process.
$\mathcal{W}$	Number of microstates.
x	Mesoscopic state (mesostate).
$\boldsymbol{x} = (x(t))$	Trajectory.
$Z\ (Z^{gc})$	Canonical (grand canonical) partition function.
β	$1/k_B T$.
γ	Rate of information-mediated transitions.
$\delta(x)$	Dirac delta.
$\delta^K_{xx'}$	Kronecker delta.
$\delta_{xx'}$	Driving from x' to x.
ΔX	Change of the state function X (energy, entropy, etc.)
ϵ_x	Mesoscopic (fluctuating) energy of mesostate x.
ϵ_ξ	Energy of a microscopic state.
ε	Small positive quantity.
ζ	Parameter in the Mandal-Jarzynski model ($\zeta = \tanh(mg\,\Delta h/2k_B T)$).
η	Stochastic efficiency ($\eta = -s^{in}/s^{out}$).
η_S	Efficiency ($\eta_S = -S^{in}/S^{out}$).
η_C	Carnot efficiency ($\eta_C = 1$).
η^{th}	Thermal efficiency ($\eta^{th} = -W/E_{hot}$).
η^{th}_C	Thermal Carnot efficiency ($\eta^{th}_C = 1 - T_{cold}/T_{hot}$).
θ	Parameter of a Poisson distribution.
$\kappa_{xx'}$	Empirical jump rate.
$\boldsymbol{\lambda} = (\lambda(t))$	Manipulation protocol.
$[\boldsymbol{\lambda}]$	Manipulation path.
λ, λ_{max}	Eigenvalue, maximal eigenvalue.
Λ	Population growth rate.
μ	Chemical potential.
μ_P	Mobility of a Brownian particle.
ν	Number of cycles in a graph.
$\nu_{xx'}$	Empirical frequency of jumps from x' to x.

ρ	Number of cell divisions in a lineage.
ξ	Microscopic state (microstate).
$\tilde{\xi}$	Time-reversed image of ξ: If $\xi = (p_r, r)$, $\tilde{\xi} = (-p_r, r)$.
$\xi(t)$	White noise.
$\sigma(x, t)$	Coefficient of the noise.
$\tilde{\sigma}_x^2$	Scaled variance of x.
τ_y, f_y	Empirical dwell time and empirical frequency in y.
$\phi_x^{(j)}(q, t)$	Generating function of the variable j conditioned to final state x.
$\phi(q)$ or $\phi^{(j)}(q)$	Generating function of variable j.
$\Phi(q)$ or $\Phi^{(j)}(q)$	Cumulant generating function of variable j.
$\psi(q)$ or $\psi^{(j)}(q)$	Scaled cumulant generating function of j.
$\omega_{xx'}$	Intrinsic rate of jumps between x and x'.
Ω_x	Phase-space region associated with mesostate x.

Stochastic Thermodynamics

CHAPTER 1

Motivation

Stochastic thermodynamics has become an established branch of nonequilibrium statistical physics. On the theoretical side, it has been discovered that the behavior of mesoscopic systems is governed by surprisingly general relations. Rapid advances in experimental techniques are leading to tests of these relations by manipulating mesoscopic physical systems.

Perhaps as a consequence of this fast development, some aspects of stochastic thermodynamics might seem obscure to noninitiates. Key results in stochastic thermodynamics, such as fluctuation relations, are so general that one might wonder what the underlying physical assumptions really are. In a broader perspective, it might be difficult to understand how the simplicity of stochastic thermodynamics relates to the daunting complexity of traditional nonequilibrium statistical physics. Preliminary answers to these questions are presented in this chapter. We conclude the chapter with an overview of the book structure.

1.1 What is stochastic thermodynamics?

In its simplest form, **stochastic thermodynamics** is a thermodynamic theory for mesoscopic, nonequilibrium physical systems interacting with equilibrium heat reservoirs.

It is useful to dissect this definition:

- *Thermodynamic theory.* As the name suggests, stochastic thermodynamics draws a correspondence between mesoscopic stochastic dynamics and macroscopic thermodynamics. This correspondence is sketched in fig. 1.1.
- *Mesoscopic, nonequilibrium physical systems.* Stochastic thermodynamics deals with **mesoscopic systems**. For our aims, a mesoscopic system is a physical system characterized by energy differences among its states on the order of the thermal energy $k_B T$, where k_B is the Boltzmann constant and T is the temperature. Prominent examples of mesoscopic systems are colloidal particles, macromolecules, nanodevices, or systems of chemical reactions at very low densities. Mesoscopic systems can be driven out of equilibrium by an external manipulation, for example, by varying in time the temperature or by controlling them with optical tweezers. More generally, all physical systems that can be described by stochastic evolution equations, where the noise

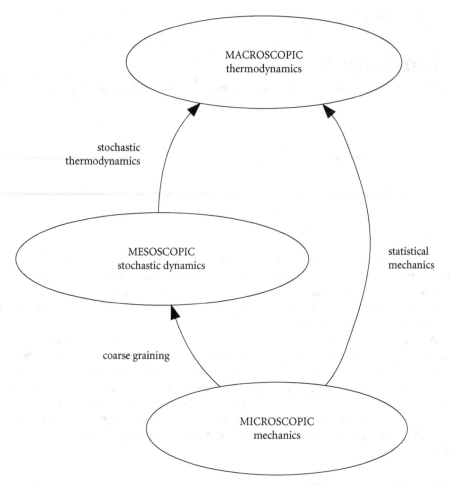

Figure 1.1. Relation between statistical mechanics, coarse-graining techniques, and stochastic thermodynamics. (Inspired by Sekimoto [153].)

models interactions with a heat reservoir, fall into the scope of stochastic thermodynamics.

- *Interacting.* The stochasticity of mesoscopic systems results from interactions with one or multiple reservoirs. Often one does not know the details of these interactions, and the functional form of the noise is dictated by general physical assumptions. Importantly, we always neglect the interaction energy between the system and the bath.
- *Equilibrium heat reservoirs.* We assume that reservoirs relax very quickly to equilibrium compared to the timescales of mesoscopic systems. Therefore, on these timescales, reservoirs are effectively always at equilibrium. This **timescale separation** is key to the simplicity of stochastic thermodynamics. In many cases, this assumption can be justified by a coarse-graining procedure.

Provocatively, one could say that Einstein's paper on Brownian motion was the first paper in history complying with our definition of stochastic thermodynamics. Indeed,

Einstein considered the stochastic dynamics of a mesoscopic colloid and used it to draw far-reaching conclusions for the thermodynamics of general nonequilibrium systems.

1.2 Why does it work and why is it useful?

Stochastic thermodynamics associates thermodynamic quantities with mesoscopic physical systems, whose evolution is described by stochastic dynamics, and predicts their properties. This task is considerably simpler than the general problem in nonequilibrium statistical physics, i.e., that of deriving macroscopic dynamics from a "fundamental" microscopic description.

In particular, several fundamental problems that arise in nonequilibrium statistical physics do not even appear in stochastic thermodynamics. One example is understanding how the irreversible nature of macroscopic thermodynamics systems emerges from microscopic dynamics. This conceptual issue is known as *Loschmidt's paradox* and has puzzled physicists since the dawn of thermodynamics. After all, macroscopic systems are made of a large number of elementary particles, and these particles evolve according to microscopic equations that are time-reversible. It is nowadays established that the solution to the Loschmidt paradox originates from the large number of degrees of freedom of thermodynamic systems. However, a rigorous derivation of irreversible macroscopic dynamics starting from reversible microscopic dynamics has proved to be rather difficult and has been carried out without simplifying assumptions only for a limited number of idealized systems. This difficulty does not arise in stochastic thermodynamics, since the stochastic equations that constitute its starting point are already irreversible.

At this point, one might wonder whether stochastic thermodynamics might be *too simple* to be really interesting. In particular, one question is, In which approximation do real mesoscopic physical systems satisfy the hypotheses of stochastic thermodynamics? Although theoretical arguments can partially answer this question, an ultimate response can only come from experiments. Stochastic thermodynamics has been successfully employed to measure equilibrium free energies from nonequilibrium measurements, and the range of mesoscopic physical systems that are experimentally controllable keeps growing. It is our hope that these experiments will clarify how safe it is to apply stochastic thermodynamics to generic mesoscopic systems and which aspects have to be treated with special care.

Due to its simplifying assumptions, stochastic thermodynamics circumvents many technical subtleties of kinetic theory. In this respect, it might appear that it provides less interesting challenges for mathematical physics. However, stochastic thermodynamics has proved to be an interesting playground for advanced mathematical tools to analyze stochastic processes, including control theory, large deviations, and probability measures in the space of trajectories. As shown in this book, these tools are precious to shed light on the nature of nonequilibrium mesoscopic processes.

1.3 Plan of the work

The main goal of this book is to provide a pedagogical introduction to stochastic thermodynamics for graduate students in physics. Our book is structured so that it can be used as a textbook for a graduate course or for independent study. To this aim, we mark those sections of the book that contain more advanced material with the notation (*). These sections can be skipped to ease a first reading or to use the book for a course covering basic concepts only.

Relevant bibliography is collected in the "Further reading" sections at the end of each chapter. These sections refer to works where results discussed in the chapter were originally presented, along with other useful references to deepen the study of specific topics.

One of the best ways to learn a subject is by problem solving. Following this philosophy, we have included an exercise section to complement most chapters. Some of these exercises are meant to be solved with paper and pen, whereas others require computer simulations. These latter exercises assume that the reader is familiar with basic computer programming (in any language). For reasons of space, we do not introduce numerical algorithms other than particularly relevant ones, such as the Gillespie algorithm. As with sections, some exercises are marked with a (*), to warn the reader that their solution is particularly challenging or that it requires concepts from a starred section.

Chapter 2 provides a brief overview of the theories upon which stochastic thermodynamics is built: thermodynamics, statistical mechanics, the theory of stochastic processes, and information theory. This overview is far from exhaustive due to reasons of space. We focus on aspects of these theories of more relevance for stochastic thermodynamics. To avoid overburdening the book with the complexities of stochastic calculus, we mainly focus on physical systems with discrete states that can be described by master equations. Throughout the book, sections requiring knowledge of stochastic processes with continuous state space are always starred.

Chapter 3 introduces the basic concepts of stochastic thermodynamics. We discuss how work, heat, and entropy can be consistently introduced at the level of single stochastic trajectories of systems described by master equations. We show that these quantities satisfy relations that are analogous to the first and second laws of traditional thermodynamics. Particular emphasis is given to the physical interpretation of these quantities.

Chapter 4 is devoted to fluctuation relations, which are perhaps the most celebrated results in stochastic thermodynamics. The connection between entropy production and statistical irreversibility is the core concept of this chapter. We exploit this connection to introduce fluctuation relations in a unified framework.

Chapter 5 discusses manipulation of information at the mesoscopic level. In the first part of the chapter, we introduce counterintuitive physical aspects of information processing, such as Maxwell demons and the Landauer cost of erasing information. In the rest of the chapter, we show how stochastic thermodynamics clarifies these concepts, both in general and in the context of concrete examples.

Chapter 6 is devoted to large deviation theory. Once confined to pure mathematics and statistics, large deviation theory has risen as a fundamental tool in statistical physics and beyond. Many current developments in stochastic thermodynamics heavily rely on large deviation theory. In this chapter, we introduce the theory and discuss its main applications in stochastic thermodynamics.

Chapter 7 presents key experimental results in stochastic thermodynamics. A focus of this chapter is how fluctuation theorems allow for estimating equilibrium free energy from nonequilibrium measurements. We also discuss other groundbreaking experiments that have tested manipulation of information at the mesoscopic level.

Chapter 8 presents a collection of developments of stochastic thermodynamics in several directions. Sections of this chapter are rather independent of each other and provide an introduction to more recent research topics. We have not marked with a (*) sections and exercises of this chapter; however, most of the material should be considered as advanced.

Chapter 9 presents perspectives on open issues and future directions.

CHAPTER 2

Basics

The prerequisites for stochastic thermodynamics are laid down in this chapter. In particular, we briefly review the thermodynamics of macroscopic systems, the statistical description of their behavior, both static and dynamic, and some basic concepts of information theory. We take advantage of this preliminary material to introduce most of the notation used in the rest of the book.

2.1 Thermodynamics

Stochastic thermodynamics describes thermodynamic processes taking place in small systems in contact with reservoirs. Before going any further, we need to specify what we mean by "thermodynamic processes" and "reservoirs," and to introduce other main concepts in traditional thermodynamics.

In its basic form, thermodynamics deals with systems that are made up of a very large number of particles and are in thermodynamic equilibrium, in the sense that their macroscopically observable properties, like density, pressure, etc., do not change with time. It is a common observation that many materials, kept isolated from the environment, reach sooner or later a state characterized by constant values of macroscopic properties. Importantly, systems that reach such an equilibrium state by different manipulations behave from then on in the same way, from the point of view of thermodynamics. *Thermodynamic equilibrium wipes out previous history.* For example, the thermodynamic behavior of two glasses of water with the same density at the same pressure does not depend on whether one of the two has been prepared by melting an ice cube in a glass of water at a higher temperature, while the other has simply kept the state it had when flowing out of the tap. There are exceptions to this rule: some systems, such as structural glasses, keep memory of past manipulations. Their thermodynamic behavior is more complex (and controversial). We do not deal with them.

The state of a thermodynamic system is characterized by a judiciously chosen, small set of macroscopically observed properties: its composition, its mass, its volume, the pressure acting on it, etc. In thermodynamics, knowledge of the values of these quantities in a given equilibrium state is sufficient to predict their values after the system undergoes a thermodynamic transformation. In this sense, traditional thermodynamics is *deterministic*.

For example, we consider a cylinder closed by a piston, holding a **simple fluid**, i.e., a fluid composed of a single chemical species. For such a system, the number n of moles of the fluid (or the number of particles $N = n N_A$, where $N_A \approx 6.02 \cdot 10^{23}$ is Avogadro's number), the volume V of the cylinder, and the pressure P applied by the piston are a complete set of thermodynamic observables.

Thermodynamics deals with two main kinds of transformations:

Adiabatic transformations. In adiabatic transformations, energy is exchanged between the system and the environment only in the form of work. This means that the system is **thermally isolated**, i.e., enclosed by walls that do not allow the exchange of heat. Values of observables like pressure, volume, etc. can be mechanically altered. In our example, this means that the walls of the cylinder do not allow any uncontrolled interaction with the surroundings, and only the position of the piston can be changed. We then let the piston rest in the new position, and we wait until the system reaches a new equilibrium state. The motion of the piston can be very slow (quasi-static transformation) or abrupt, meaning that the intermediate states are not necessarily equilibrium states. In either case, the transformation brings the system from one equilibrium state to a new equilibrium state, whose properties depend on the details of the transformation.

Heat exchange. In the case of heat exchange, the system interacts in an uncontrolled way with another thermodynamic system, for example, a similar container with nonisolating walls. We say that the two systems are put "in contact." This interaction in general involves energy exchange between the two systems. In this case, we must specify which macroscopic variables are kept constant, because it is not possible in general to keep, e.g., both the volume and the pressure constant.

Any transformation can be decomposed into a succession of (possibly infinitesimal) adiabatic transformations and heat exchanges.

We now consider two systems S_1 and S_2 that are put in contact with a much larger system S_0, one after the other, in such a way that they both reach equilibrium. We assume that S_0 is so large that its thermodynamic state is not significantly affected by being put in contact with either S_1 or S_2, whereas the thermodynamic states of S_1 or S_2 may in principle change. Experience shows that if, after this procedure, S_1 and S_2 are put in contact with each other, their thermodynamic states remain unaltered. This observation is summarized by the **zeroth law of thermodynamics**:

Two systems, each in thermodynamic equilibrium with a third one, are in equilibrium with each other.

The zeroth law of thermodynamics allows us to define a quantity Θ that assumes the same value in systems in thermodynamic equilibrium with one another. This quantity can be made observable if we put these systems in touch with a reference system $\mathfrak{T}$, small enough not to perturb their equilibrium state, and then measure a macroscopic quantity (e.g., the volume) of $\mathfrak{T}$. The system $\mathfrak{T}$ is called **thermometer**, and Θ is an "empirical temperature." We call the large system S_0 a **heat reservoir** characterized by a given value of Θ.

In general, there are multiple ways to perform a thermodynamic transformation from one equilibrium state to another. For example, we can increase the pressure of a fluid with a fixed value of V by putting it in contact with a heat reservoir at an empirical temperature Θ, or by letting an electrical current go through a resistor immersed in it, until a thermometer reads the same value of Θ. According to our postulates, the system reaches the same equilibrium state in all these cases. In particular, the energy contained in the system (its **internal energy**) is the same. In an adiabatic transformation, conservation of energy imposes that the change ΔE in the internal energy between the initial and the final state must be equal to the work W^{ad} performed on the system:

$$\Delta E = E_f - E_0 = W^{ad}. \tag{2.1}$$

We use in this book the convention that work is considered positive if it is performed *on* the system and negative if it is performed *by* the system. In macroscopic thermodynamic systems, the internal energy E is **extensive**, i.e., it is proportional to the size of the system (as measured by V or N, as long as the density is fixed).

In a nonadiabatic transformation between two equilibrium states, the work W performed on the system is in general different from W^{ad}. Therefore, a certain amount of energy exchanged by interactions between the system and its surroundings is not taken into account in W. We identify this energy Q with the heat exchanged in the transformation. We thus have

$$Q = W - \Delta E. \tag{2.2}$$

With this convention, Q is positive if it is released by the system and negative if it is provided to the system. Equation (2.2) embodies the **first law of thermodynamics**, which may be expressed as follows:

> The change of the internal energy of a system is equal to the difference between the work done on it and the heat released by it.

Our sign convention for work and heat is the opposite of that commonly used in traditional thermodynamics, but it turns out to be the most natural in stochastic thermodynamics. The reason is that, historically, thermodynamics originated from the study of thermal engines, and the emphasis was on the conversion of heat into work. Conversely, in stochastic thermodynamics, one is usually interested in describing dissipative systems, which convert work into heat.

The internal energy E is one of the macroscopic observables. For a simple fluid, knowledge of n, P, and V allows one to evaluate E. It is possible to invert this relation to express, e.g., P or Θ as a function of n, V, and E.

Given an equilibrium state $\mathcal{E}_0$ of a thermodynamic system, there are states $\mathcal{E}$ that cannot be reached from $\mathcal{E}_0$ via an adiabatic transformation, while the inverse transformation $\mathcal{E} \longrightarrow \mathcal{E}_0$ is possible. In this sense, thermodynamic systems possess an intrinsic *irreversibility*. A main goal of thermodynamics is to characterize the set of states that can be reached from a given state by an arbitrary combination of adiabatic transformations and heat exchanges.

The irreversibility of thermodynamic transformations is characterized by the concept of **entropy**. The entropy S is a function of the thermodynamic state of an equilibrium system. The irreversibility of thermodynamic transformations is captured by the **second law of thermodynamics**:

> The total entropy of a thermally isolated system cannot decrease.

The entropy S has the following properties:

Additivity. If a system is made up of several subsystems $S_1, \ldots, S_k, \ldots$, each at equilibrium, the entropy of the total system is equal to the sum of the entropies of the subsystems:

$$S\left(\bigcup_k S_k\right) = \sum_k S(S_k). \tag{2.3}$$

As a consequence of additivity, the entropy of a homogeneous system is extensive, i.e., it is proportional to the system size. This implies that, upon rescaling the other extensive observables, such as the number of particles, the volume, the internal energy, etc., by a factor $\lambda > 0$, one has

$$S(\lambda E, \lambda V, \lambda N) = \lambda S(E, V, N). \tag{2.4}$$

Monotonicity. S increases as E increases when all other variables are kept constant. This property holds for the vast majority of practical cases. There exist some intriguing systems where S can decrease with E, but we do not deal with them in this book.

Concavity. Entropy is a concave function. This means that, given two equilibrium states $\mathcal{E}_0 = (N, V_0, E_0)$ and $\mathcal{E}_1 = (N, V_1, E_1)$ of the same system and any real number α between 0 and 1, the intermediate state $\mathcal{E}_\alpha$ characterized by $(N, (1-\alpha)V_0 + \alpha V_1, (1-\alpha)E_0 + \alpha E_1)$ satisfies

$$S(\mathcal{E}_\alpha) \geq (1-\alpha)S(\mathcal{E}_0) + \alpha S(\mathcal{E}_1), \qquad 0 \leq \alpha \leq 1. \tag{2.5}$$

Properties of concave and convex functions are summarized in appendix A.1.

The second law of thermodynamics also implies several important properties of thermodynamic systems:

Temperature. Systems in thermal equilibrium with each other share the same value of the quantity

$$\frac{1}{T} = \frac{\partial S}{\partial E}, \tag{2.6}$$

if the other extensive quantities (N, V, etc.) are kept constant. To prove this result, we consider two systems S_1 and S_2 in contact with each other and adiabatically insulated from their surroundings. Their total energy $E = E_1 + E_2$ is fixed. If they are in mutual equilibrium, their total entropy cannot grow upon exchanging energy in the form of heat. Thus, at equilibrium, we have

$$\frac{\partial(S_1(E_1) + S_2(E - E_1))}{\partial E_1} = \frac{1}{T_1} - \frac{1}{T_2} = 0. \tag{2.7}$$

Therefore, T acts as the empirical temperature Θ that we defined before. The monotonicity of S implies that T cannot be negative for the cases we consider.

Heat exchange. If two systems in contact with each other do not exchange work, the system with a larger value of T provides heat to the system with a smaller value of T. We call E_1 and E_2 the initial internal energies of the two systems and T_1 and T_2 their initial temperatures. After being in contact for some time, they reach equilibrium at energy values E_1' and E_2'. The total entropy is given by

$$S = S_1(E_1') + S_2(E_2') \leq S_1(E_1) + \frac{E_1' - E_1}{T_1} + S_2(E_2) + \frac{E_2' - E_2}{T_2}$$

$$= S_1(E_1) + S_2(E_2) + \left(E_1' - E_1\right)\left(\frac{1}{T_1} - \frac{1}{T_2}\right),$$

(2.8)

where we use the concavity of S and the fact that $E_1 + E_2 = E_1' + E_2'$. The second law imposes that

$$S_1(E_1') + S_2(E_2') \geq S_1(E_1) + S_2(E_2)$$

(2.9)

and therefore

$$\left(E_1' - E_1\right)\left(\frac{1}{T_1} - \frac{1}{T_2}\right) \geq 0.$$

(2.10)

Thus, if $E_1' - E_1 > 0$, we have $1/T_1 > 1/T_2$, which corresponds to $T_1 < T_2$: the body with higher T releases energy to that with lower T. This means that T behaves as a temperature scale: hotter bodies are characterized by larger values of T, and heat flows naturally from them to colder bodies. This fact is summarized by the **Clausius statement** of the second law of thermodynamics:

> Heat can never pass from a colder to a warmer body without some other change, connected therewith, occurring at the same time.

In fact, T defined in this way coincides with the absolute temperature scale and is called the **temperature** from now on.

As an illustration of these ideas, we consider a cylinder containing n moles of an ideal gas. An **ideal gas** is a fluid in which the pressure P, the volume V, the number of particles N, and the temperature T satisfy the **equation of state**

$$P = \frac{N k_\mathrm{B} T}{V}.$$

(2.11)

We change the volume of the gas by an infinitesimal quantity $\mathrm{d}V$ by performing on it an infinitesimal amount of work

$$\mathrm{d}W = -P\,\mathrm{d}V.$$

(2.12)

If the gas is thermally isolated, we have $\mathrm{d}E = -P\,\mathrm{d}V$. It turns out that the internal energy of an ideal gas depends only on its temperature T. Thus we obtain

$$\left.\frac{\partial E}{\partial V}\right)_T = \left.\frac{\partial E}{\partial V}\right)_S + \left.\frac{\partial E}{\partial S}\right)_V \left.\frac{\partial S}{\partial V}\right)_T = 0,$$

(2.13)

where $\partial X/\partial Y)_Z$ is the partial derivative of X with respect to Y, taken at constant Z. By combining eqs. (2.12) and (2.6), we obtain

$$\left.\frac{\partial E}{\partial V}\right)_S = -P; \qquad \left.\frac{\partial E}{\partial S}\right)_V = T.$$

(2.14)

Therefore,

$$\frac{\partial S}{\partial V}\bigg)_T = \frac{P}{T}. \tag{2.15}$$

If the volume V of the system changes from V_0 to V_f at constant T, its entropy S changes by

$$\Delta S = \int_{V_0}^{V_f} dV \frac{P}{T} = N k_B \int_{V_0}^{V_f} \frac{dV}{V} = N k_B \ln \frac{V_f}{V_0}. \tag{2.16}$$

Therefore, if $V_f > V_0$, the transformation $V_0 \longrightarrow V_f$, but not its reverse, can take place in a thermally isolated system.

Another facet of these postulates is the **Kelvin-Planck statement** of the second law:

It is impossible to devise a cyclically operating heat engine, the only effect of which is to absorb energy in the form of heat from a single thermal reservoir and to deliver an equivalent amount of work.

Indeed, such a device would produce negative entropy during a cycle. To show that, we imagine enclosing the device and the reservoir with a wall, so that the entire system is thermally isolated. During the cycle, the device transfers a positive amount W of work to the environment and reduces the internal energy of the reservoir by the same amount. Since the entropy of the device does not change and the entire system is isolated, the total entropy change is $S^{res} = -W/T < 0$, in contradiction with the second law.

2.2 Thermodynamic efficiency

An important application of thermodynamics is the study of engines and their efficiency. We generally call an **engine** a physical machine that operates cyclically and converts one form of energy into another. Historically, the most important example is **heat engines**, i.e., machines that cyclically convert heat into work. The Kelvin statement of the second law of thermodynamics implies that heat engines must necessarily operate using at least two heat reservoirs at different temperatures T_{hot} and T_{cold}, with $T_{hot} > T_{cold}$.

We consider a heat engine that is alternately put in contact with two heat reservoirs. During each cycle, the system receives an amount of energy E_{hot} from the hot reservoir and releases an amount of energy E_{cold} to the cold one. At the end of the cycle, the engine returns to the same state it had at the beginning of the cycle. Therefore, the work $W = E_{cold} - E_{hot}$ performed on the engine in a cycle is equal to the net total heat released to the reservoirs (remember our sign convention!). The total change in entropy in a cycle, S^{tot}, is given by the change ΔS^{sys} of the entropy of the system plus the entropy change S^{res} of the reservoirs:

$$S^{tot} = \Delta S^{sys} + S^{res}. \tag{2.17}$$

Here and in the following we denote with ΔX the change of a state function X. The entropy change of the reservoir S^{res} is not a state function, since the internal energy and thus the energy of a reservoir can change without altering its temperature. Therefore, neither is S^{tot} a state function. In a cycle, ΔS^{sys} vanishes, whereas the entropy of the reservoirs changes by

$$S^{res} = -\frac{E_{hot}}{T_{hot}} + \frac{E_{cold}}{T_{cold}} \geq 0. \tag{2.18}$$

The extracted work attains its maximum when the entropy increase S_{cold} of the reservoir at lower temperature is the opposite of the entropy decrease S_{hot} of the reservoir at higher temperature:

$$S_{cold} = \frac{E_{cold}}{T_{cold}} \geq -S_{hot} = \frac{E_{hot}}{T_{hot}}. \tag{2.19}$$

This condition implies

$$-W \leq E_{hot} \left(1 - \frac{T_{cold}}{T_{hot}} \right). \tag{2.20}$$

Traditionally, the **thermal efficiency** η^{th} of a heat engine is defined as the ratio between the extracted work and the energy absorbed from the hot reservoir:

$$\eta^{th} = -\frac{W}{E_{hot}}. \tag{2.21}$$

Equation (2.20) implies that the maximal thermal efficiency is determined by the temperatures of the hot and cold reservoirs, independent of the amount of energy exchanged during a cycle:

$$\eta^{th} \leq \eta_C^{th} = 1 - \frac{T_{cold}}{T_{hot}}, \tag{2.22}$$

where η_C^{th} is the **thermal Carnot efficiency**.

It is interesting to consider more general engines whose energy currencies are not limited to work and heat. This is especially true in stochastic thermodynamics, where information can also be exchanged for work. To deal with these more general engines, we define the **efficiency** η_S in terms of *entropy* rather than energy. To this aim, we consider an engine operating between two arbitrary reservoirs. During a cycle, the engine extracts an amount S^{in} of entropy from a reservoir and releases an amount S^{out} of entropy to another reservoir. In this framework, we define the efficiency as

$$\eta_S = -\frac{S^{in}}{S^{out}}. \tag{2.23}$$

With this definition of efficiency, eq. (2.22) becomes

$$\eta_S \leq \eta_C, \tag{2.24}$$

where in this case the **Carnot efficiency** is simply

$$\eta_C = 1. \tag{2.25}$$

This definition of efficiency allows us to characterize engines operating between two arbitrary reservoirs.

2.3 Free energy and nonequilibrium free energy

The thermodynamic behavior of a system is identified once we know the expression of the internal energy E as a function of entropy S, particle number N, volume V, and other thermodynamically relevant observables. If the system is put in contact with a heat reservoir at temperature T, its internal energy E and entropy S are determined by the interaction with the reservoir. In this case, it is convenient to express the thermodynamic properties in terms of T rather than S. Since $T = \partial E / \partial S$ (where the other thermodynamic observables are kept constant), this change of variable can be achieved via a Legendre transformation. Properties of the Legendre transformation are summarized in appendix A.2. The **free energy** is defined as the opposite of the Legendre transform of the internal energy:

$$F(T, N, V) = E(S, N, V, \ldots) - T S, \tag{2.26}$$

where S is expressed as a function of T and of the other observables as the solution of the equation

$$\left. \frac{\partial E}{\partial S} \right)_{N,V,\ldots} = T, \tag{2.27}$$

and the dots stand for possible other thermodynamic observables. In the following, for simplicity, we limit ourselves to the basic observables S, V, and N. Given $F(T, N, V)$, the entropy S is given by

$$S(T, N, V) = - \left. \frac{\partial F}{\partial T} \right)_{N,V}, \tag{2.28}$$

whereas the pressure is given by

$$P = - \left. \frac{\partial F}{\partial V} \right)_{T,N}. \tag{2.29}$$

Therefore, the free energy fully describes the thermodynamic state, in the sense that it permits us to reconstruct all thermodynamic observables. Functions with this property are called **thermodynamic potentials**. Applying the symmetry of partial derivatives of F to relations like eqs. (2.28) and (2.29), we obtain other useful relations between thermodynamic quantities, such as

$$\left. \frac{\partial^2 F}{\partial V \partial T} \right)_N = - \left. \frac{\partial S}{\partial V} \right)_T = \left. \frac{\partial^2 F}{\partial T \partial V} \right)_N = - \left. \frac{\partial P}{\partial T} \right)_V. \tag{2.30}$$

Such equalities are known as **Maxwell relations**. In particular, one can use these relations to show that the equation of state of an ideal gas implies that its entropy depends on T and V in the form $S(T, V) = N k_B \ln V + S_0(T)$, where $S_0(T)$ does not depend on V.

We now consider a system initially at equilibrium with values S, N, and V of the thermodynamic observables that is brought to a new equilibrium by putting it in touch with a heat reservoir at temperature T. In the new equilibrium state, the free energy has the value $F(T, N, V)$ and the value of the entropy satisfies eq. (2.28). During equilibration, energy is exchanged as heat with the reservoir and as work with the external environment. We wish to characterize this exchange.

By the second law, the total entropy change of the system plus the reservoir cannot be negative:

$$S^{\text{tot}} = \Delta S^{\text{sys}} + S^{\text{res}} \geq 0. \tag{2.31}$$

We call W the work performed on the system and Q the heat released to the reservoir. The first law imposes that

$$\Delta E = W - Q = W - T S^{\text{res}}, \tag{2.32}$$

where we use the fact that the heat reservoir is always at equilibrium at temperature T. As a consequence of eqs. (2.31) and (2.32), we obtain

$$W \geq \Delta E - T \Delta S^{\text{sys}}. \tag{2.33}$$

Therefore, the equilibrium state can be identified as the one in which the expression $E - T S$ attains its minimum. Evaluating the equilibrium value S of the system entropy at this minimum, we retrieve the condition given by eq. (2.27). The value of $E - T S$ at its minimum is equal to the free energy F at temperature T. Mathematically, this result descends from the fact that Legendre transforms satisfy a variational principle (cf. eq. (A.29)). These results justify defining the **nonequilibrium free energy** by

$$F^{\text{neq}}(T, S, N, V) = E(S, N, V) - T S \tag{2.34}$$

for arbitrary values S of the entropy. The equilibrium free energy F is then obtained by the variational principle

$$F(T, N, V) = \min_{S} F^{\text{neq}}(T, S, N, V). \tag{2.35}$$

The nonequilibrium free energy is not a thermodynamic potential. One reason is that it simultaneously depends on some quantities characterizing the system (such as S) and others that characterize the reservoir (such as T). We defined F^{neq} for a system initially at equilibrium, for which S is thermodynamically defined and which is brought out of equilibrium by allowing its contact with a heat reservoir. We discuss in section 5.2 a generalization of nonequilibrium free energy to mesoscopic systems prepared in an arbitrary nonequilibrium state.

There exist thermodynamic potentials other than the free energy. They differ in the thermodynamic quantities that are kept fixed by external reservoirs. For example, the appropriate thermodynamic potential for a system in contact with a heat reservoir and kept at a fixed pressure P is the **Gibbs free energy**

$$G(T, P, N) = E + P V - T S, \tag{2.36}$$

where E, V, and S are expressed as functions of T, P, N. Beyond heat reservoirs, we can also consider particle reservoirs, which are able to exchange with the system an unlimited amount of particles of a given chemical species without changing their properties and while remaining at thermodynamic equilibrium. They are characterized by the values T of their absolute temperature and μ of their **chemical potential**. The chemical potential μ of a system with free energy F that can exchange a single chemical species is given by

$$\mu = \frac{\partial F}{\partial N}\bigg)_{T,V}, \tag{2.37}$$

where N is the number of molecules of the considered species. For a system containing a single chemical species at fixed temperature T and pressure P, the chemical potential is equal to the Gibbs free energy per particle:

$$\mu(T,P) = \frac{G(T,P,N)}{N}. \tag{2.38}$$

For systems exchanging multiple chemical species, a distinct chemical potential can be assigned to each one of them. Each chemical potential is defined by a formula similar to eq. (2.37) in which we keep constant the number of molecules of all exchangeable species but one, and take the derivative with respect to that one. By the same reasoning we followed for the free energy, it can be shown that equilibrium in the presence of particle reservoirs corresponds to the minimum of

$$\Phi(E, T, V, \mu_1, N_1, \mu_2, N_2, \ldots) = E - TS - \sum_i \mu_i N_i, \tag{2.39}$$

where the sum runs over all exchanged chemical species.

2.4 Statistical mechanics

Statistical mechanics links the microscopic description of a macroscopic system at equilibrium to its thermodynamic behavior. We consider a macroscopic system and assume for simplicity that its microscopic states ξ (also called **microstates**) are discrete, $\xi \in \{1, 2, \ldots\}$. Macroscopically, the thermodynamic equilibrium state is identified by the values of macroscopic observables, like the internal energy E, the volume V, the number of particles N, etc. At the microscopic level, the system incessantly and rapidly changes its microstate according to its dynamics. Therefore, we cannot do better than assign it a **statistical state**, i.e., a probability distribution over the microstates. We denote by p_ξ the probability of a discrete microstate ξ. If the variables ξ are continuous, the probability density is denoted by $p(\xi)$. In either case, we denote by $\langle f(\xi) \rangle$ the expectation of the function $f(\xi)$ over the probability distribution of ξ. We also use the notations p_ξ^{eq} and $\langle f(\xi) \rangle^{eq}$ whenever we want to stress that a probability distribution corresponds to thermodynamic equilibrium. We briefly review the properties of probability distributions in appendix A.3.

The **fundamental postulate of statistical mechanics** stipulates that an isolated system at thermodynamic equilibrium can be found with equal probability in any of the microstates ξ compatible with given values of the thermodynamic observables, and that the thermodynamic entropy S is related to the number $\mathcal{W}$ of the microstates that satisfy this condition by the relation

$$S = k_B \ln \mathcal{W}. \tag{2.40}$$

Here k_B is the Boltzmann constant:

$$k_B \approx 1.384 \cdot 10^{-23} \, \text{J/K}. \tag{2.41}$$

This probability distribution over the microstates is known as the **microcanonical distribution** (or **microcanonical ensemble**). The term *ensemble* is used to stress that we are effectively replacing a single system and its detailed dynamical behavior with a large collection of statistically identical systems, such that their distribution over the microstates is constant in time—in agreement with thermodynamic equilibrium at the macroscopic level. This change of description is appropriate if the system is large enough. This condition is formalized by the **thermodynamic limit**, in which one imagines the size of the system (measured by the number of particles N) to go to infinity, keeping constant the ratios V/N, E/N, ..., of extensive variables.

Starting from the microcanonical distribution, one can show that the equilibrium state of a system in equilibrium with a reservoir at temperature T is described by the **Maxwell-Boltzmann** (or **canonical**) **distribution** (or **canonical ensemble**)

$$p_\xi^{eq} = \frac{e^{-\epsilon_\xi/k_BT}}{Z},\tag{2.42}$$

where ϵ_ξ is the energy of microstate ξ and the denominator is the **partition function**

$$Z = \sum_\xi e^{-\epsilon_\xi/k_BT}.\tag{2.43}$$

The partition function is related to the free energy F by

$$F = -k_BT \ln Z.\tag{2.44}$$

This relation allows us to write the equilibrium distribution p^{eq} in the form

$$p_\xi^{eq} = e^{(F-\epsilon_\xi)/k_BT}.\tag{2.45}$$

One of the simplest thermodynamic systems is the **ideal gas**. An ideal gas is made of N point-like particles of mass m that interact weakly, so that their potential energy is negligible compared to their kinetic energy. The state of a particle i is identified by its momentum $\vec{p}_{r,i}$ and its position $\vec{r}_i$. In evaluating the partition function, we have to take into account that the particles are not distinguishable. Thus the partition function must be multiplied by a factor $1/N!$, because configurations that differ only by the exchange of particles should not be considered different. In order to make the expression of Z dimensionless, we introduce an elementary phase-space volume h for each degree of freedom, where h is Planck's constant. This value is chosen to make a connection with the behavior of quantum systems. Using Stirling's approximation for the factorial, we obtain

$$F = -k_BT \ln Z = -k_BT \ln \frac{1}{N!} \int \prod_{i=1}^N \left[\frac{d\vec{r}_i\, d\vec{p}_{r,i}}{h^3} \exp\left(-\frac{p_{r,i}^2}{2mk_BT} \right) \right]$$

$$= Nk_BT \ln \left[\frac{N}{e\,V} \left(\frac{h^2}{2\pi\, mk_BT} \right)^{3/2} \right],\tag{2.46}$$

where $e = 2.7818\ldots$ is the basis of natural logarithms. The chemical potential μ is obtained by taking the derivative of F with respect to N, as shown in eq. (2.37):

$$\mu = \frac{\partial F}{\partial N}\bigg)_{T,V} = k_{\mathrm{B}}T \ln \frac{N}{V} + \mu^{(0)}, \tag{2.47}$$

where $\mu^{(0)}$ denotes terms that are independent of the concentration.

Values of thermodynamic observables can be obtained by taking appropriate derivatives of the partition function. For example, the equation of state is obtained by taking the derivative of F, as expressed by eq. (2.46), with respect to V:

$$P = -\frac{\partial F}{\partial V}\bigg)_{T,N} = k_{\mathrm{B}}T \frac{\partial \ln Z}{\partial V}\bigg)_{T,N} = \frac{N k_{\mathrm{B}}T}{V}. \tag{2.48}$$

Similarly, defining $Z(\beta) = \sum_{\xi} e^{-\beta \epsilon_{\xi}}$, we have

$$\frac{\partial \ln Z}{\partial \beta}\bigg|_{\beta = 1/k_{\mathrm{B}}T} = -\frac{1}{Z} \sum_{\xi} \epsilon_{\xi} \, e^{-\epsilon_{\xi}/k_{\mathrm{B}}T} = -\langle \epsilon \rangle^{\mathrm{eq}}. \tag{2.49}$$

We identify the average $\langle \epsilon \rangle^{\mathrm{eq}}$ with the thermodynamic value E of the internal energy. On the one hand, taking a further derivative, we obtain

$$\frac{\partial^2 \ln Z(\beta)}{\partial \beta^2}\bigg|_{\beta = 1/k_{\mathrm{B}}T} = \langle \epsilon^2 \rangle^{\mathrm{eq}} - \left(\langle \epsilon \rangle^{\mathrm{eq}} \right)^2 \geq 0. \tag{2.50}$$

On the other hand, an explicit evaluation of the derivatives yields

$$\frac{\partial^2 \ln Z(\beta)}{\partial \beta^2}\bigg|_{\beta = 1/k_{\mathrm{B}}T} = k_{\mathrm{B}}T^2 \frac{\partial \langle \epsilon \rangle^{\mathrm{eq}}}{\partial T}\bigg)_{N,V,\dots}. \tag{2.51}$$

Comparing eqs. (2.50) and (2.51), we obtain

$$k_{\mathrm{B}}T^2 \frac{\partial \langle \epsilon \rangle^{\mathrm{eq}}}{\partial T}\bigg)_{N,V,\dots} = \langle \epsilon^2 \rangle^{\mathrm{eq}} - \left(\langle \epsilon \rangle^{\mathrm{eq}} \right)^2. \tag{2.52}$$

This relation expresses a thermodynamic derivative (on the left-hand side) in terms of a microscopic fluctuation (on the right-hand side). It is an elementary example of relations that are collectively known as **fluctuation-dissipation relations**. Importantly, eq. (2.52) tells us that the variance of the distribution of ϵ grows proportionally to its average, and therefore to the system size. Therefore, the relative uncertainty on the energy

$$\frac{\sqrt{\langle \epsilon^2 \rangle^{\mathrm{eq}} - \left(\langle \epsilon \rangle^{\mathrm{eq}} \right)^2}}{\langle \epsilon \rangle^{\mathrm{eq}}} = \sqrt{\frac{k_{\mathrm{B}}T^2}{\left(\langle \epsilon \rangle^{\mathrm{eq}} \right)^2} \frac{\partial \langle \epsilon \rangle^{\mathrm{eq}}}{\partial T}\bigg)_{N,V,\dots}} \tag{2.53}$$

scales like the inverse square root of the system size. As a consequence, also taking into account the smallness of k_{B}, energy fluctuations are negligible for macroscopic systems. Thus, although statistical mechanics describes systems with probability distributions at the microscopic level, its predictions are deterministic at the macroscopic level. This is also true for macroscopic systems out of equilibrium.

If a system is in contact with reservoirs exchanging extensive quantities other than energy, averages and fluctuations of these other quantities can be evaluated following a similar strategy. We consider, for example, a system exchanging energy and particles with a reservoir characterized by a temperature T and a chemical potential μ. The equilibrium distribution of such a system is given by the **grand canonical ensemble**

$$p_\xi^{\mathrm{eq}} = \frac{1}{Z^{\mathrm{gc}}}\, e^{-(\epsilon_\xi - \mu N_\xi)/k_B T}, \tag{2.54}$$

where we define the grand canonical partition function as a function of α and β

$$Z^{\mathrm{gc}} = \sum_\xi e^{\alpha N_\xi - \beta \epsilon_\xi}, \tag{2.55}$$

with $\beta = 1/k_B T$ and $\alpha = \mu/k_B T$. The derivatives of the logarithm of the partition function return averages and variances of thermodynamic observables in this case too:

$$\left. \frac{\partial \ln Z^{\mathrm{gc}}}{\partial \alpha} \right|_{\alpha = \mu/k_B T,\, \beta = 1/k_B T} = \langle N \rangle^{\mathrm{eq}}; \tag{2.56a}$$

$$\left. \frac{\partial \ln Z^{\mathrm{gc}}}{\partial \beta} \right|_{\alpha = \mu/k_B T,\, \beta = 1/k_B T} = - \langle \epsilon \rangle^{\mathrm{eq}}; \tag{2.56b}$$

and

$$\left. \frac{\partial^2 \ln Z^{\mathrm{gc}}}{\partial \alpha^2} \right|_{\alpha = \mu/k_B T,\, \beta = 1/k_B T} = \langle N^2 \rangle^{\mathrm{eq}} - \left(\langle N \rangle^{\mathrm{eq}} \right)^2; \tag{2.57a}$$

$$- \left. \frac{\partial^2 \ln Z^{\mathrm{gc}}}{\partial \alpha\, \partial \beta} \right|_{\alpha = \mu/k_B T,\, \beta = 1/k_B T} = \langle N \epsilon \rangle^{\mathrm{eq}} - \langle N \rangle^{\mathrm{eq}} \langle \epsilon \rangle^{\mathrm{eq}}; \tag{2.57b}$$

$$\left. \frac{\partial^2 \ln Z^{\mathrm{gc}}}{\partial \beta^2} \right|_{\alpha = \mu/k_B T,\, \beta = 1/k_B T} = \langle \epsilon^2 \rangle^{\mathrm{eq}} - \left(\langle \epsilon \rangle^{\mathrm{eq}} \right)^2. \tag{2.57c}$$

These relations allow us to estimate relative fluctuations of thermodynamic observables. The symmetry of thermodynamic derivatives yielding the Maxwell relation (2.30) corresponds to the symmetry of the covariance of fluctuations, as in the example of eq. (2.57).

In the canonical ensemble, the internal energy E is a fluctuating quantity. Its distribution can be evaluated by means of the so-called **Boltzmann-Einstein principle**. To introduce it, we first associate each value of the internal energy with the entropy of the corresponding microcanonical ensemble:

$$S(E) = k_B \ln \mathcal{W}(E) = k_B \ln \sum_\xi \delta(\epsilon_\xi - E), \tag{2.58}$$

where $\delta(x)$ is the Dirac delta function. We substitute this result in the expression for the probability of E in the canonical ensemble:

$$p^{\mathrm{eq}}(E) = \sum_{\xi} p_{\xi}^{\mathrm{eq}} \delta(\epsilon_{\xi} - E) = e^{(F-E)/k_{\mathrm{B}}T} \sum_{\xi} \delta(\epsilon_{\xi} - E)$$
$$= \exp\left[-\frac{E - TS(E) - F}{k_{\mathrm{B}}T} \right]. \tag{2.59}$$

The argument of the exponential on the right-hand side of eq. (2.59) is the difference between the nonequilibrium free energy F^{neq} with the given value of E and the equilibrium free energy F. This reasoning can be extended to multiple observables. For example, if we look for the joint probability distribution of E and an arbitrary macroscopic observable A, we obtain

$$p^{\mathrm{eq}}(E, A) = \exp\left[-\frac{E - TS(E, A) - F}{k_{\mathrm{B}}T} \right], \tag{2.60}$$

where

$$S(E, A) = k_{\mathrm{B}} \ln \sum_{\xi} \delta(\epsilon_{\xi} - E)\, \delta(a_{\xi} - A) \tag{2.61}$$

is the entropy of a constrained microcanonical ensemble in which the values of E and of A are both fixed. In this way, the evaluation of entropy can be used to estimate probabilities.

2.5 Stochastic dynamics

In stochastic thermodynamics, we study the dynamics of mesoscopic physical systems subject to random interactions with a heat reservoir. Because of this source of randomness, at a given time t a system can be found in a given discrete state x with probability $p_x(t)$. If the variable x is continuous, we denote by $p(x; t)$ the probability density of finding the system in x at time t.

The distribution $p_x(t)$ is just one way of describing the dynamics. Another way is to study trajectories of the system:

$$\boldsymbol{x} = (x(t)). \tag{2.62}$$

This notation means that $\boldsymbol{x}$ in boldface (the trajectory) identifies the whole function $x(t)$ over a given time interval. In the following, we often use this notation to distinguish the whole function $\boldsymbol{x}$ from its instantaneous value $x(t)$. Trajectories of a stochastic system are characterized by some degree of randomness. We refer to the dynamics of a stochastic system, described either in terms of time-dependent probability distributions or in terms of trajectories, as a **stochastic process**.

Many stochastic processes of physical interest possess a useful simplifying property called the **Markov property**. A Markov process is a stochastic process that has a finite memory, i.e., one in which knowledge of the recent past fully determines the statistics of the system in the present. Given an increasing sequence of time instants $(t_0, t_1, \ldots, t_n, t)$, we denote by $p_{x;t|x_n;t_n,x_{n-1};t_{n-1},\ldots,x_1;t_1,x_0;t_0}$ the conditional probability that the system is in a discrete state x at time t, given that it was in state x_n at time

t_n, in state x_{n-1} at time t_{n-1}, ..., and in state x_0 at time t_0. For any increasing sequence of time instants, a Markov process satisfies the condition

$$p_{x;t|x_n;t_n,x_{n-1};t_{n-1},\ldots,x_1;t_1,x_0;t_0} = p_{x;t|x_n;t_n}. \tag{2.63}$$

This means that, in a Markov process, given the value of x at a given time t_n, the evolution of the system at a later time is independent of events that occurred at earlier times. The same definition of a Markov process holds if x is continuous. The conditional probability density $p_{x;t|x_n;t_n}$ of a Markov process satisfies a simple relation. We pick an intermediate time t' such that $t_0 < t' < t$. By the law of total probabilities (A.35) and by the Markov property (2.63), we obtain

$$p_{x;t|x_0;t_0} = \sum_{x'} p_{x;t|x';t'} \, p_{x';t'|x_0;t_0}. \tag{2.64}$$

Equation (2.64) is called the **Chapman-Kolmogorov** equation. Therefore, the knowledge of $p_{x;t|x';t'}$ and of the distribution $p_x(t_0)$ at an initial time t_0 allows one to evaluate the distribution $p_x(t)$ at arbitrary times $t > t_0$. The conditional probability $p_{x;t|x';t'}$ is also called the **propagator** or **Green function** of the process. Conservation of probability requires

$$\sum_x p_{x;t|x';t'} = 1, \qquad \forall x' \text{ and } \forall t' \le t. \tag{2.65}$$

Applying the Chapman-Kolmogorov equation (2.64) to a time interval of infinitesimal duration dt, we obtain

$$p_x(t + dt) = \sum_{x'} p_{x;t+dt|x';t} \, p_{x'}(t), \tag{2.66}$$

where $p_{x;t+dt|x';t}$ can be written in the form

$$p_{x;t+dt|x';t} = \delta_{xx'}^{\mathrm{K}} + dt \, L_{xx'}(t), \tag{2.67}$$

where $\delta_{xx'}^{\mathrm{K}}$ is the Kronecker delta and $L_{xx'}(t)$ is a matrix that satisfies

$$L_{xx'}(t) \ge 0, \qquad\qquad \text{if } x \ne x';$$
$$L_{xx}(t) = - \sum_{x'\,(\ne x)} L_{x'x}, \qquad \forall x. \tag{2.68}$$

The matrix $L_{xx'}$ is called the **generator** of the process. Thus $p_x(t)$ satisfies an evolution equation of the form

$$\frac{dp_x(t)}{dt} = \sum_{x'} L_{xx'}(t) \, p_{x'}(t), \tag{2.69}$$

which has the solution

$$p_x(t) = \sum_{x'} G_{xx'}(t, t') \, p_{x'}(t'), \tag{2.70}$$

where the Green function is obtained by formally integrating eq. (2.69) with the initial condition $G_{xx'}(t', t') = \delta^{\mathrm{K}}_{xx'}$:

$$G_{xx'}(t, t') = p_{x;t|x';t'} = \left(\mathbb{T}\, e^{\int_{t'}^{t} dt''\, L(t'')}\right)_{xx'}. \tag{2.71}$$

Here (expression)$_{xx'}$ denotes the matrix element of expression, and $\mathbb{T}$ denotes the time-ordered product:

$$\left(\mathbb{T}\, e^{\int_{t'}^{t} dt''\, L(t'')}\right)_{xx'} = \delta^{\mathrm{K}}_{xx'} + \int_{t'}^{t} dt_0\, L_{xx'}(t_0)$$
$$+ \int_{t_0}^{t} dt_1 \int_{t'}^{t} dt_0 \sum_{x_0} L_{xx_0}(t_1) L_{x_0 x'}(t_0) + \cdots. \tag{2.72}$$

In the case where the generator does not depend on time, the Green function depends only on the time difference $(t - t')$.

Markov processes with continuous state space are also defined by the condition (2.63). In this case, we denote the propagator by $p(x; t|x'; t')$, and the Chapman-Kolmogorov equation takes the form

$$p(x; t|x_0; t_0) = \int dx'\, p(x; t|x'; t')\, p(x'; t'|x_0; t_0). \tag{2.73}$$

The evolution equation now reads

$$\frac{\partial}{\partial t} p = \mathcal{L}\, p, \tag{2.74}$$

where the generator $\mathcal{L}$ is a linear operator that in general includes derivatives with respect to x. In the next three sections, we discuss separately, and in more detail, Markov processes with discrete and continuous state space.

2.6 Master equations

Master equations describe the evolution of Markov processes in continuous time with discrete states. A master equation is defined by the **jump rates** $k_{xx'}$ from discrete state x' to x. The jump rate is proportional to the conditional probability that a jump $x' \longrightarrow x$ takes place in an infinitesimal time interval $(t, t + dt)$, given that the system is in state x' at time t. Specifically, the jump rates (or simply *rates*) are related to the propagator by

$$p_{x;t+dt|x';t} = k_{xx'}\, dt, \qquad x \neq x'. \tag{2.75}$$

The rates can in principle depend on time. Because of the normalization condition, eq. (2.65), the probability of remaining in a given state x in an infinitesimal time interval must be equal to $1 - dt \sum_{x'} k_{x'x} = 1 - dt\, k^{\mathrm{out}}_x$, where we define the **escape rate** from state x:

$$k^{\mathrm{out}}_x = \sum_{x'} k_{x'x}. \tag{2.76}$$

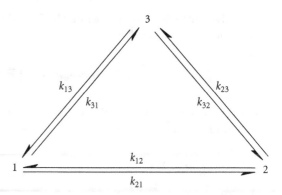

Figure 2.1. Jump network of a system with three states.

Given the rates, we construct the evolution equation for the probability $p_x(t)$ by evaluating the net probability flow reaching the state x. A state x receives an **inflow** of probability from other states at rate $\sum_{x'\,(\neq x)} k_{xx'} p_{x'}(t)$, and returns an **outflow** at rate $\sum_{x'\,(\neq x)} k_{x'x} p_x(t)$. The net probability flow is given by the inflow minus the outflow, leading to the **master equation**

$$\frac{\mathrm{d}}{\mathrm{d}t} p_x(t) = \sum_{x'\,(\neq x)} \left[k_{xx'}\, p_{x'}(t) - k_{x'x}\, p_x(t) \right]. \tag{2.77}$$

We now assume that the rates $k_{xx'}$ are independent of time, and look at the behavior of the probability distribution $p_x(t)$ for $t \to \infty$. A **stationary** distribution p_x^{st} is a probability distribution that satisfies

$$\sum_{x'\,(\neq x)} \left[k_{xx'}\, p_{x'}^{\mathrm{st}} - k_{x'x}\, p_x^{\mathrm{st}} \right] = 0, \qquad \forall x, \tag{2.78}$$

and that is therefore a solution of the master equation (2.77) constant in time.

It is useful to represent a master equation via a **jump network**, where the nodes represent the states x and the arrows $x' \longrightarrow x'$ represent possible jumps, i.e., jumps with nonvanishing rates. We draw two opposite arrows between a pair of states if jumps in both directions are possible. An example of a jump network with three states is shown in fig. 2.1, where each arrow $x' \longrightarrow x$ is associated with a nonzero rate $k_{xx'}$.

We say that a jump network is **connected** if any states x can be reached from any other state x' by means of a sequence of jumps with nonvanishing rates. In the following, we always assume that this property holds, since disconnected master equations represent multiple noninteracting physical systems that can be studied independently. Under such assumptions, and provided that the number of states is finite, the **Perron-Frobenius theorem** asserts that the leading eigenvalue, i.e., the eigenvalue with the largest real part, is purely real and nondegenerate. Moreover, its associated eigenvector can be chosen to have strictly positive entries. In the case of master equations with a finite number of states, the leading eigenvalue must be zero, otherwise the equation would not preserve normalization. The normalized eigenvector can therefore be interpreted as the stationary probability distribution. If the number of states is infinite, the stationary distribution may not exist: this is the case, for example, of a particle diffusing on an infinite line, when the rates of jumps to the left or to the right are independent of its position. If the number of states is finite, the real parts of all eigenvalues except the leading one are negative, and therefore the stationary distribution is approached exponentially fast in time:

$$\lim_{t \to \infty} p_x(t) = p_x^{st}, \quad \forall x. \tag{2.79}$$

The Perron-Frobenius theorem is not limited to generators of Markov processes. In particular, it does not require the generator to preserve normalization, i.e., that its columns sum to one. An elementary proof of these properties that does not explicitly rely on the Perron-Frobenius theorem is reported in appendix A.5.

The master equation can be seen as a continuity equation for the probability. This interpretation becomes more transparent by introducing the **probability current**

$$J_{xx'}(t) = k_{xx'} p_{x'}(t) - k_{x'x} p_x(t), \quad x \neq x', \tag{2.80}$$

which quantifies the net probability flow from state x' to state x at time t. In terms of the probability currents, the master equation can be rewritten in the compact form

$$\frac{d}{dt} p_x(t) = \sum_{x' (\neq x)} J_{xx'}(t). \tag{2.81}$$

Equation (2.81) states that, at any given time, the rate change of the probability of being in a state x is given by the total net flow to state x from all other states x'. If the probability distribution p is the stationary one, we have

$$\sum_{x' (\neq x)} J_{xx'} = 0, \quad \forall x. \tag{2.82}$$

If for any allowed jump $x \longrightarrow x'$ (i.e., such that $k_{x'x} > 0$) the reverse jump is also allowed ($k_{xx'} > 0$), the system exhibits **microscopic reversibility**. Most of the systems studied in stochastic thermodynamics possess this property. In this case, we often represent the jump network by a graph in which each possible jump between two states is represented by an undirected **edge** between the corresponding nodes. We discuss in section 8.5 how to handle microscopically irreversible systems.

If the stationary distribution satisfies, beyond eq. (2.78), the stronger condition

$$k_{xx'} p_{x'}^{st} = k_{x'x} p_x^{st}, \quad \forall x \neq x', \tag{2.83}$$

then in the stationary state the probability current $J_{xx'}$ vanishes for each pair (x, x') of different states. The condition in eq. (2.83) is known as the **detailed balance condition**. If it is satisfied, the stationary distribution is called the **equilibrium distribution** and we denote it by p^{eq}. A master equation admits an equilibrium distribution if, for any sequence $(x_0, x_1, \ldots, x_n)$ of states all different from one another, we have

$$k_{x_0 x_1} k_{x_1 x_2} \cdots k_{x_n x_0} = k_{x_0 x_n} k_{x_n x_{n-1}} \cdots k_{x_1 x_0}. \tag{2.84}$$

This condition requires in particular that microscopic reversibility is satisfied. The condition of eq. (2.84) might seem obscure at first, but it becomes clearer in the context of a specific jump network as in fig. 2.1. Because of the conservation of probability, the sum of currents arriving at each node must vanish in the stationary state; see eq. (2.82). This implies that nonvanishing stationary currents can only flow in loops. This requires in particular that the jump network contains a sequence of distinct nodes, each of them connected to the previous one by an edge, and where an edge connects the last one to the first. This sequence defines a **cycle** in the network. A nonvanishing current in a cycle can only be sustained if the rates "pushing" the currents clockwise and counterclockwise do not balance. The balance conditions for the rates along each cycle of the network are indeed given by eq. (2.84). This also means that, if there are no cycles (e.g., if the network is linear or treelike), then the stationary solution always satisfies detailed balance conditions, provided that microscopic reversibility holds. When the condition (2.84) is not satisfied for some cycles, the probability currents along these cycles do not vanish in the stationary state. In this case, the stationary distribution is also called a **nonequilibrium steady state**.

In thermodynamic systems, the condition of detailed balance is often associated with thermodynamic equilibrium, and the stationary distribution p_x^{eq} appearing in eq. (2.83) is the Boltzmann distribution, eq. (2.42). We further discuss this point in section 3.1. Then, assuming that $k_{x'x} > 0$, eq. (2.83) implies that

$$\frac{k_{xx'}}{k_{x'x}} = \frac{p_x^{eq}}{p_{x'}^{eq}} = e^{-(\epsilon_x - \epsilon_{x'})/k_B T}. \tag{2.85}$$

Although the master equation (2.77) is linear, solving it explicitly can be difficult when the number of states is large. To tackle it numerically, it is convenient to simulate an ensemble of random trajectories rather than integrating the master equation itself. Trajectories of a master equation can be very efficiently simulated by means of the **Gillespie algorithm**, which is briefly described in appendix A.6.

2.7 Trajectories of master equations

During its evolution in a time interval $[t_0, t_f]$, a system described by a master equation visits a sequence $x_0, x_1, \ldots, x_f$ of states (fig. 2.2). We call t_k the random time at which the system jumps from state x_{k-1} to state $x_k \neq x_{k-1}$. The system is in state x_k during a time interval $t_k \leq t < t_{k+1}$. Thus the trajectory x in the given time interval is made of a sequence of **dwells**, where the system remains in the same state, separated by **jumps**, where the system changes state, as shown in fig. 2.2. The whole trajectory x is then encoded in the sequence

$$x = ((x_0, t_0), (x_1, t_1), \ldots, (x_f, t_n), t_f). \tag{2.86}$$

We wish to evaluate the probability density $\mathcal{P}_x$ of the trajectory x. To this aim, we discretize the time interval into $\mathcal{N}$ intervals of short duration $\Delta t = \mathcal{T}/\mathcal{N}$, where $\mathcal{T} = t_f - t_0$ is the duration of the whole time interval. We then approximate the trajectory by the sequence $x^{dsc} = (x_0, x_1, \ldots, x_{\mathcal{N}})$ of states at the time t_ℓ, where $t_\ell = t_0 + \ell \Delta t$. For each small interval $(t_\ell, t_\ell + \Delta t)$, the conditional probability that $x(t_\ell + \Delta t) = x$, given that $x(t_\ell) = x'$, is given by $p_{x;t+\Delta t|x';t}$. The probability of the discrete trajectory is then

$$\mathcal{P}_{x^{dsc}|x_0} = \prod_{\ell=1}^{\mathcal{N}} p_{x_\ell;t_{\ell-1}+\Delta t|x_{\ell-1};t_{\ell-1}}. \tag{2.87}$$

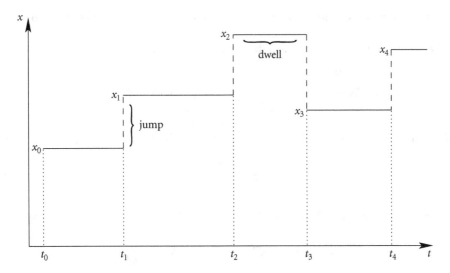

Figure 2.2. Example of a trajectory of a master equation.

From the master equation, we approximate the conditional probability by

$$p_{x;t+\Delta t|x';t} \approx \delta^{K}_{xx'} + \Delta t\, L_{xx'}(t), \tag{2.88}$$

where $L_{xx'}(t)$ is the generator. This expression yields conditional probabilities that are properly normalized. Therefore, the probability of the trajectory x is approximately equal to

$$\mathcal{P}_{x|x_0} \approx \mathcal{P}_{x^{\mathrm{dsc}}|x_0} \approx \prod_{\ell=1}^{\mathcal{N}} \left(\delta^{K}_{x_\ell x_{\ell-1}} + \Delta t\, L_{x_\ell x_{\ell-1}}(t_{\ell-1}) \right). \tag{2.89}$$

On the one hand, we can explicitly evaluate the products over each dwell:

$$\prod_{\ell \in \mathrm{dwell}} p_{x_\ell;t_{\ell-1}+dt|x_\ell;t_{\ell-1}} = \prod_{\ell} \left(1 - dt_\ell\, k^{\mathrm{out}}_{x_\ell} \right) \approx e^{-\sum_\ell k^{\mathrm{out}}_{x_\ell}(t)\, dt_\ell} \approx e^{-\int dt\, k^{\mathrm{out}}_{x(t)}(t)}. \tag{2.90}$$

On the other hand, the probability that the system undergoes a jump from x' to x in the short time interval $[t_\ell, t_\ell + dt]$ is given by $k_{xx'}\, dt\, p_{x'}(t_\ell)$ to first order in dt. Therefore, the factors contributing to the probability of a trajectory x due to the sequence of dwells and jumps are

$$\mathcal{P}_x = e^{-\int_{t_n}^{t_f} dt'\, k^{\mathrm{out}}_{x_n}(t')} k_{x_n x_{n-1}}(t_n)\, e^{-\int_{t_n}^{t_{n-1}} dt'\, k^{\mathrm{out}}_{x_{n-1}}(t')} \cdots$$
$$\times\, e^{-\int_{t_2}^{t_1} dt'\, k^{\mathrm{out}}_{x_1}(t')} k_{x_1 x_0}(t_1)\, e^{-\int_{t_1}^{t_0} dt'\, k^{\mathrm{out}}_{x_0}(t')} p_{x_0}(t_0). \tag{2.91}$$

This expression includes the probability of the initial state $p_{x_0}(t_0)$. The choice of writing the factors on the right-hand side of eq. (2.91) in temporal order from right to left might seem awkward for a reader used to writing in the Latin alphabet from left to right. However, it becomes quite natural when thinking of the probability density as a product of matrix elements, one for each infinitesimal time step, acting on the initial probability $p_{x_0}(t_0)$.

We define the integral over trajectories in the following way:

$$\int \mathcal{D}x \cdots = \sum_{n=0}^{\infty} \sum_{x_0, x_1, \ldots, x_n} \int_{t_0}^{t_2} dt_1 \int_{t_2}^{t_3} dt_2 \cdots \int_{t_{n-1}}^{t_f} dt_n \cdots . \tag{2.92}$$

With this definition, the probability density $\mathcal{P}_x$ satisfies the normalization condition

$$\int \mathcal{D}x \, \mathcal{P}_x = 1. \tag{2.93}$$

Indeed, comparing eqs. (2.91) and (2.92) with eqs. (2.71) and (2.72), we obtain

$$\int \mathcal{D}x \, \mathcal{P}_x = \sum_{x_f x_0} \left(\mathbb{T} \, e^{\int_{t_0}^{t_f} dt \, L(t)} \right)_{x_f x_0} p_{x_0}(t_0)$$

$$= \sum_{x_f x_0} P_{x_f; t_f | x_0; t_0} \, p_{x_0}(t_0) = 1. \tag{2.94}$$

2.8 Fokker-Planck equation (*)

The stochastic dynamics of systems with continuous state space and continuous trajectories is described by the Fokker-Planck equation and by stochastic differential equations. These tools were introduced in physics to investigate **Brownian motion**. Einstein described Brownian motion as the phenomenon by which "bodies of microscopically visible size suspended in a liquid perform [random] movements of such magnitude that they can be easily observed in a microscope." We follow his line of reasoning to derive the Fokker-Planck equation.

We consider a particle that moves along a one-dimensional continuous coordinate x. We call $p(x; t)$ the probability density of the position x of a particle at time t. We first assume that the particle is immersed in a uniform fluid and is not subject to external applied forces. Due to random interactions with the fluid particles, during a short time interval of duration Δt, the particle experiences a random displacement Δx. If Δt is very small, the displacement Δx is largely in the direction of the initial velocity of the particle due to inertia. However, interactions with the particles of the fluid soon wipe out this dependence. Therefore, if Δt is large enough but still small, we expect this displacement to be independent in nonoverlapping time intervals. We call $\psi(\Delta x)$ the probability distribution of Δx over these time intervals. In the absence of externally applied forces, $\psi(\Delta x)$ must be an even function of Δx due to symmetry. We express $p(x; t + \Delta t)$ in terms of $p(x; t)$ and the displacement distribution

$$p(x; t + \Delta t) = \int d\Delta x \, \psi(\Delta x) \, p(x - \Delta x; t). \tag{2.95}$$

Since Δt is small, Δx is also small. We therefore expand $p(x - \Delta x; t)$ in a Taylor series to second order:

$$p(x; t + \Delta t) \approx \int d\Delta x \, \psi(\Delta x) \left[p(x; t) - \Delta x \frac{\partial}{\partial x} p(x; t) + \frac{1}{2} \Delta x^2 \frac{\partial^2}{\partial x^2} p(x; t) \right]. \quad (2.96)$$

We then perform the integration over Δx. Since $\psi(\Delta x)$ is an even function, the term proportional to Δx vanishes upon integration. We thus obtain

$$p(x; t + \Delta t) - p(x; t) \approx \frac{1}{2} \langle \Delta x^2 \rangle \frac{\partial^2}{\partial x^2} p(x; t). \quad (2.97)$$

We now divide eq. (2.97) by Δt and take the limit $\Delta t \to 0$. In taking this limit, we make the crucial assumption that

$$\lim_{\Delta t \to 0} \frac{\langle \Delta x^2 \rangle}{2 \Delta t} = D, \quad (2.98)$$

where D is a finite quantity called the **diffusion constant**. If this assumption holds, we obtain the **diffusion equation**:

$$\frac{\partial}{\partial t} p(x; t) = D \frac{\partial^2}{\partial x^2} p(x; t). \quad (2.99)$$

Starting from a localized initial condition $p(x; t_0) = \delta(x - x_0)$, the solution of the diffusion equation is a Gaussian distribution,

$$p(x; t) = \frac{1}{\sqrt{4\pi \, D\mathcal{T}}} \exp\left(-\frac{(x - x_0)^2}{4 \, D\mathcal{T}} \right), \quad (2.100)$$

where $\mathcal{T} = t - t_0$.

In a more general case, for instance when the particle is subject to forces, the distribution of the displacements Δx might depend on x, t, and might not necessarily be even. We express it as

$$p(\Delta x; \Delta t, x, t) = p(x + \Delta x, t + \Delta t | x, t). \quad (2.101)$$

We assume that the following limits exist and define the **drift** and **diffusion coefficients**, respectively, by

$$v(x, t) = \lim_{\Delta t \to 0} \frac{\langle \Delta x \rangle_{x,t}}{\Delta t}; \quad (2.102a)$$

$$D(x, t) = \lim_{\Delta t \to 0} \frac{\langle \Delta x^2 \rangle_{x,t}}{2 \, \Delta t}. \quad (2.102b)$$

The averages $\langle \ldots \rangle_x$ are taken over the distribution of displacements. Then the dynamics is described by the **Fokker-Planck equation** (also called the **Kolmogorov forward equation**):

$$\frac{\partial}{\partial t} p(x; t) = -\frac{\partial}{\partial x} \left[v(x, t) \, p(x; t) \right] + \frac{\partial^2}{\partial x^2} \left[D(x, t) \, p(x; t) \right]. \quad (2.103)$$

Equation (2.103) is derived in appendix A.7. The first term in eq. (2.103) is the drift and is associated with the local mean velocity of particles. The second term is the diffusion, which represents the "random," unbiased component of the motion.

The Fokker-Planck equation can also be written as a continuity equation,

$$\frac{\partial}{\partial t} p(x; t) = -\frac{\partial}{\partial x} J(x; t), \tag{2.104}$$

where the probability current is defined by

$$J(x, t) = v(x, t)\, p(x; t) - \frac{\partial}{\partial x}\left[D(x, t)\, p(x; t) \right]. \tag{2.105}$$

The Fokker-Planck equation is a second-order partial differential equation. To solve it, we need to specify the initial condition, i.e., the distribution $p(x; t_0)$, and the **boundary conditions**. Often the Fokker-Planck equation is studied in the infinite interval $x \in (-\infty, \infty)$, for which the boundary conditions are simply that the probability vanishes as $|x| \to \infty$.

The solution $p(x; t|x_0; t_0)$ of the Fokker-Planck equation with initial condition $\delta(x - x_0)$ at $t = t_0$ is the **propagator**, which expresses the conditional probability density of finding the particle close to x at time $t > t_0$, given that it was at x_0 at time t_0. In some cases, one is interested in the dependence of this probability density on the earlier state x_0 and the earlier time t_0. This dependence is governed by the **Kolmogorov backward equation**:

$$-\frac{\partial}{\partial t_0} p(x; t|x_0; t_0) = v(x_0, t_0)\, \frac{\partial}{\partial x_0} p + D(x_0, t_0)\, \frac{\partial^2}{\partial x_0^2} p. \tag{2.106}$$

We derive this equation in appendix A.7.

We now consider cases where the coefficients $v(x, t)$ and $D(x, t)$ of the Fokker-Planck equation do not depend on time. The stationarity condition reads

$$\frac{\partial}{\partial x} J^{st}(x) = 0, \tag{2.107}$$

where the stationary current is

$$J^{st}(x) = v(x)\, p^{st}(x) - \frac{\partial}{\partial x}\left[D(x)\, p^{st}(x) \right]. \tag{2.108}$$

When considering an infinite interval, the stationary solution does not necessarily exist. A prominent example is the diffusion equation (2.99), whose solution (2.100) does not tend to a limiting form as $\mathcal{T} \to \infty$.

In some cases, the Fokker-Planck equation also admits a stationary solution with a vanishing stationary current:

$$J^{st}(x) = 0, \qquad \forall x. \tag{2.109}$$

Equation (2.109) is the detailed balance condition for Fokker-Planck equations. In general, detailed balance is a rather restrictive condition. However, there are physically relevant scenarios where this property holds, such as many one-dimensional systems and most systems at thermodynamic equilibrium, as discussed at the end of section 2.6. In the one-dimensional case, when detailed balance holds, the stationary solution is given by

$$p^{st}(x) \propto \frac{1}{D(x)} \exp\left(\int_{x_0}^{x} dx' \frac{v(x')}{D(x')}\right), \tag{2.110}$$

where x_0 is arbitrary and the proportionality constant is determined by the normalization condition $\int_{-\infty}^{\infty} dx\, p^{st}(x) = 1$.

2.9 Langevin equation (*)

Trajectories of a Brownian particle with drift coefficient $v(x, t)$ and diffusion coefficient $D(x, t)$ are solutions of the **Langevin equation**

$$\frac{dx}{dt} = v(x, t) + \sigma(x, t)\xi(t), \tag{2.111}$$

where $\sigma(x, t)$ is a function related to $D(x, t)$ and $\xi(t)$ is a random quantity. We impose that this random quantity is unbiased, $\langle \xi(t) \rangle = 0$, $\forall t$. We also assume that displacements in nonoverlapping time intervals are uncorrelated. This amounts to requiring $\langle \xi(t)\xi(t') \rangle = 0$ for t and t' sufficiently far apart. Assuming that this holds whenever $t \neq t'$, we obtain $\langle \xi(t)\xi(t') \rangle \propto \delta(t - t')$. We set the proportionality constant equal to 1 by suitably defining $\sigma(x, t)$. A variable $\xi(t)$ satisfying these properties is called **white noise**. In the mathematical literature, Langevin equations are an example of **stochastic differential equations**.

Equation (2.111) is ill defined as it stands, since the variance of the random quantity $\xi(t)$ is a delta function in time. The mathematical tools to deal with stochastic differential equations were developed well after the work of Langevin and are the subject of the theory of **stochastic calculus**. Stochastic calculus is nowadays a rather developed field of mathematics. In this section, we concentrate on the most important and useful results without going too deep into the mathematical details.

We formally evaluate the time integral of $\xi(t)$:

$$W(t) = \int_{t_0}^{t} dt'\, \xi(t') = \int_{t_0}^{t} dW. \tag{2.112}$$

The random process $W(t)$ is called the **Wiener process**. It has vanishing average

$$\langle W(t) \rangle = \left\langle \int_{t_0}^{t} dt'\, \xi(t') \right\rangle = \int_{t_0}^{t} dt'\, \langle \xi(t') \rangle = 0. \tag{2.113}$$

Since $\xi(t)$ is unbiased on average, so is its time integral $W(t)$. The second moment of $W(t)$ is given by

$$\begin{aligned}
\langle W^2(t) \rangle &= \left\langle \int_{t_0}^{t} dt' \int_{t_0}^{t} dt''\, \xi(t')\xi(t'') \right\rangle = \int_{t_0}^{t} dt' \int_{t_0}^{t} dt''\, \langle \xi(t')\xi(t'') \rangle \\
&= \int_{t_0}^{t} dt' \int_{t_0}^{t} dt''\, \delta(t' - t'') = \mathcal{T},
\end{aligned} \tag{2.114}$$

where $\mathcal{T} = t - t_0$. In general, it can be shown that the distribution of $W(t)$ is Gaussian:

$$p(W(t) = w) = \frac{1}{\sqrt{2\pi\,\mathcal{T}}}\,e^{-w^2/2\mathcal{T}}. \tag{2.115}$$

Mathematicians prefer to formally define $W(t)$ as a process whose increments over a time interval of duration $\mathcal{T}$ are Gaussian random variables with zero mean and variance $\mathcal{T}$. They allow physicists to use $\xi(t)$ as an ill-defined "derivative" of $W(t)$.

We now consider a Langevin equation without drift:

$$\frac{dx}{dt} = \sigma\,\xi(t), \tag{2.116}$$

where we assume that the coefficient σ is constant. Integrating from $t = t_0$ to a generic time t, we obtain

$$x(t) = x_0 + \sigma\,W(t). \tag{2.117}$$

Thus the distribution of $x(t)$ is a Gaussian centered in x_0 and with variance $\sigma^2\left\langle W^2(t)\right\rangle = \sigma^2\mathcal{T}$. By comparing with the diffusion equation, eq. (2.99) and its solution, we conclude that this Langevin equation describes the trajectories of a diffusion equation with a diffusion coefficient

$$D = \sigma^2/2. \tag{2.118}$$

By going through the derivation of the Fokker-Planck equation in section 2.8, one sees that the solution of the Langevin equation (2.111) with constant σ describes a process satisfying the Fokker-Planck equation (2.103) with the same drift $v(x, t)$ and with diffusion coefficient $D = \sigma^2/2$.

Processes where σ depends on x are often called Langevin equations with **multiplicative noise** in the physics literature. This term is somewhat misleading as it appears to be limited to the case in which $\sigma \propto x$. In the presence of multiplicative noise, the interpretation of eq. (2.111) is mathematically subtle. Depending on its interpretation, it may or may not represent the behavior of the solutions of eq. (2.103). This is due to the fact that, for short time intervals dt, the typical increments of W are much larger than dt, and thus the second term of the Langevin equation dominates over the first one.

Formally, solving a Langevin equation requires the evaluation of **stochastic integrals** of the form

$$\int_{t_0}^{t_f} dt\,\xi(t)\,f(x(t), t) = \int_{t_0}^{t_f} dW(t)\,f(x(t), t), \tag{2.119}$$

where $f(x, t)$ is a given function. The ambiguity in the interpretation of the Langevin equation stems from the ill-defined nature of stochastic integrals. There are two major conventions to resolve this ambiguity and therefore to assign a precise interpretation to the corresponding Langevin equation:

Ito convention. The Ito convention is defined by the prescription

$$I_I = \lim_{dt \to 0} \sum_{k=0}^{\mathcal{N}} [W(t_k + dt) - W(t_k)] f(x(t_k), t_k), \tag{2.120}$$

where $t_i = t_0 + i\,dt$ and the sum runs over the $\mathcal{N} = (t_f - t_0)/dt$ intervals of duration dt in which we divide the interval $[t_0, t_f]$. In the Ito convention, the function $f(x(t), t)$ is evaluated at the beginning of each infinitesimal time interval $[t_k, t_k + dt]$. We denote the Ito convention by a dot product symbol:

$$I_I = \int_{t_0}^{t_f} dW \cdot f(x(t'), t'). \tag{2.121}$$

Stratonovich convention. The Stratonovich convention is defined by

$$I_S = \lim_{dt \to 0} \sum_{k=0}^{\mathcal{N}} [W(t_k + dt) - W(t_k)] f\left(\frac{x(t_k + dt) + x(t_k)}{2}, t_k + \frac{dt}{2}\right)$$

$$= \lim_{dt \to 0} \sum_{k=0}^{N} [W(t_k + dt) - W(t_k)] \frac{1}{2} \left[f(x(t_k + dt), t_k + dt) + f(x(t_k), t_k)\right]. \tag{2.122}$$

In the Stratonovich convention, $f(x, t)$ is evaluated at the midpoint of each infinitesimal interval $[t_k, t_k + dt]$. We can equivalently evaluate the function $f(x, t)$ as the average of its values at the boundaries of the interval: the difference between these two prescriptions vanishes as $dt \to 0$. We denote the Stratonovich convention by a circle product symbol:

$$I_S = \int_{t_0}^{t_f} dW \circ f(x(t'), t'). \tag{2.123}$$

In conventional (nonstochastic) integrals, the choices (2.120) and (2.122) yield the same result as $dt \to 0$. This is not necessarily the case for stochastic integrals. Therefore, a Langevin equation is not fully defined unless one declares whether stochastic integrals are interpreted according to the discretization of eq. (2.120) or (2.122).

The Langevin equation (2.111) is equivalent to the Fokker-Planck equation (2.103) under the Ito convention. As shown in appendix A.8, the Fokker-Planck equation corresponding to the Stratonovich convention looks slightly different:

$$\frac{\partial}{\partial t} p(x; t) = -\frac{\partial}{\partial x}\left(v(x, t)\, p(x; t)\right) + \frac{1}{2}\frac{\partial}{\partial x}\left[\sigma(x, t)\frac{\partial}{\partial x}\left(\sigma(x, t)\, p(x; t)\right)\right]. \tag{2.124}$$

In this interpretation, the current is defined by

$$J(x, t) = w(x, t)\, p(x; t) - D(x, t)\frac{\partial}{\partial x} p(x; t), \tag{2.125}$$

where the diffusion coefficient is given by $D(x, t) = \sigma^2(x, t)/2$ as in the Ito convention, and

$$w(x, t) = v(x, t) - \frac{1}{2}\sigma(x, t)\frac{\partial}{\partial x}\sigma(x, t) = v(x, t) - \frac{1}{2}\frac{\partial}{\partial x}D(x, t). \tag{2.126}$$

When σ (or D) does not depend on x, eq. (2.124) reduces to eq. (2.103), confirming the equivalence of the two interpretations in this special case.

There are rules for transforming a Langevin equation interpreted in the Ito convention to an equivalent one interpreted in the Stratonovich convention and back. As shown in appendix A.8, one obtains the rule that

$$\frac{dx}{dt} = v(x,t) + \frac{1}{2}\sigma(x,t)\frac{\partial}{\partial x}\sigma(x,t) + \sigma(x,t)\xi \quad \text{(Ito)}$$

is equivalent to

$$\frac{dx}{dt} = v(x,t) + \sigma(x,t)\xi \quad \text{(Stratonovich)}. \tag{2.127}$$

The extra term necessary to change convention is often called the **noise-induced drift**.

Which convention is the most appropriate to describe a given natural phenomenon? The short answer is "it depends." One can show that, considering a noise source with a finite correlation time and then taking the limit of vanishing correlation time, the resulting Langevin equation is of the Stratonovich type. Therefore, Stratonovich is the correct interpretation for many systems affected by noise characterized by a negligible but finite correlation time. On the other hand, the evaluation of Stratonovich integrals requires knowledge of the "future," i.e., of the value of functions after the initial instant of each time step; see eq. (2.122). This aspect is unrealistic when the noise is generated by the physical process itself. In general, when modeling a system with multiplicative noise, it is safer to derive rigorously the Langevin equation and therefore to make sure that the interpretation is correct.

Traditionally, most physicists tend to prefer to work with the Stratonovich convention, whereas mathematicians tend to prefer the Ito convention. It is useful to be aware of the pros and cons of this choice. The Stratonovich convention has the advantage of respecting the chain rule, i.e., $d[f(x)]/dt = f'(x)\,dx/dt$, even if x follows a Langevin equation. In the case of Ito, assuming that $x(t)$ is the solution of eq. (2.111), the derivative df/dt of $f(x(t))$ is instead given by

$$\frac{df}{dt} = v(x,t)f'(x) + \frac{1}{2}\sigma^2(x,t)f''(x) + \sigma(x,t)f'(x)\xi. \tag{2.128}$$

This relation is known as the **Ito formula**. Therefore, the Ito approach introduces a complication that is not present in the Stratonovich one. It has however the advantage that expectations of stochastic integrals, such as

$$\left\langle \int dW \cdot f(x) \right\rangle,$$

always vanish in the Ito convention. This is in general not true for the Stratonovich convention, where such integrals must be evaluated on a case-by-case basis.

2.10 Information

It is often useful to quantify the "uncertainty" associated with a certain probability distribution over the states of a system $\mathcal{S}$. A measure of uncertainty must satisfy three reasonable properties: it must vanish when the system is known to be in one state, it must be maximal when the distribution is uniform over all possible states, and, if the system is made up of two independent systems, it must be the sum of the two uncertainties. Claude Shannon showed that the only function of the probabilities that satisfies these three requirements (up to a multiplicative constant) is the **Shannon entropy**

$$H(\mathcal{S}) = -\sum_x p_x \ln p_x, \tag{2.129}$$

where the sum runs over all the states of system $\mathcal{S}$. The larger $H(\mathcal{S})$, the larger the uncertainty about $\mathcal{S}$, up to a maximum $H(\mathcal{S}) = \ln \mathcal{N}$ when the probability is uniformly distributed over $\mathcal{N}$ states. With a slight abuse of notation, we interchangeably use $H(\mathcal{S})$ or $H(p)$ to denote the Shannon entropy of a system $\mathcal{S}$ having distribution $p = (p_x)$ over its states. In the latter case, we use the notation $p = (p_x)$ to denote by the symbol p the entire vector p_x. In the simple case of a binary variable ($x \in \{0, 1\}$) with equal probabilities $p_0 = p_1 = 1/2$, one has

$$H(\mathcal{S}) = \ln 2. \tag{2.130}$$

This quantity is called a **bit** of information.

If the system $\mathcal{S}$ is at thermal equilibrium, its thermodynamic entropy S is proportional to the Shannon entropy of its distribution over the microstates

$$S = k_B H(\mathcal{S}) = -k_B \sum_\xi p_\xi^{eq} \ln p_\xi^{eq}, \tag{2.131}$$

where k_B is the Boltzmann constant. Indeed, evaluating $H(\mathcal{S})$ with p^{eq} given by (2.45) leads to

$$H(\mathcal{S}) = -\sum_\xi p_\xi^{eq} \frac{F - \epsilon_\xi}{k_B T} = \frac{1}{k_B T} \left(\langle \epsilon \rangle^{eq} - F \right) = \frac{S}{k_B}, \tag{2.132}$$

since $F = \langle \epsilon \rangle^{eq} - TS$. Equation (2.131) is known as the **Gibbs relation**.

The dissimilarity between two probability distributions $p = (p_x)$ and $q = (q_x)$ over the states of the same system is measured by the **Kullback-Leibler divergence** (also called the **relative entropy**),

$$D_{KL}(p\|q) = \sum_x p_x \ln \frac{p_x}{q_x}. \tag{2.133}$$

The Kullback-Leibler divergence has the following properties:

- $D_{KL}(p\|q) \geq 0$;
- $D_{KL}(p\|q) = 0$ only if $p_x = q_x, \forall x$.

To prove these results, we consider that $-\ln x$ is a convex function of x and therefore satisfies the **Jensen inequality** $\langle -\ln x \rangle \geq -\ln \langle x \rangle$; see appendix A.1. We therefore obtain

$$D_{KL}(p\|q) = -\sum_x p_x \ln \frac{q_x}{p_x} \geq -\ln \sum_x p_x \frac{q_x}{p_x} = -\ln \sum_x q_x = -\ln 1 = 0. \tag{2.134}$$

In eq. (2.134), equality holds only if $q_x = p_x$ for all x, since $-\ln x$ is a strictly convex function. One has in general $D_{\mathrm{KL}}(p\|q) \neq D_{\mathrm{KL}}(q\|p)$. For instance, if $q_x > 0$, $\forall x$, while $p_x = 0$ for some x, $D_{\mathrm{KL}}(p\|q)$ is finite, but $D_{\mathrm{KL}}(q\|p)$ diverges.

We now consider two systems $\mathcal{S}_1$ and $\mathcal{S}_2$ whose states are respectively denoted by x and y. The information shared by the two systems is measured by the **mutual information**

$$I(\mathcal{S}_1 : \mathcal{S}_2) = \sum_{x,y} p_{x,y} \ln \frac{p_{x,y}}{p_x p_y}, \tag{2.135}$$

where $p_{x,y}$ is the joint probability distribution of the two systems and p_x, p_y are their **marginal distributions**; see appendix A.3. The mutual information can be seen as the Kullback-Leibler divergence between the joint distribution $p_{x,y}$ and the product of the marginal distributions p_x and p_y. Therefore, it is nonnegative and vanishes only if the two variables x and y are independent. The mutual information is symmetric: $I(\mathcal{S}_1 : \mathcal{S}_2) = I(\mathcal{S}_2 : \mathcal{S}_1)$. It can also be expressed in terms of the conditional probability distribution $p_{x|y}$ of the state of $\mathcal{S}_1$, given that of $\mathcal{S}_2$, by

$$I(\mathcal{S}_1 : \mathcal{S}_2) = \sum_{x,y} p_{x,y} \ln p_{x|y} - \sum_{x} p_x \ln p_x. \tag{2.136}$$

The second term on the right-hand side is the Shannon entropy $H(\mathcal{S}_1)$ of $\mathcal{S}_1$. The first term is minus the **conditional entropy** of $\mathcal{S}_1$, given $\mathcal{S}_2$:

$$H(\mathcal{S}_1|\mathcal{S}_2) = -\sum_{x,y} p_{x,y} \ln p_{x|y}. \tag{2.137}$$

When the two systems are statistically independent, i.e., $p_{x,y} = p_x p_y$, we have $H(\mathcal{S}_1|\mathcal{S}_2) = H(\mathcal{S}_1)$, and the mutual information vanishes. We moreover have

$$I(\mathcal{S}_1 : \mathcal{S}_2) = H(\mathcal{S}_1) - H(\mathcal{S}_1|\mathcal{S}_2) = H(\mathcal{S}_2) - H(\mathcal{S}_2|\mathcal{S}_1). \tag{2.138}$$

The Shannon entropy $H(\mathcal{S}_1, \mathcal{S}_2)$ of the joint distribution of the states of $\mathcal{S}_1$ and $\mathcal{S}_2$ is called the **joint entropy** of the two variables. It satisfies the relation

$$H(\mathcal{S}_1, \mathcal{S}_2) = H(\mathcal{S}_1) + H(\mathcal{S}_2|\mathcal{S}_1), \tag{2.139}$$

which is known as the **chain rule**. By exploiting (2.138), we also have

$$H(\mathcal{S}_1, \mathcal{S}_2) = H(\mathcal{S}_1) + H(\mathcal{S}_2) - I(\mathcal{S}_1 : \mathcal{S}_2). \tag{2.140}$$

Equation (2.140) clarifies that the mutual information quantifies the reduction of the uncertainty of the pair $(\mathcal{S}_1, \mathcal{S}_2)$ due to their mutual dependence. A similar relation holds for the Kullback-Leibler divergence:

$$D_{\mathrm{KL}}(p(\mathcal{S}_1, \mathcal{S}_2)\|q(\mathcal{S}_1, \mathcal{S}_2)) = D_{\mathrm{KL}}(p(\mathcal{S}_1)\|q(\mathcal{S}_1)) + D_{\mathrm{KL}}(p(\mathcal{S}_2|\mathcal{S}_1)\|q(\mathcal{S}_2\|\mathcal{S}_1)), \tag{2.141}$$

where

$$D_{\mathrm{KL}}(p(\mathcal{S}_2|\mathcal{S}_1)\|q(\mathcal{S}_2\|\mathcal{S}_1)) = \sum_{x,y} p_{x,y} \ln \frac{p_{x|y}}{q_{x|y}}. \tag{2.142}$$

This result can be verified by a direct calculation. An immediate consequence of eq. (2.141) is that the Kullback-Leibler divergence between two systems cannot increase if we average out some variables.

Given three systems $\mathcal{S}_1$, $\mathcal{S}_2$, and $\mathcal{S}_3$, whose states are respectively denoted by x, y, and z, the **conditional mutual information** between $\mathcal{S}_1$ and $\mathcal{S}_2$, given $\mathcal{S}_3$, is defined by

$$I(\mathcal{S}_1 : \mathcal{S}_2 | \mathcal{S}_3) = \sum_{x,y,z} p_{xyz} \ln \frac{p_{xy|z}}{p_{x|z} p_{y|z}}, \tag{2.143}$$

where $p_{x|z}$ and $p_{y|z}$ are conditional marginal distributions, i.e.,

$$p_{y|z} = \sum_x p_{xy|z}; \qquad p_{x|z} = \sum_y p_{x,y|z}. \tag{2.144}$$

The conditional mutual information $I(\mathcal{S}_1 : \mathcal{S}_2 | \mathcal{S}_3)$ is nonnegative and vanishes only if the joint conditional probability distribution of x and y, given z, factorizes into the product of the conditional marginal distributions. This happens, for instance, if $\mathcal{S}_1$ depends on $\mathcal{S}_2$ only via $\mathcal{S}_3$. The mutual information also satisfies a chain rule, which reads

$$I(\mathcal{S}_1, \mathcal{S}_2 : \mathcal{S}_3) = I(\mathcal{S}_1 : \mathcal{S}_3) + I(\mathcal{S}_2 : \mathcal{S}_3 | \mathcal{S}_1), \tag{2.145}$$

where $I(\mathcal{S}_1, \mathcal{S}_2 : \mathcal{S}_3)$ is the mutual information between $\mathcal{S}_1 \cup \mathcal{S}_2$ and $\mathcal{S}_3$. Derivation of eq. (2.145) is left as exercise 2.7.

2.11 Further reading

The aim of this chapter is limited to giving a bird's-eye perspective on theories that lie at the foundation of stochastic thermodynamics. Several books explain much more extensively the main concepts presented in each of the sections of this chapter, and sometimes choosing among the many classic references might be a matter of personal taste. Callen [28] and Pippard [130] are established textbooks in thermodynamics. Feynman's lectures in physics [54] provide an original introduction to thermodynamics and statistical mechanics, including ideas that, in hindsight, were seminal for stochastic thermodynamics. De Groot and Mazur [37] is a good reference on "nonstochastic" nonequilibrium thermodynamics. Landau et al. [95], Chandler [29], and Peliti [125] are classic references on statistical mechanics.

Equation (2.52) is a basic example of a fluctuation-dissipation relation. These relations are of paramount importance in nonequilibrium statistical physics, as discussed throughout this book. Marini Bettolo Marconi et al. [110] provide a comprehensive review on fluctuation-dissipation relations.

The basic concepts of stochastic processes were introduced into physics by the pioneering work of Einstein [44] on Brownian motion (English translation in [46]). Nowadays the theory of stochastic processes is rather well developed, even if it is still playing a relatively marginal role in many physics curricula. There are many excellent books on stochastic processes that present the theory at different mathematical levels and with slightly different angles. Berg [15] is a basic introduction that focuses on the physical concepts rather than on the mathematics. Gardiner [59], van Kampen [171], and Risken [139] are references of a more mathematical nature. Øksendal

[120] is an excellent, though even more advanced book. Other books focus on specific aspects of the theory of stochastic processes. One important example is first-passage time problems, discussed in Redner [138].

The seminal paper by Shannon [154] already contains the main ideas in information theory. Khinchin [87] provides a more systematic introduction to Shannon's theory. Cover and Thomas [32] and MacKay [105] present information theory in a more modern and extended way.

2.12 Exercises

2.1   Consider a master equation with three states, $x \in \{0, 1, 2\}$, and with constant rates $k_{xx'}$ $(x \neq x')$, none of which vanishes. Write down the explicit form of the master equation. Evaluate the steady-state probability distribution p_x^{st}, $x \in \{0, 1, 2\}$ and the corresponding probability current $J^{st} = k_{x'x}p_x^{st} - k_{xx'}p_{x'}^{st}$ with $x' = x + 1 \mod 3$. Derive the conditions on the rates k such that detailed balance is satisfied, i.e., such that J^{st} vanishes.

2.2   Consider a master equation with four states, $x \in \{0, 1, 2, 3\}$, and with rates $k_{xx'}$ $(\forall x \neq x')$, none of which vanishes. Show that if $k_{xx'} = k_{x'x}$, then the unique stationary distribution is $p_0 = p_1 = p_2 = p_3 = 1/4$. Show with an example that this is not necessarily the case if some of the rates vanish.

2.3   A collection of N white balls and N black balls, with $N \geq 3$, are randomly distributed in two urns, so that each contains N balls. At each step t, one ball is extracted from each urn; the two balls are swapped and put back in the urns. Denote by x_t the number of white balls in the first urn. Show that x_t is a Markov process and express its jump rates $k_{xx'} = p(x_{t+1} = x | x_t = x')$. Discuss whether the process satisfies detailed balance. Evaluate the stationary distribution p_x^{st} and the decay rate of the correlation function $\langle x_t x_0 \rangle - \langle x_0 \rangle^2$, where x_0 is drawn from the stationary distribution.

2.4   Consider an infinite sequence of independent, identically distributed real-valued random variables $y = (y_0, y_1, \ldots)$. Another sequence $x = (x_0, x_{1,2}, \ldots)$ is recursively defined by

$$x_0 = y_0; \qquad x_n = y_n + a\,x_{n-1},$$

where $0 < a < 1$.

a. Show that, if the distribution of y_ℓ is Gaussian, the distribution of x_ℓ approaches for $\ell \to \infty$ a stationary Gaussian distribution and evaluate its parameters.

b. Show that, if y_ℓ is *not* Gaussian distributed, the stationary distribution of x_ℓ, if it exists, is not Gaussian.

2.5   A Brownian particle is confined in a one-dimensional potential. Its position $x(t)$ satisfies the following Langevin equation in the Stratonovich representation:

$$\frac{dx}{dt} = -\Gamma kx + \sigma\,\xi(t),$$

where $\sigma > 0$ and $\xi(t)$ is Gaussian random white noise, satisfying $\langle\xi(t)\rangle = 0$ and $\langle\xi(t)\,\xi(t')\rangle = \delta(t-t')\;\forall t, t'$. Write the formal solution of the equation for $t \geq 0$ as a function of the initial condition $x(0) = x_0$ and of the realization $\xi(t')$ ($0 \leq t' \leq t$) of the noise. Evaluate $\langle x(t)\rangle$ and $\langle(x(t) - \langle x(t)\rangle)^2\rangle$ for $t \to \infty$. Assuming that the potential is harmonic, $U(x) = \frac{1}{2}kx^2$, find the relation that Γ and σ must satisfy for the particle to reach the equilibrium distribution $p^{eq}(x) \propto \exp(-U(x)/k_BT)$ for $t \to \infty$. Write down the Fokker-Planck equation associated with the Langevin equation and show that the equilibrium distribution is a solution.

2.6   Prove eq. (2.110) for a system satisfying a one-dimensional Fokker-Planck equation (2.103), assuming that the detailed balance condition eq. (2.109) is satisfied.

2.7   Prove the chain rule for the mutual information, eq. (2.145).

CHAPTER 3

Stochastic Thermodynamics

The fundamental concepts of stochastic thermodynamics are introduced in this chapter. We focus on an out-of-equilibrium mesoscopic physical system described by a master equation and discuss how stochastic thermodynamic observables can be defined at the level of individual trajectories.

3.1 The system

We consider a mesoscopic physical system that can be found in discrete **mesostates**, identified by a variable x. Mesostates can represent, for example, the number of molecules participating in a chemical reaction or different conformations of a protein. Physically, a mesostate corresponds to a collection of microscopic configurations, from which it is obtained by a suitable coarse-graining procedure. We discuss in more detail the concept of coarse graining in section 3.12. For the time being, we assume that this procedure has been successfully carried out, and that the dynamics at the mesoscopic level has been properly determined. Since stochastic thermodynamics focuses on the mesoscopic dynamics, unless some ambiguity is possible, we refer to mesostates simply as *states*.

Our mesoscopic system interacts with a heat reservoir. Such interactions are uncontrolled, making the dynamics of the system intrinsically stochastic. This means that we expect to observe a different trajectory every time we repeat an experiment, even if we carefully prepare the system in the same state at the start of each realization. We therefore introduce the probability distribution $p_x(t)$ of finding the system in state x at time t. This probability distribution evolves according to a master equation of the form given in eq. (2.77):

$$\frac{\mathrm{d}}{\mathrm{d}t}p_x(t) = \sum_{x' \, (\neq x)} \left[k_{xx'} p_{x'}(t) - k_{x'x} p_x(t) \right]. \tag{3.1}$$

Physical details of the system and the nature of its interactions with the reservoir are all encoded into the jump rates, which can in principle be time dependent: $k_{xx'} = k_{xx'}(t)$. The first step to establish a connection with thermodynamics is to determine the nature of this encoding.

We begin with the simple case where the system is in contact with a single heat reservoir. A crucial assumption in stochastic thermodynamics is that this reservoir is very large and relaxes quickly enough to be considered at thermodynamic equilibrium at all times. In the absence of external forces, the jump rates satisfy the detailed balance condition (2.85):

$$\frac{k_{xx'}}{k_{x'x}} = \exp\left(\frac{\epsilon_{x'} - \epsilon_x}{k_B T}\right), \tag{3.2}$$

where ϵ_x is the energy of the system in state x. The condition (3.2) ensures that the canonical equilibrium distribution $p_x^{eq} = e^{(F-\epsilon_x)/k_B T}$ is the unique stationary distribution of the process and therefore that, independent of the initial condition $p_x(t_0)$, the system eventually relaxes to it. It also implies that the net probability current between any pair of states vanishes at stationarity. These two conditions are the hallmarks of a system at equilibrium with a heat reservoir.

It is often easier to estimate energies and the temperature T rather than the jump rates directly. In such cases, we can use eq. (3.2) to express rates in terms of known quantities. However, eq. (3.2) contains the two unknowns $k_{xx'}$ and $k_{x'x}$ and is therefore underdetermined. This means that there is some freedom in expressing the two jump rates. One choice that we often make is to write them in a symmetric form:

$$k_{xx'} = \omega_{xx'}\, e^{-(\epsilon_x - \epsilon_{x'})/2k_B T}, \qquad k_{x'x} = \omega_{xx'}\, e^{-(\epsilon_{x'} - \epsilon_x)/2k_B T}, \tag{3.3}$$

where $\omega_{xx'} = \omega_{x'x}$ is the **intrinsic rate** of jumps between states x and x' in the absence of an energy difference. The intrinsic rate vanishes if the two states are not connected by jumps. Another possibility is

$$k_{xx'} = \omega_{xx'}\, e^{-\epsilon_x/k_B T}, \qquad k_{x'x} = \omega_{xx'}\, e^{-\epsilon_{x'}/k_B T}. \tag{3.4}$$

Alternatively, it might be the case that only one jump rate depends on the energies:

$$k_{xx'} = \omega_{xx'}\, e^{-(\epsilon_x - \epsilon_{x'})/k_B T}, \qquad k_{x'x} = \omega_{xx'}. \tag{3.5}$$

These possibilities are not exhaustive, and the choice is often dictated by physical considerations. For example, if we know that the rate $k_{x'x}$ cannot depend on the values of the energy, then eq. (3.5) is the appropriate choice. Intrinsic rates may incorporate the effect of activation barriers, as discussed in section 3.12.

We can perturb the system away from equilibrium in two different, but not mutually exclusive, ways:

Manipulation. A manipulation is an external, time-dependent control of the system. In a manipulation, the energies of states depend on an external control parameter $\epsilon_x = \epsilon_x(\lambda)$, which is made to change with time: $\lambda = \lambda(t)$. When we need to stress this dependence, we use the short notation $\epsilon_x(t)$ instead of $\epsilon_x(\lambda(t))$. Experiments carried out with optical tweezers provide paradigmatic examples of manipulated systems. In such experiments, a laser creates an energy potential that traps a mesoscopic physical system, be it a colloidal particle or a macromolecule. The external manipulation modulates the stiffness and/or the position of the optical trap.

Driving. A driving is a coupling of the system to an external agent that exchanges with it a certain amount of energy when the system performs specific jumps. In this case, we denote by $\delta_{xx'}$ the energy provided by the external agent during a jump from x' to x. The energy thus provided breaks detailed balance and is immediately released to the reservoir. For example, in the case of biological enzymes fueled by chemical energy, the driving represents the energetic contribution due to ATP hydrolysis.

Our first step is to understand how the detailed balance condition (2.84) is modified in systems brought out of equilibrium by means of manipulations and/or drivings. Because of the conservation of energy, during a jump from state x' to state x, the heat reservoir instantaneously receives an amount of energy equal to

$$q_{xx'} = \epsilon_{x'}(\lambda) - \epsilon_x(\lambda) + \delta_{xx'}, \tag{3.6}$$

which is the sum of the energy $\epsilon_{x'}(\lambda) - \epsilon_x(\lambda)$ lost by the system and of the energy $\delta_{xx'}$ instantaneously provided by the external driving.

In a mesoscopic system, jumps occur due to sudden interactions with the heat reservoir at equilibrium at temperature T. On this fast timescale, the manipulated parameter can be considered to be constant. The effect of a driving can be directly included in the energy budget of the jump (see eq. (3.6)) and therefore just modifies the amount of heat instantaneously released to the reservoir. We conclude that, on the timescale of a single jump, the heat reservoir cannot distinguish between an equilibrium system and a nonequilibrium system that provides it with the same amount of heat. This argument implies that, even out of equilibrium, the ratio between the forward and reverse jump rates between pairs of states x and x' must be equal to $\exp(-q_{xx'}/k_BT)$, exactly as in the equilibrium case:

$$\frac{k_{xx'}}{k_{x'x}} = \exp\left(\frac{q_{xx'}}{k_BT}\right) = \exp\left(\frac{\epsilon_{x'}(\lambda) - \epsilon_x(\lambda) + \delta_{xx'}}{k_BT}\right). \tag{3.7}$$

Equation (3.7) is the **generalized detailed balance condition**. It implies in particular that the jump $x \longrightarrow x'$ is allowed if and only if the reverse jump $x' \longrightarrow x$ is also allowed, i.e., that the system exhibits microscopic reversibility; see section 2.6. Physically, this means that it is always possible to observe the reverse of any observed jump at a finite temperature, although it can be very unlikely. We discuss how this hypothesis can be relaxed in section 8.5.

Equation (3.7) can be used to express jump rates in terms of energies, temperatures, and driving. Also this problem is underdetermined, and similar considerations to those made in the context of eq. (3.2) apply here. It is often the case that the driving acts only on one jump direction. In this case, we write the rates in the form

$$k_{xx'} = \omega_{xx'}\, e^{-(\epsilon_x - \epsilon_{x'})/(2k_BT) + \delta_{xx'}/k_BT}, \qquad k_{x'x} = \omega_{xx'}\, e^{-(\epsilon_{x'} - \epsilon_x)/2k_BT}. \tag{3.8}$$

A model of a mesoscopic system whose jump rates satisfy eq. (3.7) is said to be **thermodynamically consistent**.

In general, the energy of a nonequilibrium system exhibits a slow modulation due to the time-dependent manipulation, interrupted by sudden changes due to jumps caused by drivings and interactions with the heat reservoir, as shown in fig. 3.1.

3.2   Work and heat in stochastic thermodynamics

In stochastic thermodynamics, work and heat are defined at the level of individual trajectories and are therefore stochastic quantities.

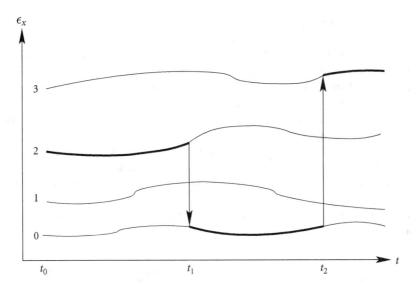

Figure 3.1. Trajectory of a manipulated system. Thin lines represent energy levels of different mesostates, modulated in time as an effect of the manipulation. Bold lines mark the states occupied by the system; arrows represent jumps. The system is initialized in state 2. Its energy changes as time goes from $t = t_0$ to $t = t_1$. At $t = t_1$ the system jumps to state 0, releasing a quantity of heat $q_{02}(t_1) = \epsilon_2(t_1) - \epsilon_0(t_1)$ to the reservoir. The energy of state 0 changes between $t = t_1$ and $t = t_2$. Then a jump occurs at $t = t_2$ from 0 to 3, exchanging with the reservoir a quantity of heat given by $q_{30}(t_2) = \epsilon_0(t_2) - \epsilon_3(t_2)$. The fact that $q_{30}(t_2) < 0$ in this case means that heat is absorbed from the reservoir. In general, the energy of the system can change either because the energy of the occupied state changes (work by the manipulating device) or because of jumps to a state of different energy (due to heat exchanged with the reservoir or work by the driving or both).

We consider a trajectory x of our mesoscopic system, starting from time t_0 with the system in state x_0 and ending at time t_f with the system in state x_f. At the times t_k, the system jumps from state x_{k-1} to state x_k, with $k = 1, \ldots, n$ and $x_n = x_f$; see section 2.7. Along the trajectory, energy is exchanged among the system, the external forces, and the reservoir. The **stochastic work** w is the energy provided by the external forces to the system. The **stochastic heat** q is the energy released by the system to the reservoir. In this book, we use lowercase letters, like w, q, and ϵ, for stochastic, trajectory-dependent thermodynamic observables and uppercase ones for their averaged counterparts, like the standard thermodynamic work W, heat Q, and energy E.

We have seen how a system can be perturbed out of equilibrium by a time-dependent manipulation or by external drivings. Both of these factors contribute in general to the stochastic work. Their combined effect can be expressed by

$$w(x) = \underbrace{\sum_{k=0}^{n} \int_{t_k}^{t_{k+1}} dt \, \frac{d\lambda}{dt} \frac{d\epsilon_{x_k}}{d\lambda}}_{\text{manipulated work}} + \underbrace{\sum_{k=1}^{n} \delta_{x_k x_{k-1}}}_{\text{driven work}} . \tag{3.9}$$

The first term on the right-hand side of eq. (3.9) is the work made by the manipulation. Since $(d\lambda/dt)\,d\epsilon_{x_k}/d\lambda = d\epsilon_{x_k}/dt$, this term represents the total energy change between consecutive jumps. As the reservoir does not exchange energy outside jumps, this energy change necessarily originates from work by the manipulation. The second term on the right-hand side of eq. (3.9) is the total work performed by the driving along each jump. Because of microscopic reversibility, the driven work in a jump $x \longrightarrow x'$ is equal to $\delta_{x'x}$ and that in the opposite jump is $\delta_{xx'} = -\delta_{x'x}$, independent of how the rates $k_{xx'}$ and $k_{x'x}$ have been defined.

We now turn our attention to the stochastic heat. During a jump from state x' to state x at time t, the reservoir receives an amount of heat $q_{xx'}$ given by eq. (3.6), with $\lambda = \lambda(t)$. As we discussed, heat is exchanged with the reservoir only during jumps—the external manipulation is too slow to contribute to the energy budget on the timescale of a single jump. Therefore, the total heat exchanged with the heat reservoir along the trajectory $\boldsymbol{x}$ is

$$q(\boldsymbol{x}) = \sum_{k=1}^{n} q_{x_k x_{k-1}}(t_k) = \sum_{k=1}^{n} \left[-\epsilon_{x_k}(t_k) + \epsilon_{x_{k-1}}(t_k) + \delta_{x_k x_{k-1}} \right]. \qquad (3.10)$$

From eqs. (3.9) and (3.10), it follows that

$$w(\boldsymbol{x}) - q(\boldsymbol{x}) = \epsilon_{x_f}(t_f) - \epsilon_{x_0}(t_0). \qquad (3.11)$$

We call eq. (3.11) the **first law of stochastic thermodynamics**, since it formally resembles the first law of thermodynamics, eq. (2.2). The reason for using the cautionary word *formally* is explained in the next section.

3.3 Mesoscopic and calorimetric heat (*)

To understand the physical interpretation of mesoscopic thermodynamic observables, it is useful to scrutinize in more detail each term in eq. (3.11) and compare them with those appearing in the first law of thermodynamics, eq. (2.2).

We start by reflecting on the actual meaning of the energy of a mesostate ϵ_x. For an object such as a rigid colloid without any relevant hidden degrees of freedom, this energy can be interpreted without ambiguity. A more subtle case is that of a macromolecule, where each mesostate represents a collection of microscopic states ξ, each characterized by its energy ϵ_ξ. To understand this scenario, we assume that the timescales of the fast microscopic dynamics and the slow mesoscopic dynamics are well separated. If the manipulation is slow enough, and the driving connects different mesostates, then jumps between states in the same mesostate satisfy the detailed balance condition. Therefore, the occupation probabilities of the microstates within a given mesostate are proportional to their Boltzmann factors $e^{-\epsilon_\xi/k_BT}$, as in equilibrium. On the other hand, the relative occupation probabilities of different mesostates can in general be different from those in equilibrium.

We evaluate the equilibrium probability p_x^{eq} of a mesostate x under these assumptions. From the law of total probabilities, we obtain

$$p_x^{eq} = \sum_{\xi \in x} p_\xi^{eq} \propto \sum_{\xi \in x} e^{-\epsilon_\xi/k_BT}, \qquad (3.12)$$

where the sum runs over all microstates ξ contained in the mesostate x. The occupation probability of the mesostate x can be recast in the form of a Boltzmann factor $e^{-\epsilon_x/k_BT}$ by defining ϵ_x via

$$\epsilon_x = -k_BT \ln \left(\sum_{\xi \in x} e^{-\epsilon_\xi/k_BT} \right). \tag{3.13}$$

Equation (3.13) reveals that, if the mesostate x has an internal structure, ϵ_x represents its *free energy* rather than its energy. In principle, mesostates with different values of ϵ_x might contain microstates with the same average energies but different entropies. As a consequence, eq. (3.11) does not really correspond to the first law of thermodynamics: in eq. (2.2), the energies are proper energies, not free energies.

To solve this conundrum, we inspect more closely the other fluctuating thermodynamic quantities in eq. (3.11). The stochastic work in eq. (3.9) is the sum of a driven and a manipulated contribution. The driven work is equal to the total energy provided by the external agent and is therefore unambiguous. The definition of manipulated work requires more care. In a manipulated system, we have

$$\begin{aligned}
w(\boldsymbol{x}) &= \sum_{k=0}^{n} \int_{t_k}^{t_{k+1}} dt \, \frac{d\lambda}{dt} \frac{\partial \epsilon_{x_k}}{\partial \lambda} \\
&= \sum_{k=0}^{n} \int_{t_k}^{t_{k+1}} dt \, \frac{d\lambda}{dt} \frac{\sum_{\xi \in x} (\partial \epsilon_{\xi_k}/\partial \lambda) \, e^{-\epsilon_{\xi_k}/k_BT}}{\sum_{\xi \in x} e^{-\epsilon_\xi/k_BT}} \\
&= \sum_{k=0}^{n} \int_{t_k}^{t_{k+1}} dt \, \frac{d\lambda}{dt} \left\langle \frac{\partial \epsilon_{\xi_k}}{\partial \lambda} \right\rangle_x^{\text{eq}},
\end{aligned} \tag{3.14}$$

where we denote by $\langle \ldots \rangle_x^{\text{eq}}$ the equilibrium average over the microscopic states included in the mesoscopic state x. This means that, under our assumptions, the definition of work in stochastic thermodynamics is consistent with the microscopic one.

The conclusion is that, to save the energy balance, we have to reconsider the physical interpretation of the stochastic heat, eq. (3.10). In particular, it is necessary to distinguish between two definitions of heat:

Mesoscopic heat. The mesoscopic heat is defined by eq. (3.10) and is associated with changes in the free energies of the mesoscopic states. This is the natural definition of heat in stochastic thermodynamics. It however neglects the heat exchanges necessary to alter the entropy of the mesostates.

Calorimetric heat. The calorimetric heat is defined as the heat exchanged with the reservoir that would be measured by an infinitely sensitive calorimeter. Such a calorimeter would appreciate the amount of heat required to rearrange the distribution of microscopic states inside each mesostate. Following this definition, we express the calorimetric heat q^{cal} by subtracting from the mesoscopic heat the entropic contribution

$$q^{\text{cal}}(\boldsymbol{x}) = q(\boldsymbol{x}) - T \sum_{k=1}^{n} \left(S_{x_k}(t_k) - S_{x_{k-1}}(t_k) \right), \tag{3.15}$$

where $S_x = -k_B \sum_{\xi \in x} p_\xi \ln p_\xi$ is the entropy of the internal states of a mesostate.

Summarizing, in stochastic thermodynamics "energy" actually means "free energy" if we take into account the internal structure of each mesostate. However, the definition of mesoscopic heat in stochastic thermodynamics also neglects the heat released in the reservoir (or absorbed by the system) due to the change of entropy of these internal degrees of freedom. These two contributions compensate each other in the energy budget, and this is the reason why eq. (3.11) formally looks like the first law of thermodynamics. In other words, eq. (3.11) is a coarse-grained first law of thermodynamics that overlooks the part of the heat exchange that does not affect the probability distribution of mesostates. This shows once more that the assumption of separation of scales between the microscopic and the mesoscopic dynamics is crucial in stochastic thermodynamics. It is thanks to this assumption that microscopic energy transactions are always thermodynamically reversible, so that we can practically forget about them.

3.4 ATP hydrolysis by myosin

Many proteins in cells act as molecular motors, meaning that they consume chemical energy to perform work. This work may result in transport of cargos to different parts of the cell or in the assembly of molecular structures such as filaments. Myosins are a broad class of such proteins. They consume energy in the form of ATP to generate forces on organic filaments made of the actin protein. This mechanism is the basis for the contraction of muscle cells. Myosins consume ATP both when attached to actin filaments or detached from them. We focus on the latter case, where the reaction kinetics is substantially slower.

After ATP is bound to myosin, it is hydrolyzed, i.e., converted into ADP plus a phosphate group P. In physiological conditions, this reaction provides to the motor protein about $25.5\,k_BT$ of free energy. After this reaction, the phosphate group and ADP, in that order, are released by the protein. At this point, the myosin is free to bind to another ATP molecule.

We describe the cycle of a single myosin protein in the cell, while detached from an actin filament, using the tools developed in this chapter. We represent the protein cycle via the jump network in fig. 3.2a. We denote the four states where myosin is unbound, ATP-bound, ADP+P-bound, and ADP-bound by $\emptyset$, ATP, ADP+P, and ADP, respectively. We consider as "external driving" the ATP hydrolysis and assign the corresponding work to the jump ATP $\longrightarrow$ ADP + P. The myosin cycle is rather well characterized biochemically, so that all the jump rates have been experimentally estimated.

The stochastic heat q is given by eq. (3.10), with $q_{xx'}$ defined in eq. (3.6) and the energies ϵ_x stated in the caption to fig. 3.2. We model hydrolysis of ATP by the nonequilibrium driving

$$\delta_{\mathrm{ADP+P,ATP}} = -\delta_{\mathrm{ATP,ADP+P}} = \Delta\mu = 25.5\,k_BT, \qquad (3.16)$$

where $\Delta\mu$ is the chemical potential imbalance between ATP and the products of hydrolysis. The other reactions are not driven. The stochastic work w is expressed by eq. (3.9).

The stochastic work w and heat q are shown in fig. 3.2b as a function of time for a sample trajectory. The difference between the stochastic work and heat is the internal energy change; see eq. (3.11). The motor spends most of its time performing the hydrolysis reaction, due to the slow release of the phosphate group. During this reaction,

(a)

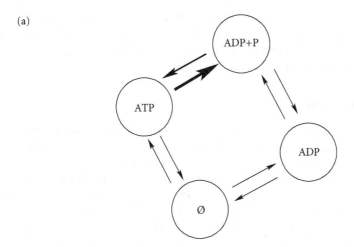

(b)

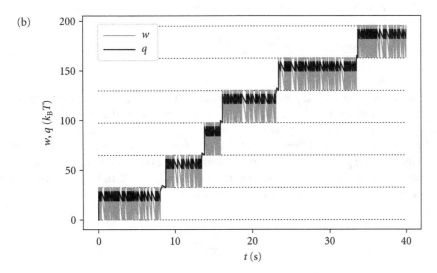

Figure 3.2. (a) Jump network for the myosin model (from Howard [80, ch. 14]). Parameters are based on measurements of myosin S1 in rabbit skeletal muscle. The energy of unbound myosin is taken as a reference, $\epsilon_\emptyset = 0$. The other energies are $\epsilon_{ATP} \approx -19\,k_BT$, $\epsilon_{ADP+P} \approx 4\,k_BT$, $\epsilon_{ADP} \approx -2\,k_BT$ with $\Delta\mu \approx 25.5\,k_BT$. Thicker arrows represent jumps where work (positive or negative) is performed by external driving. Jump rates: $k_{ATP,\emptyset} = 2 \cdot 10^4\,s^{-1}$, $k_{\emptyset,ATP} = 10^{-4}\,s^{-1}$, $k_{ADP+P,ATP} = 100\,s^{-1}$, $k_{ATP,ADP+P} = 10\,s^{-1}$, $k_{ADP,ADP+P} = 0.1\,s^{-1}$, $k_{ADP+P,ADP} = 2 \cdot 10^{-4}\,s^{-1}$, $k_{\emptyset,ADP} = 2\,s^{-1}$, $k_{ADP,\emptyset} = 15\,s^{-1}$. (b) Stochastic work ($w$) and heat ($q$) as a function of time t in a simulation of the process with the above parameters. The system spends most of the time vacillating between the ATP and ADP+P states, receiving or releasing energy equal to $\Delta\mu$, leading to large fluctuations in w, while the fluctuations in q are much smaller. Lines are at intervals of $25.5\,k_BT$, corresponding to the total energy released to the reservoir in one cycle.

the driving exchanges an amount of work $\Delta\mu = 25.5\,k_BT$, whereas the exchanged heat is $q_{\mathrm{ATP,ADP+P}} = \epsilon_{\mathrm{ADP+P}} - \epsilon_{\mathrm{ATP}} - \Delta\mu \approx -2\,k_BT$. Since, for this motor, the hydrolysis reaction occurs rather close to equilibrium, fluctuations of the stochastic heat during hydrolysis are much less pronounced than those of the stochastic work.

3.5 General reservoirs

So far, we have considered a nonequilibrium mesoscopic system in contact with a single heat reservoir. In this section, we discuss more general reservoirs, focusing on how the generalized detailed balance condition (3.7) is modified.

We start by considering a single reservoir that, besides providing and absorbing heat, can also exchange particles with the system. This generalization is particularly useful for applying stochastic thermodynamics to systems of chemical reactions, where a particle-exchanging reservoir is called a **chemostat**. We extensively discuss chemical reactions in section 3.10. For a system coupled to a reservoir exchanging particles and heat, the generalized detailed balance condition becomes

$$\frac{k_{xx'}}{k_{x'x}} = \exp\left(\frac{\epsilon_{x'}(\lambda) - \epsilon_x(\lambda) + \sum_i \mu^i(\lambda)\Delta n^i_{xx'} + \delta_{xx'}}{k_BT}\right), \qquad (3.17)$$

where $\mu^i(\lambda)$ is the chemical potential of species i and $\Delta n^i_{xx'}$ is the number of molecules of species i exchanged with the reservoir during a jump from state x' to state x. The quantity $\epsilon_{x'}(\lambda) - \epsilon_x(\lambda) + \sum_i \mu^i(\lambda)\Delta n^i_{xx'}$ appearing on the right-hand side of eq. (3.17) can be interpreted as the difference in the Gibbs free energy between states x' and x.

Another useful generalization is to cases where the system is simultaneously in contact with multiple reservoirs, labeled by an index $r = 1, \ldots, R$. In this case, two generic states x' and x can be connected by multiple jumps,

$$k_{xx'} = \sum_{r=1}^{R} k^{(r)}_{xx'}, \qquad (3.18)$$

each caused by an interaction with a different reservoir. Equation (3.17) applies to each reservoir individually:

$$\frac{k^{(r)}_{xx'}}{k^{(r)}_{x'x}} = \exp\left[\frac{1}{k_BT^{(r)}}\left(\epsilon_{x'}(\lambda) - \epsilon_x(\lambda) + \delta_{xx'} + \sum_i \mu^{i,(r)}(\lambda)\Delta n^i_{xx'}\right)\right], \qquad (3.19)$$

where $T^{(r)}$ and $\mu^{i,(r)}$ are, respectively, the temperature and the chemical potential of species i for reservoir r.

In some cases, the external driving in the generalized detailed balance relation (3.7) is implicitly related to particle reservoirs. To show that relationship, we consider a system S in contact with two particle reservoirs $\mathcal{R}_1$ and $\mathcal{R}_2$, both at temperature T but characterized by different values, μ_1 and μ_2, of the chemical potential of particles that they exchange with the system. Because of this difference, the composite system, i.e., the system plus the two reservoirs, is out of equilibrium. We focus on a jump $x' \longrightarrow x$ that is always accompanied by the transfer of one particle from $\mathcal{R}_1$ to $\mathcal{R}_2$. The net energy change of the composite system in this jump is given by

$$\Delta \epsilon_{xx'}^{\text{tot}} = \epsilon_x - \epsilon_{x'} - \mu_1 + \mu_2. \tag{3.20}$$

The jump rates must satisfy the generalized detailed balance condition at the level of the composite system:

$$\frac{k_{xx'}}{k_{x'x}} = \exp\left(-\frac{\Delta \epsilon_{xx'}^{\text{tot}}}{k_B T}\right) = \exp\left(\frac{\epsilon_{x'} - \epsilon_x + \mu_1 - \mu_2}{k_B T}\right). \tag{3.21}$$

The generalized detailed balance relation expresses the thermodynamic consistency of the process. This expression can be cast in the form of eq. (3.7) by associating with the particle exchange an effective driving,

$$\delta_{xx'} = \Delta \mu = \mu_1 - \mu_2. \tag{3.22}$$

This reasoning illustrates that the driving $\delta_{xx'}$ can be interpreted as the effect of an external force facilitating the jump or, equivalently, as the chemical potential imbalance associated with particle transfer between reservoirs. For this reason, in the following we use interchangeably $\delta_{xx'}$ or $\Delta \mu$ to denote a chemical driving originating from a chemical potential imbalance.

Equations (3.7), (3.17), and (3.19) also apply if we manipulate the temperature, $T = T(\lambda)$. To formalize this case, one can imagine many heat reservoirs in equilibrium at different temperatures, and a manipulation protocol that continuously "disconnects" the system from a reservoir at temperature $T(\lambda(t))$ and connects it to a different one at a slightly different temperature $T(\lambda(t + dt))$. Also this procedure requires a separation between the timescales at which T is manipulated and those of the individual jumps. This hypothesis ensures that, on the timescales of individual jumps, the system is always coupled to one equilibrium heat reservoir at a well-defined temperature.

3.6 Stochastic entropy

The uncertainty associated with the distribution $p_x(t)$ over the mesostates is quantified by the Shannon entropy

$$H[\mathcal{S}(t)] = -\sum_x p_x(t) \ln p_x(t). \tag{3.23}$$

In keeping with equilibrium thermodynamics, we assign to the system a **nonequilibrium entropy**

$$S^{\text{sys}}(t) = k_B H[\mathcal{S}(t)]. \tag{3.24}$$

This entropy pertains to an *ensemble* of systems undergoing the same process, while each individual system follows a different trajectory, due to differences in the initial condition and in the interactions with the heat reservoir. To characterize individual systems, we define the **stochastic entropy**

$$s_x^{\text{sys}}(t) = -k_B \ln p_x(t). \tag{3.25}$$

For an individual trajectory $x = (x(t))$, the value of the stochastic entropy at time t is $s_{x(t)}(t) = -k_B \ln p_{x(t)}(t)$. Although the stochastic entropy is defined for individual trajectories, computing it requires knowledge of the probabilities $p_x(t)$ that characterize the whole ensemble of trajectories. The average and stochastic entropies are related by

$$S^{\text{sys}}(t) = \langle s^{\text{sys}}(t)\rangle_t, \tag{3.26}$$

where the average is defined by $\langle \cdots \rangle_t = \sum_x \cdots p_x(t)$. The entropy change of the system between the initial and final time is

$$\Delta s^{\text{sys}} = s_{x_f}^{\text{sys}}(t_f) - s_{x_0}^{\text{sys}}(t_0). \tag{3.27}$$

We are now in a position to evaluate the total entropy change of the system plus the heat reservoir along a given trajectory x:

$$s^{\text{tot}}(x) = \Delta s^{\text{sys}} + s^{\text{res}}(x), \tag{3.28}$$

where $s^{\text{res}}(x)$ is the trajectory-dependent change in entropy of the reservoir due to the heat released by the system. Equation (3.28) is the stochastic counterpart of eq. (2.31). Since the reservoir is always at equilibrium, its entropy change is

$$s^{\text{res}}(x) = \frac{q(x)}{T}, \tag{3.29}$$

where the stochastic heat $q(x)$ is given by eq. (3.10).

3.7 Stochastic entropy and entropy production in a manipulated two-level system

Lasers provide a versatile and powerful way to manipulate mesoscopic physical and biological systems. We consider a system consisting of a single defect center in natural IIa-type diamond simultaneously excited by a red laser of wavelength 680 nm and a green laser of wavelength 514 nm. The red laser can bring the system to a high-energy state from which it decays, emitting fluorescence. The green laser also excites the system to a state from which it decays nonradiatively. The state usually reached by this decay is slightly different from that reached after absorbing red photons. However, with a smaller rate the state excited by green light can decay in the ground state reached after absorption of red light and vice versa.

The jump network for this system is represented in fig. 3.3a. The system can be found in four states, which form two sets, corresponding respectively to the left (0) and right (1) part of the figure. The jump rates between the two sets are on the order of a few inverse milliseconds, whereas those within each of them are on the order of a few inverse nanoseconds. Since jumps within each set are hard to resolve experimentally, we describe the process as a single effective two-level system with two states, (0) and (1). If the system is in (0) it appears bright, whereas in (1) it appears dark. The jump rates between these two states can be manipulated by varying the power of the applied lasers.

For a two-level system, the detailed balance condition for the steady-state probability distribution always holds:

$$k^\uparrow p_0^{\text{st}} = k^\downarrow p_1^{\text{st}}. \tag{3.30}$$

(a)

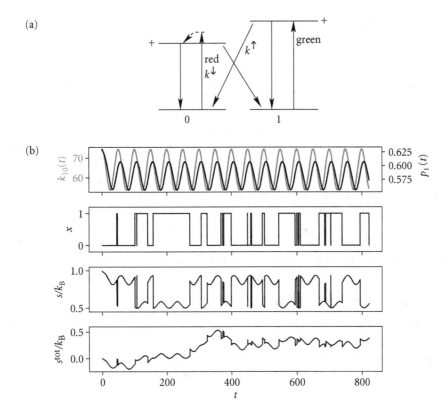

(b)

Figure 3.3. (a) Energy levels of the diamond-defect system. The system is described by four states, forming two sets. Transitions within each set are too fast to be resolved. The much smaller jump rates $k^{\downarrow}$ and $k^{\uparrow}$ can be manipulated by changing the intensity of the applied laser radiation. (Adapted from Schuler et al. [147], with permission; see also Tietz et al. [166].) (b) Simulated trajectories. Top to bottom: Escape rate $k^{\uparrow}(t)$ in kHz, as a function of time (gray) and probability of occupation $p_1(1)$ of the dark state (black); a trajectory of the system; stochastic entropy $s = s_{x(t)}(t)$ for the given trajectory; and the corresponding total produced entropy $s^{\text{tot}}(t)$. Parameters: $a_0^{-1} = 26$ ms, $b_0^{-1} = 31$ ms, $\gamma = 0.16$, $t_{\text{m}} = 50$ ms. Time is measured in ms.

The jump rates at a given intensity of laser radiation are expressed as

$$k^{\uparrow}(t) = a_0[1 + \lambda(t)], \qquad k^{\downarrow}(t) = b_0, \qquad (3.31)$$

where $\lambda(t) = \gamma \sin(2\pi t/t_{\text{m}})$ is the experimental control parameter.

It is important to clarify that jumps in this system are caused by quantum effects that do not necessarily require interactions with a heat reservoir. Nevertheless, we can formally study the system using stochastic thermodynamics. We consider a trajectory

$$\boldsymbol{x} = ((x_0, t_0), (x_1, t_1), \dots, (x_n, t_n), t_f),$$

such that

$$x = x_k, \qquad \text{for } t_k \leq t \leq t_{k+1}, \qquad (3.32)$$

where the last jump takes place at t_n and we set $t_{n+1} = t_f$. The total entropy produced by the system in the trajectory x is given by

$$s^{\text{tot}}(x) = \Delta s^{\text{sys}} + s^{\text{res}}(x) = s_{x_f}^{\text{sys}}(t_f) - s_{x_0}^{\text{sys}}(t_0) + k_B \sum_{k=1}^{n} \ln \frac{k_{x_k x_{k-1}}(\lambda(t_k))}{k_{x_{k-1} x_k}(\lambda(t_k))}. \qquad (3.33)$$

In the second equality, we have used the generalized detailed balance condition (3.7).

In this case, we expressed the entropy production in terms of the rates and the probabilities only: since jumps in this system are not caused by interactions with a heat reservoir, we cannot invoke eq. (3.7) to interpret rates in terms of energies and temperature. This also means that the stochastic work w and heat q do not have a direct physical interpretation. On the other hand, entropy production is still a useful measure of the irreversibility of the dynamics (fig. 3.3b). The system entropy remains bounded, whereas the total produced entropy appears to be negative at short times for the given trajectory and then tends to increase, as shown in the bottom plot.

3.8 Average entropy production rate

We now discuss the average rate at which entropy is produced. It is convenient to separate the contribution due to jumps from the contribution during dwells. During each jump from state x' to state x at a time t, the reservoir absorbs an amount of heat $q_{xx'}$ given by eq. (3.6). This corresponds to a change of reservoir entropy equal to

$$s^{\text{res}} = \frac{q_{xx'}}{T} = k_B \ln \frac{k_{xx'}}{k_{x'x}}. \qquad (3.34)$$

During the same jump, the system entropy changes by

$$\Delta s^{\text{sys}} = k_B \ln \frac{p_{x'}}{p_x}. \qquad (3.35)$$

Therefore, the total entropy change during a jump is

$$s^{\text{tot}} = k_B \ln \frac{k_{xx'} p_{x'}}{k_{x'x} p_x}. \qquad (3.36)$$

We now consider the entropy change between jumps. Because of the separation of timescales, there is no heat exchange with the reservoir between jumps, and therefore the reservoir entropy does not change. When the system is in state x, its entropy changes at a rate

$$\frac{ds^{\text{sys}}}{dt} = -k_B \frac{d\lambda}{dt} \frac{\partial}{\partial \lambda} \ln p_x = \frac{k_B}{p_x} \frac{d\lambda}{dt} \frac{\partial p_x}{\partial \lambda}. \qquad (3.37)$$

Because of conservation of probability, this change vanishes on average:

$$\left\langle \frac{ds^{\text{sys}}}{dt} \right\rangle = k_B \sum_x p_x \frac{ds_x^{\text{sys}}}{dt} = k_B \frac{d\lambda}{dt} \sum_x \frac{\partial p_x}{\partial \lambda} = 0. \qquad (3.38)$$

This means that, on average, the entropy of both the system and the reservoir can only change during jumps. On average, the net jump rate from state x' to state x is given by the probability current $J_{xx'}(t)$; see eq. (2.80). Therefore, the net average entropy production rate is

$$\dot{S}^{\text{tot}}(t) = \frac{k_B}{2} \sum_{xx'} J_{xx'}(t) \ln \frac{k_{xx'}(t)p_{x'}(t)}{k_{x'x}(t)p_x(t)}, \tag{3.39}$$

where the factor $1/2$ compensates for counting the current between each pair of states twice. Equation (3.39) is a useful result, often referred to as the **Schnakenberg formula**. In this expression, we denote the derivative by a dot to stress that the total entropy production is not a state function, as discussed in section 2.2. We reserve the notation d/dt for derivatives of state functions. Since the sign of $\ln[k_{xx'}(t)p_{x'}(t)/k_{x'x}(t)p_x(t)]$ is the same as the sign of $J_{xx'}(t) = k_{xx'}(t)p_{x'}(t) - k_{x'x}(t)p_x(t)$, each term in the sum in eq. (3.39) is nonnegative. This implies that the average entropy production rate is nonnegative as well, and that it vanishes if and only if the current between each pair of states vanishes. This latter case corresponds to detailed balance and therefore to thermodynamic equilibrium.

It follows from our discussion that the average entropy production rate can be split into the production rate of the reservoir entropy and that of the system entropy, $\dot{S}^{\text{tot}}(t) = \dot{S}^{\text{res}}(t) + dS^{\text{sys}}/dt$. We express these two contributions by

$$\dot{S}^{\text{res}}(t) = \frac{k_B}{2} \sum_{xx'} J_{xx'}(t) \ln \frac{k_{xx'}(t)}{k_{x'x}(t)}; \tag{3.40}$$

$$\frac{dS^{\text{sys}}}{dt} = \frac{k_B}{2} \sum_{xx'} J_{xx'}(t) \ln \frac{p_{x'}(t)}{p_x(t)}. \tag{3.41}$$

In the particular case of a system in a steady state, the internal entropy is constant on average, and one has $dS^{\text{sys}}/dt = 0$ and $\dot{S}^{\text{tot}} = \dot{S}^{\text{res}}$.

3.9 Network theory of nonequilibrium steady states (*)

In this section, we analyze nonequilibrium steady states of master equations based on an analogy with the Kirchhoff theory of electric circuits. We consider a master equation that satisfies microscopic reversibility. We say that two nodes x and x' are connected by an edge if the jump $x \longrightarrow x'$ and its reverse have nonvanishing rates; see section 2.6. We represent the jump network associated with the master equation as shown in fig. 2.1, where edges are represented as segments.

In the steady state, the probability current $J_{xx'} = k_{xx'}p_{x'}^{\text{st}} - k_{x'x}p_x^{\text{st}}$ satisfies

$$\sum_{x'} J_{xx'} = 0, \tag{3.42}$$

for each node x. Equation (3.42) is analogous to Kirchhoff's law for currents, with $J_{xx'}$ playing the role of the electric current.

We define a **cycle** $\mathcal{C}$ in the network as a sequence of distinct states such that each is connected by an edge to the previous one and to the next, and the last to the first. Equation (3.42) implies that $J_{xx'} = 0$ if the edge between x and x' does not belong to any

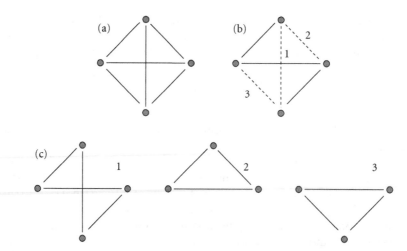

Figure 3.4. Example of construction of a fundamental set of cycles from the jump network of a four-state system in which all jumps are allowed. The panels show (a) the original network, (b) the spanning tree, where the dashed lines are the chords that have been removed, (c) the resulting set of fundamental cycles.

cycle in the network. If the jump network is made up of a single cycle, it follows from eq. (3.42) that the current is constant, i.e., independent of the node. The key idea of the Kirchhoff theory is to decompose an arbitrary network into a set of cycles. To this aim, we need to introduce a few concepts from graph theory.

Our starting point is a connected jump network; see section 2.6. In principle, two nodes could be connected by more than one edge due to coupling with different reservoirs; see section 3.5. We neglect this complication here. From the original network, we construct its **core network** by removing all edges that do not belong to any cycle. Currents can run only in the core network in the steady state. We now decompose the core network into fundamental cycles. We construct a treelike network by sequentially removing an arbitrary edge from an arbitrary cycle until there are no cycles left (fig. 3.4). In this way, we obtain a **spanning tree** of the core network. The spanning tree is connected by construction, it includes all nodes of the core network, and it does not contain any cycles. If the core network has N states, the spanning tree has $N - 1$ edges. Since the edge to be removed at each step is arbitrary, the spanning tree is not unique.

The edges removed to construct the spanning tree are called **chords** and are labeled by an index $\alpha \in \{1, \ldots, \nu\}$. By adding one chord α to the spanning tree, we obtain a network with exactly one cycle. We call this cycle C_α. The set of cycles (C_α) with $\alpha = 1, \ldots, \nu$ is called the set of **fundamental cycles**; see fig. 3.4. They are called fundamental since they are independent, their union reproduces the core network, and any linear function defined on the cycles in the original network can be represented by a linear combination of functions defined on the cycles in the set (C_α). A proof of this nontrivial fact is sketched in appendix A.9.

Thanks to this result, we can express the probability current flowing through any edge in the steady state as a sum of the currents flowing in the fundamental cycles. To this aim, we assign an arbitrary orientation to each edge xx', either $x \to x'$ or $x' \to x$:

$$J_{xx'} = \sum_{\alpha=1}^{\nu} s_{xx'}(\alpha) J_\alpha, \tag{3.43}$$

where the sum runs over the fundamental cycles. Here J_α is the current of the cycle, which can be identified with the current running through the chord unique to that cycle aligned with its orientation; and $s_{xx'}(\alpha) = 1$ if the edge xx' has the same orientation of C_α, $s_{xx'}(\alpha) = -1$ if it has the opposite orientation, and it vanishes if it does not belong to C_α.

Similar to the current, the average entropy production rate is also a linear function of the cycles in the jump network. Indeed, eq. (3.39) shows that the average entropy production rate is a linear function of the currents and therefore does not receive contributions from jumps that are not part of a cycle. Therefore, the Kirchhoff theory tells us that we can linearly decompose the average entropy production rates into contributions from the cycles in the fundamental set. The contribution of each cycle is equal to its associated average entropy production rate according to eq. (3.39). Following this reasoning, we introduce, for each oriented cycle α, its **affinity**:

$$A_\alpha = k_B T \ln \left(\prod_{j=1}^{n} \frac{k_{x_{j+1}x_j}}{k_{x_j x_{j+1}}} \right) = \sum_{j=1}^{n} \delta_{x_{j+1}x_j}, \tag{3.44}$$

where we set $x_{n+1} = x_1$. In the second equality, we have used the generalized detailed balance condition (3.7). We thus obtain the decomposition

$$\dot{S}^{\mathrm{tot}} = \frac{1}{T} \sum_\alpha A_\alpha J_\alpha. \tag{3.45}$$

Therefore, the entropy production rate in arbitrary networks can be decomposed into a sum of the products of currents times affinities, where each product pertains to a cycle in the fundamental set.

3.10 Stochastic chemical reactions

Stochastic chemical reactions have gained importance in recent years thanks to the advancement of cell biology. Such advancements have revealed that many proteins are present in concentrations as little as a few units per cell, so that representing their density as a continuous quantity is not appropriate. At such low concentrations, protein numbers are intrinsically stochastic quantities that can be described quite naturally with the tools of stochastic thermodynamics.

In stochastic chemical reactions, the state of a system is represented by a vector $n_A, n_B, n_C, \ldots$, whose entries are the number of molecules of each chemical species A, B, C, ..., present at a given time. By *species*, we mean either individual molecules or bound states of more molecules. For example, in a dimerization reaction $A + A \rightleftharpoons AA$, the state of the system is identified by the two-dimensional vector (n_A, n_{AA}), where n_A is the number of monomers and n_{AA} is the number of dimers. In a well-stirred container, the **law of mass action** stipulates that the rate of each reaction is proportional to the concentration of each reactant, raised to the power of the corresponding stoichiometric coefficient. We assume that this is the case. The reaction rates are then simple functions of the number of particles.

It is convenient to classify the possible reactions as follows:

One-body reactions. In one-body reactions, the reactant is a single molecule. A decay reaction $A \to \emptyset$ is an example of a one-body reaction. Since each molecule participates individually in a simple reaction, the total reaction rate is given by an individual rate times the number of molecules present at time t. In the case of the decay reaction, one has $k_{n_A-1,n_A} = k\,n_A$, where k is the individual rate.

Binary reactions. In binary reactions, two molecules react together. For example, the formation of a bound state $A + B \to AB$ is a binary reaction, where AB is the bound state of a molecule A with a molecule B. On the contrary, the breaking of a bound state, such as $AB \to A + B$, is a one-body reaction. The total rate of binary reactions is proportional to the number of molecules of either species participating in it and is inversely proportional to the available volume. In the case of the bound state formation, we have $k_{(n_A-1,n_B-1,n_{AB}+1),(n_A,n_B,n_{AB})} = k\,n_A\,n_B$, where k is a rate constant. There is an exception when the two molecules entering the reaction belong to the same species A, like in the dimerization reaction $A + A \to AA$. In this case, the rate is equal to $k\,n_A(n_A - 1)$. This rate duly vanishes when there is a single A molecule available.

We can similarly define higher-order reactions with three or more reactants, although in practice they are less common. The probability $p_{n_A,n_B,n_C,\ldots}(t)$ of the numbers of species evolves according to a master equation, whose rates can be determined using the above approach. The link with thermodynamics is obtained by expressing the jump rates in terms of the energy of states via the generalized detailed balance condition, eq. (3.7) or (3.17).

As an illustration, we consider an isomerization reaction, in which a chemical A is transformed into a chemical B:

$$A \underset{k^-}{\overset{k^+}{\rightleftharpoons}} B. \tag{3.46}$$

We denote by $[A] = \langle n_A \rangle / V$ and $[B] = \langle n_B \rangle / V$ the concentrations of chemicals A and B, respectively, where V is the volume of the system. We also introduce the individual rates $k^\pm$ of the reactions $A \rightleftharpoons B$ and write the master equation for $p_{n_A,n_B}(t)$:

$$\frac{dp_{n_A,n_B}}{dt} = k^+ (n_A + 1)p_{n_A+1,n_B-1}(t) + k^- (n_B + 1)p_{n_A-1,n_B+1}(t)$$
$$- \left(k^+ n_A + k^- n_B\right) p_{n_A,n_B}(t). \tag{3.47}$$

If the system is not in contact with a particle reservoir, the total number of particles $N = n_A + n_B$ is fixed. We therefore express eq. (3.47) in terms of a single variable by substituting $n_B = N - n_A$. In this representation, the jump rates $k_{n'_A n_A}$ read

$$k_{n'_A n_A} = k^+ n_A\, \delta_{n'_A,n_A-1} + k^- (N - n_A)\, \delta_{n'_A,n_A+1}. \tag{3.48}$$

Multiplying eq. (3.47) by n_A and summing over n_A, we obtain an equation for the average $\langle n_A \rangle = \sum_{n_A} n_A p_{n_A}$:

$$\frac{d}{dt} \langle n_A \rangle = k^+ \langle n_A(n_A - 1) \rangle + k^- \langle (N - n_A)(n_A + 1) \rangle$$

$$- k^+ \langle n_A^2 \rangle - k^- \langle (N - n_A) n_A \rangle \qquad (3.49)$$

$$= -k^+ \langle n_A \rangle + k^- \langle N - n_A \rangle.$$

Dividing by the volume V of the container, we obtain the rate equation in terms of the concentrations of A and B:

$$\frac{d[A]}{dt} = -k^+ [A] + k^- [B], \qquad (3.50)$$

and therefore, at equilibrium,

$$\frac{[B]}{[A]} = \frac{k^+}{k^-}. \qquad (3.51)$$

If the energy difference between a B molecule and an A molecule is equal to $\Delta \epsilon$, we must have, at temperature T,

$$\frac{[B]}{[A]} = e^{-\Delta \epsilon / k_B T}. \qquad (3.52)$$

Therefore, the reaction (3.46) is compatible with thermodynamic equilibrium if

$$\frac{k^+}{k^-} = e^{-\Delta \epsilon / k_B T}. \qquad (3.53)$$

Equation (3.53) is known in chemical kinetics as the **de Donder relation**, providing a connection between the forward and backward reaction steps of a chemical reaction and the affinity, i.e., the change in free energy associated with the reaction.

From (3.48), provided n_A, $(N - n_A) \gg 1$, we obtain

$$\frac{k_{n_A+1,n_A}}{k_{n_A,n_A+1}} = \frac{k^+ n_A}{k^- (N - n_A - 1)} \approx e^{-\Delta \epsilon_{n_A} / k_B T}, \qquad (3.54)$$

where $\Delta \epsilon_{n_A} = \epsilon_{n_A+1} - \epsilon_{n_A}$, in which

$$\epsilon_{n_A} = n_A k_B T \ln n_A + (N - n_A) [\epsilon + k_B T \ln(N - n_A)], \qquad (3.55)$$

is the free energy of a state with n_A A particles and $(N - n_A)$ B particles, taking into account the entropy of mixing. Equation (3.54) connects the "macroscopic" rates with the "microscopic" ones appearing in (3.53).

3.11 Linear response theory (*)

We consider a mesoscopic system that is brought out of equilibrium by a weak time-dependent manipulation. As a consequence of this manipulation, average values of observables might deviate from their equilibrium values and in principle depend on time. It is often useful to evaluate these deviations at first order in the intensity of the manipulation, as measured by the control parameter λ. This approach is known as **linear response theory**. Although we introduce it in the context of stochastic thermodynamics, linear response theory can be applied to a broader range of thermodynamic systems close to equilibrium.

We assume that the energy of the state x can be manipulated in r different ways, and we denote by $\lambda = (\lambda_\alpha)$, $\alpha = 1, \ldots, r$ the corresponding control parameters. Such parameters act linearly and independently on the energies, so that

$$\epsilon_x(\lambda) = \epsilon_x^{(0)} - \sum_{\alpha=1}^{r} \lambda_\alpha X_{\alpha,x}, \tag{3.56}$$

where the observables $X_{\alpha,x}$, $\alpha = 1, \ldots, r$ are given functions of x. At fixed values of the control parameters, the rates $k_{xx'}$ satisfy the detailed balance condition (2.85). We denote by $\langle \ldots \rangle_0$ the equilibrium average in the unperturbed case $\lambda = 0$. At equilibrium, the dynamics is invariant under time translations. We wish to evaluate the average values $\langle X_\beta(t) \rangle$ of the observables $X_{\beta,x}$ at time t as a function of the manipulation protocol $\lambda = (\lambda(t))$:

$$\langle X_\beta(t) \rangle_\lambda = \sum_x X_{\beta,x} p_x(t; \lambda), \tag{3.57}$$

where $p_x(t; \lambda)$ is the probability distribution of the states in the presence of the manipulation λ. We define the observables so that $\langle X_\alpha \rangle_0 = 0$ for all α. We then expand $\langle X_\alpha(t) \rangle$ as a functional power series,

$$\langle X_\beta(t) \rangle_\lambda = \int_{-\infty}^{\infty} dt' \sum_\alpha \mathcal{K}_{\beta\alpha}(t, t') \lambda_\alpha(t') + \text{higher-order terms}, \tag{3.58}$$

where we introduce the **linear response function** $\mathcal{K}_{\beta\alpha}(t, t')$.

We now express the linear response function in terms of equilibrium averages only. The function $\mathcal{K}_{\beta\alpha}(t, t')$ must vanish whenever $t' > t$, since the value of $\langle X_\beta(t) \rangle_\lambda$ cannot depend on future values of $\lambda_\alpha(t')$. Moreover, the effects of the manipulation must be invariant under time translation. This implies

$$\mathcal{K}_{\beta\alpha}(t, t') = K_{\beta\alpha}(t - t'), \qquad \forall \alpha, \beta. \tag{3.59}$$

We consider a manipulation $\lambda(t) = \lambda^{(0)} \theta(-t) e^{\varepsilon t}$, where $\varepsilon > 0$ is arbitrarily small and $\theta(t)$ is the Heaviside step function. Such a manipulation grows gradually from 0 for $t \to -\infty$ to a small constant value $\lambda^{(0)}$ and is then switched off at $t = 0$. Since ε is small, at $t = 0$ the probability distribution of the system is the equilibrium probability distribution $p^{eq}(\lambda)$ in the presence of a constant $\lambda^{(0)}$:

$$p_x(0; \lambda) = p_x^{eq}(\lambda) = e^{(F(\lambda) - \epsilon_x^{(0)} + \sum_\alpha \lambda_\alpha^{(0)} X_{\alpha,x})/k_B T}$$

$$\approx p^{eq}(0) \left(1 + \frac{1}{k_B T} \sum_\alpha \lambda_\alpha^{(0)} X_{\alpha,x} \right), \tag{3.60}$$

where we expand to linear order in $\lambda^{(0)}$ and use the fact that

$$\frac{\partial F(\lambda)}{\partial \lambda_\alpha} = \langle X_\alpha \rangle_0 = 0, \qquad \forall \alpha. \tag{3.61}$$

The average $\langle X_\beta(t) \rangle_\lambda$ for $t > 0$ is equal to the average of A_β over a probability distribution $p(t)$ that satisfies the unperturbed master equation (2.77), but with the initial condition $p^{eq}(\lambda)$. Given the initial condition $p_{x'}(t = 0; \lambda)$, the solution of this equation is given by $p_x(t) = \sum_{x'} G_{xx'}(t) p_{x'}(0; \lambda)$, where the matrix $G_{xx'}(t)$ is the Green function introduced in eq. (2.71). We then have, to first order in λ,

$$p_x(t) \approx p_x^{eq}(0) + \frac{1}{k_B T} \sum_{x'} G_{xx'}(t) \sum_\alpha \lambda_\alpha^{(0)} X_\alpha p_{x'}^{eq}(0) \tag{3.62}$$

and therefore, taking into account that $\langle X_\beta \rangle_0 = 0$,

$$\langle X_\beta(t) \rangle_\lambda \approx \frac{1}{k_B T} \sum_{xx'} X_{\beta,x} G_{xx'}(t) \sum_\alpha \lambda_\alpha^{(0)} X_\alpha p_{x'}^{eq}(0)$$

$$= \frac{1}{k_B T} \sum_\alpha \langle X_\beta(t) X_\alpha(0) \rangle_0 \, \lambda_\alpha^{(0)}. \tag{3.63}$$

Comparing with (3.58) and taking into account the expression of our protocol λ, we conclude that

$$\int_{-\infty}^0 \mathrm{d}t' \, K_{\beta\alpha}(t - t') = \frac{1}{k_B T} \langle X_\beta(t) X_\alpha(0) \rangle_0 = \frac{1}{k_B T} C_{\beta\alpha}(t), \tag{3.64}$$

where $C_{\beta\alpha}(t) = \langle X_\beta(t) X_\alpha(0) \rangle_0$ is the equilibrium time-dependent correlation function. We have, on the other hand,

$$\int_{-\infty}^0 \mathrm{d}t' \, K_{\beta\alpha}(t - t') = \int_t^{+\infty} \mathrm{d}t' \, K_{\beta\alpha}(t'). \tag{3.65}$$

Taking the derivative with respect to t, and taking into account that $K(t)$ vanishes for $t < 0$, we obtain the expression of the linear response function:

$$K_{\beta\alpha}(t) = -\frac{\theta(t)}{k_B T} \frac{\mathrm{d}}{\mathrm{d}t} \langle X_\beta(t) X_\alpha(0) \rangle_0 = -\frac{\theta(t)}{k_B T} \frac{\mathrm{d}}{\mathrm{d}t} C_{\beta\alpha}(t). \tag{3.66}$$

This result extends the **fluctuation-dissipation relation** from static to time-dependent quantities.

Fluctuation-dissipation relations connect equilibrium fluctuations of observables, measured by the correlation function on the right-hand side of eq. (3.66), with the response to a small external perturbation, given by the coefficient $K_{\beta\alpha}$ on the left-hand side. These relations played a fundamental role in the development on nonequilibrium statistical physics. We return to them in chapter 4 in the broader context of fluctuation relations.

3.12 More on coarse graining (*)

Throughout this chapter, we have emphasized that stochastic thermodynamics deals with a coarse-grained, mesoscopic description of a physical system. The separation between the slow timescales of the mesoscopic dynamics and the fast timescales of the heat reservoir (and of the microscopic degrees of freedom of the system) is crucial to achieve a simple closed description, where we can essentially forget about the fast microscopic degrees of freedom. In practice, there are different techniques to eliminate the fast dynamics and obtain an effective mesoscopic description. These techniques can be classified into two broad categories. **Coarse-graining procedures** group states together in such a way that the internal dynamics among states in a group is fast, whereas dynamics among different groups is slower. **Decimation procedures** eliminate states characterized by fast dynamics. With a slight abuse of language, we often refer to both categories as *coarse graining*. Both categories include several methods, whose effectiveness depends on the problem at hand.

In many practical situations, the choice of a coarse-graining scale is not unique. There can be different mesoscopic scales, corresponding to different levels of coarse graining. If these levels are well separated, one should be able, in principle, to formulate stochastic thermodynamics at each mesoscopic level, and these descriptions should be consistent among them. For example, we study how rates are transformed by changing coarse-graining levels. We consider an equilibrium physical system with two states, 1 and 2, separated by a high-energy state f (where f stands for "fast"). The master equation reads

$$\frac{d}{dt}p_1 = k_{1f}p_f - k_{f1}p_1;$$

$$\frac{d}{dt}p_f = k_{f1}p_1 + k_{f2}p_2 - (k_{1f} + k_{2f})p_f; \qquad (3.67)$$

$$\frac{d}{dt}p_2 = k_{2f}p_f - k_{f2}p_2.$$

The rates are expressed as in eq. (3.3). We now perform an **adiabatic elimination** of the fast state f. The adiabatic elimination is one of the most common decimation procedures and is schematized in fig. 3.5 for the case at hand. Since the dynamics of state f is fast, it always stays very close to a steady state on the timescales of the slow states. We therefore eliminate this state by setting $dp_f/dt = 0$. This implies $p_f = (k_{f1}p_1 + k_{f2}p_2)/(k_{1f} + k_{2f})$. Substituting into the remaining two equations yields

$$\frac{d}{dt}p_1 = k_{12}^{\text{eff}}p_2 - k_{21}^{\text{eff}}p_1,$$

$$\qquad (3.68)$$

$$\frac{d}{dt}p_2 = k_{21}^{\text{eff}}p_1 - k_{12}^{\text{eff}}p_2,$$

where

$$k_{21}^{\text{eff}} = \frac{k_{2f}k_{f1}}{k_{1f} + k_{2f}} = \frac{\omega_{1f}\omega_{2f}\,e^{(\epsilon_1 - \epsilon_2)/(2k_{\mathrm{B}}T)}e^{-\epsilon_f/(2k_{\mathrm{B}}T)}}{\omega_{1f}\,e^{-\epsilon_1/(2k_{\mathrm{B}}T)} + \omega_{2f}\,e^{-\epsilon_2/(2k_{\mathrm{B}}T)}};$$

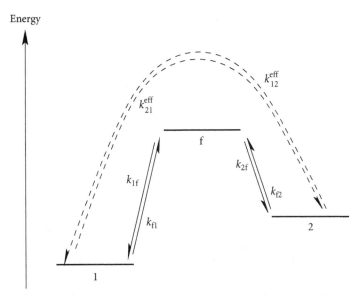

Figure 3.5. Adiabatic elimination of an intermediate high-energy state. Continuous arrows denote the original rates; dashed arrows denote coarse-grained rates. See eq. (3.69).

$$k_{12}^{\text{eff}} = \frac{k_{1f}k_{f2}}{k_{1f} + k_{2f}} = \frac{\omega_{1f}\omega_{2f}\, e^{(\epsilon_2-\epsilon_1)/(2k_BT)} e^{-\epsilon_f/(2k_BT)}}{\omega_{1f}\, e^{-\epsilon_1/(2k_BT)} + \omega_{2f}\, e^{-\epsilon_2/(2k_BT)}}. \tag{3.69}$$

The effective rates k_{21}^{eff} and k_{12}^{eff} in eq. (3.69) always satisfy detailed balance. This means that the system remains compatible with equilibrium also after the coarse-graining procedure. However, the precise expressions of the rates are significantly more complicated than the original ones. These expressions considerably simplify when k_{1f} and k_{2f} are equal, i.e., when

$$\omega_{1f} = \omega\, e^{(\epsilon_1-\epsilon_f)/(2k_BT)};$$
$$\omega_{2f} = \omega\, e^{(\epsilon_2-\epsilon_f)/(2k_BT)}. \tag{3.70}$$

Here ω is a rate constant. In this case, the effective rates become

$$k_{21}^{\text{eff}} = \frac{\omega}{2}\, e^{(\epsilon_1-\epsilon_f)/k_BT};$$
$$k_{12}^{\text{eff}} = \frac{\omega}{2}\, e^{(\epsilon_2-\epsilon_f)/k_BT}. \tag{3.71}$$

Such an exponential dependence of the rates on the "activation barriers" $\epsilon_f - \epsilon_1$ and $\epsilon_f - \epsilon_2$ is known in physical chemistry as the **Arrhenius law**. This dependence holds qualitatively also for other choices of ω_{1f} and ω_{2f}, as long as k_{1f} and k_{2f} do not differ too much.

We have seen that the mesoscopic energies usually treated in stochastic thermo-dynamics are, strictly speaking, free energies. Such free energies are automatically consistent across coarse-graining levels. Suppose, e.g., that a mesostate x contains different microstates y. We thus obtain

$$\epsilon_x = -k_{\mathrm{B}}T \ln \left(\sum_{y \in x} e^{-\epsilon_y/k_{\mathrm{B}}T} \right); \qquad \epsilon_y = -k_{\mathrm{B}}T \ln \left(\sum_{\xi \in y} e^{-\epsilon_\xi/k_{\mathrm{B}}T} \right). \qquad (3.72)$$

One then also has

$$\epsilon_x = -k_{\mathrm{B}}T \ln \left(\sum_{y \in x} \sum_{\xi \in y} e^{-\epsilon_\xi/k_{\mathrm{B}}T} \right) = -k_{\mathrm{B}}T \ln \left(\sum_{y \in x} e^{-\epsilon_y/k_{\mathrm{B}}T} \right). \qquad (3.73)$$

This property is general and does not require particular assumptions about how the states y are assigned to the states x.

We now turn our attention to work. For simplicity, we consider a purely manipulated system, without external driving. We follow the same line of eq. (3.14) and compare the definitions of work $w^{(x)}$ and $w^{(y)}$ at the two coarse-graining levels x and y.

$$w^{(y)} = \sum_{k=0}^{n} \int_{t_k}^{t_{k+1}} \mathrm{d}t \, \frac{\mathrm{d}\lambda}{\mathrm{d}t} \, \frac{\partial \epsilon_{y_k}}{\partial \lambda}; \qquad (3.74)$$

$$w^{(x)} = \sum_{k=0}^{n'} \int_{t_k}^{t_{k+1}} \mathrm{d}t \, \frac{\mathrm{d}\lambda}{\mathrm{d}t} \, \frac{\partial \epsilon_{x_k}}{\partial \lambda} = \sum_{k=0}^{n'} \int_{t_k}^{t_{k+1}} \mathrm{d}t \, \frac{\mathrm{d}\lambda}{\mathrm{d}t} \, \frac{\sum_{y \in x} (\partial \epsilon_y/\partial \lambda) \, e^{-\epsilon_y/k_{\mathrm{B}}T}}{\sum_{y \in x} e^{-\epsilon_y/k_{\mathrm{B}}T}}$$

$$= \sum_{k=0}^{n'} \int_{t_k}^{t_{k+1}} \mathrm{d}t \, \frac{\mathrm{d}\lambda}{\mathrm{d}t} \, \left\langle \frac{\partial \epsilon_y}{\partial \lambda} \right\rangle_{x,\mathrm{eq}}; \qquad (3.75)$$

where $\langle \ldots \rangle_{x,\mathrm{eq}}$ denotes an equilibrium average over the states y included in the coarse-grained state x. The two definitions of work are therefore equivalent when the states y within a state x can be considered at equilibrium among themselves, i.e., when the dynamics among the states within x is much faster than that among the mesostates x.

We now discuss the behavior of entropy production at different coarse-graining levels. We should distinguish between two cases. In the first case, the mesostates y within a state x can be considered always at equilibrium as we assumed before. In this case, we can follow the same logic of section 3.3 and conclude that, since heat exchanges between such mesostates and the heat reservoir are reversible, they do not contribute to entropy production. The conclusion in this case therefore is that entropy production does not change at different coarse-graining levels.

The situation is different when considering cases where the internal dynamics of the mesostates y inside a coarser mesostate x is fast but out of equilibrium. In these cases, entropy production tends to become smaller at coarser levels of description. One main reason is that increasing the coarse-graining level can "destroy" loops in the jump network that contribute to the entropy production. For example, if the three states 1, 2, 3 represented in fig. 3.6 are collapsed in the mesostate A, the entropy production associated with the current circulating in the loop does not appear anymore in the description of the system.

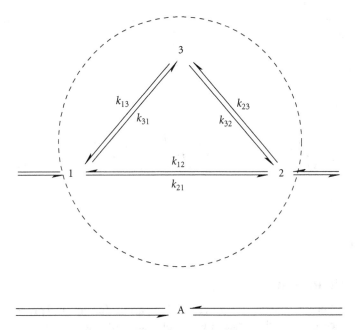

Figure 3.6. The three states 1, 2, 3 are collapsed into the state A upon coarse graining. As a consequence, the entropy production associated with the current circulating in the 1, 2, 3 loop disappears.

In some cases, due to coarse graining, there might be multiple microscopic pathways among pairs of states. This means that the effective rate $k_{xx'}$ represents the sum of rates $k_{xx'}^{(r)}$ associated with physically distinct possible jumps indexed by r, with $k_{xx'} = \sum_r k_{xx'}^{(r)}$. In this case, the correct expression of the average entropy production rate is

$$\dot{S}^{\text{tot}} = \frac{k_B}{2} \sum_{xx'r} J_{xx'}^{(r)} \ln \frac{k_{xx'}^{(r)} p_{x'}^{\text{st}}}{k_{x'x}^{(r)} p_{x}^{\text{st}}}, \tag{3.76}$$

where $J^{(r)}$ is defined by

$$J_{xx'}^{(r)} = k_{xx'}^{(r)} p_{x'}^{\text{st}} - k_{x'x}^{(r)} p_{x}^{\text{st}}, \qquad \forall x \neq x', \forall r. \tag{3.77}$$

We also define the effective current

$$J_{xx'} = \sum_r J_{xx'}^{(r)}, \qquad \forall x \neq x'. \tag{3.78}$$

The effective entropy production $\dot{S}^{\text{cg}}$ is the entropy production rate associated with the effective rates

$$\dot{S}^{\text{tot,eff}} = \frac{k_B}{2} \sum_{xx'} J_{xx'} \ln \frac{k_{xx'} p_{x'}^{\text{st}}}{k_{x'x} p_{x}^{\text{st}}}. \tag{3.79}$$

The effective average entropy production rate provides a lower bound to the actual average entropy production rate

$$0 \leq \dot{S}^{\text{tot,eff}} \leq \dot{S}. \tag{3.80}$$

One has indeed

$$\dot{S}^{\text{tot}} = \frac{k_B}{2} \sum_{xx'r} \left(k_{xx'}^{(r)} p_{x'}^{\text{st}} - k_{x'x}^{(r)} p_x^{\text{st}} \right) \ln \frac{k_{xx'}^{(r)} p_{x'}^{\text{st}}}{k_{x'x}^{(r)} p_x^{\text{st}}} = k_B \sum_{xx'r} k_{xx'}^{(r)} p_{x'}^{\text{st}} \ln \frac{k_{xx'}^{(r)} p_{x'}^{\text{st}}}{k_{x'x}^{(r)} p_x^{\text{st}}}$$

$$\geq k_B \sum_{xx'} k_{xx'} p_{x'}^{\text{st}} \ln \frac{k_{xx'} p_{x'}^{\text{st}}}{k_{x'x} p_x^{\text{st}}} = \dot{S}^{\text{tot,eff}}, \tag{3.81}$$

where we apply the logsum inequality (A.18).

3.13 Continuous systems (*)

Stochastic thermodynamics has developed in parallel for physical processes with discrete state space, described by master equations, and continuous state space, described by Langevin equations. The main results in stochastic thermodynamics, such as the fluctuation theorems discussed in the next chapter, can be equivalently derived in either case, although the formalism is quite different. In this section, we briefly describe the fundamental ideas of stochastic thermodynamics based on Langevin equations.

We consider a one-dimensional physical system with negligible inertia, described by the Langevin equation

$$\frac{dx}{dt} = \mu_P \mathcal{F}(x,t) + \sqrt{2D}\,\xi(t) = \mu_P \left[-\frac{\partial}{\partial x}\epsilon(x,\lambda) + f(x,t) \right] + \sqrt{2D}\,\xi(t), \tag{3.82}$$

where we split the total force $\mathcal{F}(x,t)$ into a term $-\partial\epsilon(x,\lambda)/\partial x$ originating from a conservative energy field $\epsilon(x,\lambda)$ and a nonconservative term $f(x,t)$ representing the nonequilibrium driving. We assume that the diffusion coefficient D is constant, which implies that the equation can be interpreted with either the Ito or the Stratonovich convention with identical results. Then the Fokker-Planck equation corresponding to eq. (3.82) reads

$$\frac{\partial p(x;t)}{\partial t} = \frac{\partial}{\partial x} \left[\mu_P \left(\frac{\partial}{\partial x}\epsilon(x,\lambda) - f(x,t) \right) p(x;t) + D\frac{\partial}{\partial x} p(x;t) \right]. \tag{3.83}$$

This corresponds to the probability current $J = \mu_P \mathcal{F} p - D\,\partial p/\partial x$ (cf. eq. (2.105)). We assume that the **mobility** μ_P and the diffusion coefficient D are connected by the **Einstein relation**,

$$D = k_B T \mu_P. \tag{3.84}$$

The Einstein relation is the condition for the Langevin equation (3.82) to be thermo-dynamically consistent. Indeed, if this condition is satisfied, the driving is absent, and λ is constant, the equation satisfies detailed balance and the probability current vanishes at stationarity. In this case, the Fokker-Planck equation (3.83) associated with the Langevin equation (3.82) admits as a solution the equilibrium distribution

$$p^{eq}(x) = e^{(F(\lambda) - \epsilon(x,\lambda))/k_B T}, \tag{3.85}$$

where $F(\lambda) = -k_B T \ln \int dx \, e^{-\epsilon(x,\lambda)/k_B T}$ is the free energy. If the nonconservative driving f is present and the boundary conditions are suitable, the equation admits a nonequilibrium stationary solution in which the work performed by the driving is released as heat to the reservoir.

The energy change associated with the Langevin equation is

$$d\epsilon = \frac{\partial \epsilon(x,\lambda)}{\partial x} \circ dx + \frac{\partial \epsilon(x,\lambda)}{\partial \lambda} \frac{d\lambda}{dt} dt. \tag{3.86}$$

The $\circ$ symbol reminds us that the product is interpreted in the Stratonovich sense. If we decided to adopt the Ito convention, eq. (3.86) would have contained an extra term; see section 2.9.

Similar to (3.9), the infinitesimal work is

$$dw = \frac{\partial \epsilon(x,\lambda)}{\partial \lambda} \frac{d\lambda}{dt} dt + f(x,t) \circ dx, \tag{3.87}$$

where the products are interpreted in the Stratonovich sense for the same reason as in eq. (3.86). Combining eqs. (3.86) and (3.87), we find a simple expression for the infinitesimal heat exchange with the reservoir:

$$dq = dw - d\epsilon = \left[-\frac{\partial \epsilon(x,\lambda)}{\partial x} + f(x,t) \right] \circ dx = \mathcal{F}(x,t) \circ dx, \tag{3.88}$$

where we recall the expression of $\mathcal{F}(x,t)$ given by eq. (3.82).

We now look at the total entropy production rate. Like in the case of discrete states, the stochastic entropy $s^{sys}(x,t)$ is defined by

$$s^{sys}(x,t) = -k_B \ln p(x;t), \tag{3.89}$$

where $p(x;t)$ is the probability distribution of the ensemble at time t. Using the rules of Stratonovich calculus, the rate of increase of the system entropy along the trajectory x is

$$\begin{aligned} \frac{d}{dt} s^{sys}(x(t),t) &= -k_B \frac{d}{dt} \ln p(x(t),t) \\ &= -k_B \left[\frac{1}{p(x;t)} \frac{\partial}{\partial x} p(x;t) \circ \frac{dx}{dt} + \frac{1}{p(x;t)} \frac{\partial}{\partial t} p(x;t) \right] \\ &= k_B \left[\frac{J(x,t)}{D p(x;t)} \circ \frac{dx}{dt} - \frac{\mathcal{F}(x,t)}{k_B T} \circ \frac{dx}{dt} - \frac{1}{p(x;t)} \frac{\partial}{\partial t} p(x;t) \right]. \end{aligned} \tag{3.90}$$

In the third line, we used the Einstein relation and the definition (2.105) of the current for the Fokker-Planck equation. Since $s^{res} = q/T$, we obtain

$$\dot{s}^{tot} = \frac{ds^{sys}}{dt} + \dot{s}^{res} = k_B \left[\frac{J(x,t)}{D p(x;t)} \circ \frac{dx}{dt} - \frac{1}{p(x;t)} \frac{\partial p(x(t),t)}{\partial t} \right]. \tag{3.91}$$

We now use this expression to evaluate the average entropy production rate. The average of the last term on the right-hand side vanishes:

$$\left\langle \frac{1}{p(x;t)} \frac{\partial}{\partial t} p(x;t) \right\rangle = \int dx \, \frac{\partial}{\partial t} p(x;t) = \frac{\partial}{\partial t} \int dx \, p(x;t) = 0. \tag{3.92}$$

To evaluate the average of the first term on the right-hand side of eq. (3.91), it is easier to apply the Ito convention, which has the advantage that the average of the noise term vanishes. We therefore switch from the Stratonovich to the Ito convention using the rule presented in eq. (2.127), obtaining

$$
\begin{aligned}
\langle ds^{tot} \rangle &= k_B \left\langle \frac{J(x,t)}{D \, p(x;t)} \circ dx \right\rangle \\
&= k_B \left\langle \frac{J(x,t) \mu_P \mathcal{F}(x,t)}{D \, p(x;t)} \, dt + \sqrt{\frac{2}{D}} \frac{J(x,t)}{p(x;t)} \circ dW \right\rangle \\
&= k_B \left\langle \frac{J(x,t) \mu_P \mathcal{F}(x,t)}{D \, p(x;t)} \, dt + \sqrt{\frac{2}{D}} \frac{J(x,t)}{p(x;t)} \cdot dW \right. \\
&\quad \left. + \frac{1}{p(x;t)} \frac{\partial}{\partial x} J(x,t) \, dt - \frac{J(x,t)}{p^2(x,t)} \frac{\partial}{\partial x} p(x;t) \, dt \right\rangle.
\end{aligned}
\tag{3.93}
$$

Here dW is the increment of the Wiener process, defined in eq. (2.112). The average of the second term vanishes in the Ito convention, as we anticipated. The average of the third term also vanishes since $\langle (\partial J(x,t)/\partial x)/p(x;t) \rangle = \int dx \, \partial J(x,t)/\partial x$, which can be transformed into a boundary term. We are therefore left with

$$
\begin{aligned}
\langle \dot{s}^{tot} \rangle &= k_B \left\langle \frac{J(x,t) \mu_P \mathcal{F}(x,t)}{D \, p(x;t)} - \frac{J(x,t)}{p^2(x,t)} \frac{\partial p(x;t)}{\partial x} \right\rangle \\
&= k_B \left\langle \frac{J^2(x,t)}{D \, p^2(x,t)} \right\rangle = k_B \int dx \, \frac{J^2(x,t)}{D \, p(x;t)},
\end{aligned}
\tag{3.94}
$$

where we again take advantage of eq. (2.105). This formula generalizes the expression for the average entropy production rate for master equations (3.39), to continuous systems.

3.14 Further reading

The ideas developed in this chapter constitute the founding core of stochastic thermodynamics. We discuss here a few references complementing our presentation, although there are many other seminal contributions that we do not review for reasons of space.

Sekimoto [151, 152] and Crooks [33] define stochastic work and heat for physical systems described by master equations. In our exposition, we loosely follow the last reference. The book by Sekimoto [153] is a classic reference that, among other things, highlights the conceptual difference between mesoscopic and calorimetric heat (ch. 6). Qian [134], Seifert [148], and other classic works discuss the definition and interpretation of stochastic entropy production.

The example of the myosin molecular motor is inspired by the book by Howard [80]. Stochastic chemical reactions have been both a fruitful playground for the development of stochastic thermodynamics and extremely relevant for applications; see Schnakenberg [146] and Rao and Esposito [135] for a more modern treatment.

The Schnakenberg formula (3.39) expresses the average entropy production rate. Schnakenberg's work develops ideas introduced by Bergmann and Lebowitz [16] in the context of the nonequilibrium statistical mechanics of Hamiltonian systems interacting with thermal reservoirs—a work that, in retrospect, anticipates several ideas of stochastic thermodynamics.

Bo and Celani [18] review in a comprehensive way coarse-graining and decimation techniques for stochastic processes, including potential pitfalls. See also Pigolotti and Vulpiani [129] for a discussion of adiabatic elimination in master equations. Puglisi et al. [133] proposes the idea that disappearence of loops under coarse graining leads to a reduced entropy production rate. Kawaguchi and Nakayama [85] point out that, when variables, even and odd, coexist under time reversal, the entropy production may increase upon coarse graining. Rao and Esposito [136] derive a general procedure to identify the conservative and the minimal set of nonconservative contributions to the entropy production.

Esposito [48] discusses more generally stochastic thermodynamics under coarse graining. Seifert [148] obtains the expression for the entropy production rate for continuous systems. The book by Sekimoto [153] and the review by Seifert [149] provide an extensive introduction to stochastic thermodynamics focused on Langevin equations. Expressions of observables for continuous systems can be also obtained from their discrete counterparts by taking a continuum limit; see, e.g., Gingrich et al. [70].

3.15 Exercises

3.1   Consider a system with three states, $x = 0, 1, 2$. The jump rates $k_{x'x}$ are all equal to 1, except the jumps $0 \leftrightarrow 1$, which are driven:

$$k_{10} = e^{\delta/2k_B T}; \qquad k_{01} = e^{-\delta/2k_B T}.$$

Simulate the process by the Gillespie algorithm for different values of δ and show that the total entropy produced up to time t can become negative, at least for small δ and short times.

3.2   Consider a system with two states $x = 0$ and $x = 1$ with energies $\epsilon_0 = 0$ and $\epsilon_1 = \epsilon(t)$, where $\epsilon(t) = \epsilon_f t/\mathcal{T}$ and ϵ_f is a constant. The system is at equilibrium at time $t_0 = 0$ and is manipulated in the time interval $[t_0, t_f]$ with $t_f = \mathcal{T}$. Simulate the stochastic dynamics of the two-state system for a given value of ϵ and different values of the duration $\mathcal{T}$, implementing a discrete-step simulation. From the simulations, compute the average total entropy production as a function of $\mathcal{T}$. Estimate analytically the total entropy production in the limiting cases of large and small $\mathcal{T}$ and verify that the result agrees with the simulations.

3.3   A reversible dimerization reaction $A + A \rightleftharpoons AA$ is characterized by a dimerization rate $k^+ = 1$ and a dissociation rate of dimers $k^- = 1$. Simulate the stochastic chemical reaction from an initial state with $n_A = 1000$ free monomers and no dimers, for a long enough time until equilibrium is reached. Estimate analytically the average total entropy production and compare the estimate with simulation results.

3.4   A three-state system is manipulated in a periodic way, $\lambda(t + \mathcal{T}) = \lambda(t)$. Its jump rates are given by

$$k_{xx'} = \omega_{xx'} \, e^{\epsilon_{x'}(\lambda)/k_B T}, \qquad x \in \{0, 1, 2\}.$$

Show that, with this choice of rates, the probability current averaged over a period $J_\mathcal{T} = \int_0^\mathcal{T} dt \, J_{xx'}(t)/\mathcal{T}$ always vanishes.

3.5   A particle is dragged by a force $f = 2$ in a periodic potential $\epsilon(x) = \sin(2\pi x)$, where parameters are in dimensionless units $\mu = k_B T = D = 1$. Simulate the Langevin equation corresponding to this physical system. Use simulations to show that the average total entropy production at steady state is linear in the duration of the time interval $\mathcal{T}$.

CHAPTER 4

Fluctuation Relations

Fluctuation relations are remarkable constraints on distributions of fluctuating quantities like stochastic heat, work, and entropy. They hold in rather general settings and find their origin in the assumptions of microscopic reversibility and thermodynamic consistency. Mathematically, fluctuation relations are based on the asymmetry between the probability of an observed trajectory and that of its time-reversed counterpart. This asymmetry is a hallmark of nonequilibrium systems and, in stochastic thermodynamics, is directly related to entropy production. In this chapter, we discuss how fluctuation theorems descend from this asymmetry and their main consequences for stochastic thermodynamics.

4.1 Irreversibility and entropy production

Nonequilibrium processes produce entropy because of their irreversible nature. The connection between irreversibility and entropy production can be made very explicit in stochastic thermodynamics. Our first step in this direction is to assign a precise meaning to the concept of **irreversibility**.

We consider a mesoscopic system satisfying generalized detailed balance, eq. (3.7). We imagine performing an experiment in which we prepare the system in an initial statistical state and then manipulate it during a time interval $[t_0, t_f]$. We assign to the control parameter λ the time-dependent value $\lambda(t)$. We call this procedure the **forward protocol** and the corresponding trajectories **forward trajectories**. With a slight abuse of language, we often use the expression *forward protocol* for the function $\lambda = (\lambda(t))$ as well.

We now imagine preparing the system, e.g., in the final state of the forward protocol, and performing the experiment with a **backward protocol**, i.e., by reversing the experimental procedure in time. Mathematically, this means that the control parameter is varied from $\lambda(t_f)$ to $\lambda(t_0)$, tracing back the values it assumed in the forward protocol. We call a stochastic trajectory generated by this procedure a **backward trajectory**. How likely is it to observe backward trajectories coinciding with typical forward trajectories, but in the reverse temporal order? If the probabilities of the most probable trajectories with the forward protocol are larger than those of their time-reversed images with the backward protocol, we can conclude that the underlying dynamics is irreversible.

We express the probability density $\mathcal{P}_x(\lambda)$ of a forward trajectory x with the forward protocol λ by

$$\mathcal{P}_x(\lambda) = \mathcal{P}_{x|x_0}(\lambda)\, p_{x_0}(t_0), \tag{4.1}$$

where $\mathcal{P}_{x|x_0}(\lambda)$ is the conditional probability of the trajectory x, given its initial state x_0 at time t_0:

$$
\begin{aligned}
\mathcal{P}_{x|x_0}(\lambda) = {}& e^{-\int_{t_n}^{t_f} k_{x_f}^{\mathrm{out}}(t)\, dt}\, k_{x_f x_{n-1}}(t_m)\, e^{-\int_{t_{n-1}}^{t_n} k_{n-1}^{\mathrm{out}}(t)\, dt}\, \cdots \\
& \times e^{-\int_{t_2}^{t_3} k_{x_2}^{\mathrm{out}}(t)\, dt}\, k_{x_2 x_1}(t_2)\, e^{-\int_{t_1}^{t_2} k_{x_1}^{\mathrm{out}}(t)\, dt}\, k_{x_1 x_0}(t_1)\, e^{-\int_{t_0}^{t_1} k_{x_0}^{\mathrm{out}}(t)\, dt};
\end{aligned}
\tag{4.2}
$$

see section 2.7. We now associate with each trajectory x a backward trajectory $\widehat{x}$. The backward trajectory traces all the states visited by x in reverse order, dwelling in each state the same time as the forward trajectory. It is obtained from the original trajectory by an operation of **time reversal**:

$$\widehat{t} = t_f - (t - t_0), \qquad t_0 \leq t \leq t_f. \tag{4.3}$$

We assume here for simplicity that the operation of time reversal leaves the states x unaffected. We discuss systems in which this is not the case in sections 4.10 and 4.14. We now express the backward trajectory using time reversal:

$$\widehat{x}(t) = x(\widehat{t}). \tag{4.4}$$

The initial state and time, respectively, of the backward trajectory are the final state and time of the forward one.

We also introduce a backward dynamics. The jump rates of the backward dynamics are the same as the forward one, but with the **backward protocol**

$$\widehat{\lambda}(t) = \lambda(\widehat{t}), \tag{4.5}$$

where $\widehat{t}$ is defined in eq. (4.3). The probability density of the backward trajectory $\widehat{x}$ with the backward protocol $\widehat{\lambda}$ is

$$\mathcal{P}_{\widehat{x}}(\widehat{\lambda}) = \mathcal{P}_{\widehat{x}|x_f}(\widehat{\lambda})\, p_{x_f}(t_f), \tag{4.6}$$

where $\mathcal{P}_{\widehat{x}|x_f}(\widehat{\lambda})$ is the conditional probability of the backward trajectory given its initial state, which is the final state x_f of the forward trajectory. This conditional probability is expressed by

$$
\begin{aligned}
\mathcal{P}_{\widehat{x}|x_f}(\widehat{\lambda}) = {}& e^{-\int_{t_0}^{t_1} k_{x_0}^{\mathrm{out}}(t)\, dt}\, k_{x_0 x_1}(t_1)\, e^{-\int_{t_1}^{t_2} k_{x_1}^{\mathrm{out}}(t)\, dt}\, \cdots \\
& \times e^{-\int_{t_{n-1}}^{t_n} k_{x_{n-1}}^{\mathrm{out}}(t)\, dt}\, k_{x_{n-2} x_{n-1}}(t_{n-1})\, e^{-\int_{t_{n-1}}^{t_n} k_{x_{n-1}}^{\mathrm{out}}(t)\, dt} \\
& \times k_{x_{n-1} x_f}(t_n)\, e^{-\int_{t_n}^{t_f} k_{x_f}^{\mathrm{out}}(t)\, dt},
\end{aligned}
\tag{4.7}
$$

where we exploit the relation

$$k_{xx'}(\widehat{\lambda}(\widehat{t})) = k_{xx'}(\lambda(t)) = k_{xx'}(t). \tag{4.8}$$

We now evaluate the ratio between conditional probabilities of forward and backward trajectories. The expression considerably simplifies since the exponential factors associated with the dwells cancel out:

$$\frac{\mathcal{P}_{\boldsymbol{x}|x_0}(\boldsymbol{\lambda})}{\mathcal{P}_{\widehat{\boldsymbol{x}}|x_f}(\widehat{\boldsymbol{\lambda}})} = \prod_{i=1}^{n} \frac{k_{x_i x_{i-1}}}{k_{x_{i-1} x_i}}. \tag{4.9}$$

We further assume that our system is coupled to a heat reservoir at temperature T, so that its jump rates satisfy the generalized detailed balance condition, eq. (3.7). Substituting this condition into eq. (4.9), we obtain

$$\frac{\mathcal{P}_{\boldsymbol{x}|x_0}(\boldsymbol{\lambda})}{\mathcal{P}_{\widehat{\boldsymbol{x}}|x_f}(\widehat{\boldsymbol{\lambda}})} = \exp\left[\frac{1}{k_B T} \sum_{k=1}^{n-1} \left(\epsilon_{x_{k-1}} - \epsilon_{x_k} + \delta_{x_k x_{k-1}}\right)\right] = e^{q(\boldsymbol{x})/k_B T}, \tag{4.10}$$

where the second equality follows from the definition of stochastic heat, eq. (3.10). Equation (4.10) expresses the imbalance between conditional probabilities of forward and backward trajectories in terms of the heat released to the reservoir. We now express the ratio of unconditioned probabilities of trajectories by including the probabilities $p_{x_0}(t_0)$ and $p_{x_f}(t_f)$ of the initial and final states, respectively:

$$\frac{\mathcal{P}_{\boldsymbol{x}}(\boldsymbol{\lambda})}{\mathcal{P}_{\widehat{\boldsymbol{x}}}(\widehat{\boldsymbol{\lambda}})} = \frac{p_{x_0}(t_0)\mathcal{P}_{\boldsymbol{x}|x_0}(\boldsymbol{\lambda})}{p_{x_f}(t_f)\mathcal{P}_{\widehat{\boldsymbol{x}}|x_f}(\widehat{\boldsymbol{\lambda}})} = \exp\left[\frac{s_{x_f}^{\text{sys}}(t_f) - s_{x_0}^{\text{sys}}(t_0)}{k_B} + \frac{q(\boldsymbol{x})}{k_B T}\right] = e^{s^{\text{tot}}(\boldsymbol{x})/k_B}. \tag{4.11}$$

And, solving for the total entropy production $s^{\text{tot}}(\boldsymbol{x})$,

$$s^{\text{tot}}(\boldsymbol{x}) = k_B \ln\left[\frac{\mathcal{P}_{\boldsymbol{x}}(\boldsymbol{\lambda})}{\mathcal{P}_{\widehat{\boldsymbol{x}}}(\widehat{\boldsymbol{\lambda}})}\right]. \tag{4.12}$$

Equation (4.12) is central in stochastic thermodynamics. It provides the announced connection between the entropy production and irreversibility, quantified as the asymmetry between the probabilities of forward and backward trajectories. The consequences of this relation are the focus of this chapter.

At equilibrium, where the system is not manipulated and the jump rates satisfy the detailed balance relation (2.85), one has

$$\mathcal{P}_{\boldsymbol{x}} = \mathcal{P}_{\widehat{\boldsymbol{x}}}, \qquad \forall \boldsymbol{x}. \tag{4.13}$$

Equation (4.13) implies that the entropy production $s^{\text{tot}}(\boldsymbol{x})$ vanishes for all trajectories $\boldsymbol{x}$ at equilibrium.

4.2 Integral fluctuation relation

Irreversibility and entropy production are related by eq. (4.12) at the level of individual trajectories. We now consider an ensemble of trajectories starting from an initial distribution of states $p_0(x_0)$. Because of microscopic reversibility, the same set of states is accessible to both dynamics, hence there is a one-to-one correspondence between trajectories of the forward and backward dynamics. We introduce averages over the forward and backward ensembles of trajectories:

$$\langle\ldots\rangle_F = \int \mathcal{D}x\, P_x(\lambda)\ldots\,, \qquad \langle\ldots\rangle_B = \int \mathcal{D}\widehat{x}\, P_{\widehat{x}}(\widehat{\lambda})\ldots\,, \tag{4.14}$$

where we use $\int \mathcal{D}\widehat{x} = \int \mathcal{D}x$, a property that can be verified by changing the signs of the infinitesimal time increments in eq. (2.92). Using eq. (4.12), we directly obtain

$$\left\langle e^{-s^{\mathrm{tot}}(x)/k_B}\right\rangle_F = \int \mathcal{D}x\, P_{\widehat{x}}(\widehat{\lambda}) = 1. \tag{4.15}$$

Equation (4.15) is the **integral fluctuation relation**, a relation that has many important consequences. In particular, since the exponential is a convex function, the Jensen inequality entails $\ln\langle e^{-x}\rangle \geq \ln e^{-\langle x\rangle} = -\langle x\rangle$; see appendix A.1. Therefore, eq. (4.15) implies

$$\left\langle s^{\mathrm{tot}}(x)\right\rangle_F \geq 0, \tag{4.16}$$

which is the **second law of stochastic thermodynamics**. Like the traditional second law of thermodynamics, eq. (4.16) reduces to an equality for equilibrium processes. In this case, the equality is not only valid on average but also trajectory-wise: because of detailed balance, the total entropy production vanishes for any equilibrium trajectory; see section 4.1.

On the other hand, out of equilibrium, each trajectory is characterized by its own entropy production, and the second law is only valid on average. Since $e^{-s^{\mathrm{tot}}(x)/k_B} < 1$ for $s^{\mathrm{tot}}(x) > 0$, positive values of the entropy production must be compensated by negative values for eq. (4.15) to hold. This means that, out of equilibrium, there is always a nonvanishing probability of observing trajectories characterized by a negative total entropy production. This fact is one of the most important physical consequences of the integral fluctuation relation.

There are situations in which the probability distribution of s^{tot} is Gaussian. For example, this is the case in the **linear response regime**, i.e., in the regime close to thermodynamic equilibrium where linear response theory as described in section 3.11 applies. We prove this fact in section 6.2. For Gaussian distributions, the integral fluctuation relation (4.15) implies a relation between the mean and the variance of the distribution of s^{tot}:

$$2k_B\langle s^{\mathrm{tot}}\rangle = \sigma^2_{s^{\mathrm{tot}}}. \tag{4.17}$$

This equation is another example of a fluctuation-dissipation relation. Its derivation is left as exercise 4.1.

4.3   Dragged particle on a ring

We now consider a particle on a ring of discrete states, $i = 1, \ldots, N$, with periodic boundary conditions. All states have the same energy. An external agent drags the particle in the counterclockwise direction with a constant driving δ. The jump rates are given by

$$k^+ = k_{x+1,x} = \omega\, e^{\delta/k_B T}; \qquad k^- = k_{x-1,x} = \omega, \tag{4.18}$$

(a)

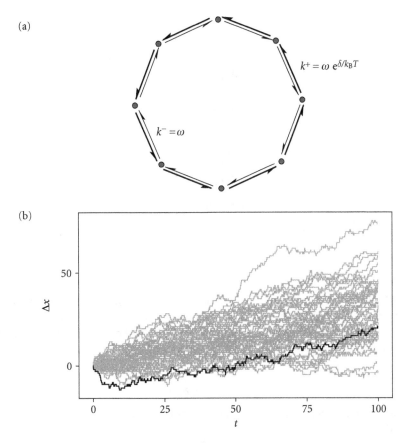

$k^+ = \omega \, e^{\delta/k_B T}$

$k^- = \omega$

(b)

Figure 4.1. (a) Jump network of the dragged particle on a ring. The driven jumps, $x \longrightarrow x+1$, are shown with bold arrows. (b) Examples of trajectories for $\omega = 1$ and $\delta = 0.2 \, k_B T$. The y-axis represents the total displacement Δx since the initial time. A random trajectory is highlighted.

where $+, -$ stand for counterclockwise and clockwise, respectively. The jump network is shown in fig. 4.1a. We want to verify that the integral fluctuation relation holds. We consider a time interval $[t_0, t_f]$. During this time interval, the particle performs a number n^+ of counterclockwise jumps and a number n^- of clockwise jumps. Since all states are identical, clockwise and counterclockwise jumps occur at constant rates. This implies that n^+ and n^- are Poisson distributed:

$$
\begin{aligned}
p_{n^+} &= \frac{1}{n^+!} \left[\omega \, e^{\delta/k_B T} \mathcal{T} \right]^{n^+} \exp\left(-\omega \, e^{\delta/k_B T} \mathcal{T} \right), \\
p_{n^-} &= \frac{1}{n^-!} \left[\omega \mathcal{T} \right]^{n^-} \exp\left(-\omega \mathcal{T} \right),
\end{aligned}
$$

(4.19)

where $\mathcal{T} = t_f - t_0$. Examples of trajectories are shown in fig. 4.1b.

The external agent performs work equal to δ for every counterclockwise jump. The same amount of work is returned to the external agent at every clockwise jump. The total work is therefore

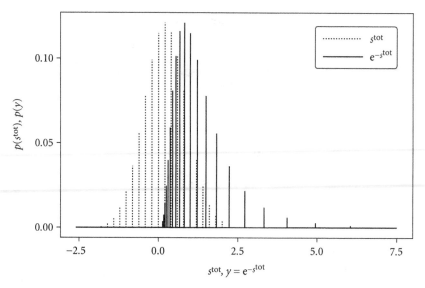

Figure 4.2. Probability distribution of s^{tot} and $y = \exp^{-s^{tot}}$ for the dragged particle over a duration $\mathcal{T} = 5\,\text{s}$, with $\omega = 1\,\text{s}^{-1}$ and $\delta = 0.2\,k_B T$. For these parameters, the average entropy production is $\omega \mathcal{T} \delta\,[\exp(\delta/k_B T) - 1]/T \approx 0.224\,k_B$. Entropy production is plotted in units of k_B.

$$w = (n^+ - n^-)\,\delta. \tag{4.20}$$

Since the internal energy is constant, all the work is released into the heat reservoir. This implies that

$$s^{tot} = \frac{w}{T} = \frac{\delta}{T}\,(n^+ - n^-) \tag{4.21}$$

and therefore

$$p(s^{tot}) = \sum_{n^+,n^-=0}^{\infty} p_{n^+} p_{n^-}\,\delta^{K}_{T\,s^{tot}/\delta,(n^+-n^-)}. \tag{4.22}$$

For our parameter choice, the distribution of s^{tot} is close to a Gaussian, as shown in fig. 4.2. The distribution of e^{-s^{tot}/k_B} is instead markedly skewed and exhibits a number of peaks at large values, corresponding to the events in which s^{tot} is large and negative.

It is possible to analytically compute the sums in eq. (4.22). However, it is enough for our purposes to substitute eq. (4.22) directly in the integral fluctuation relation, obtaining

$$\left\langle e^{-s^{tot}/k_B} \right\rangle = \sum_{s^{tot}} e^{-s^{tot}/k_B} p(s^{tot}) = \sum_{n^+,n^-=0}^{\infty} p_{n^+} p_{n^-}\,e^{-\delta\,(n^+-n^-)/k_B T}$$

$$= \left[\sum_{n^+=0}^{\infty} p_{n^+} e^{-\delta\,n^+/k_B T} \right] \left[\sum_{n^-=0}^{\infty} p_{n^-} e^{\delta\,n^-/k_B T} \right] \tag{4.23}$$

$$= \exp\left[\omega \mathcal{T} \left(1 - e^{\delta/k_B T} \right) \right] \exp\left[\omega \mathcal{T} \left(e^{\delta/k_B T} - 1 \right) \right] = 1.$$

This calculation illustrates how the integral fluctuation relation results from a delicate balance between the contribution of more likely jumps that produce entropy and less likely jumps characterized by a negative entropy production.

4.4 Back to linear response theory (*)

The fluctuation-dissipation relation (4.17) descends from the integral fluctuation relation (4.15) close to equilibrium. However, it is historically more appropriate to say that fluctuation relations are the generalization of fluctuation-dissipation relations far from equilibrium, since the latter were discovered earlier. In this section, we return to the linear response theory developed in section 3.11 and connect it with time-reversal symmetry.

We consider the fluctuation-dissipation relation (3.66) and assume that the observables X, as well as the dynamics, are invariant under time reversal. At equilibrium, the probability of a trajectory x is equal to the probability of its backward trajectory $\widehat{x}$. This probability is also invariant under time translations. As a consequence, the correlation functions introduced in section 3.11 are symmetric:

$$
C_{\alpha\beta}(t) = \int \mathcal{D}x \, \mathcal{P}_x \, X_{\alpha,x(t)} \, X_{\beta,x(0)} = \int \mathcal{D}x \, \mathcal{P}_x \, X_{\alpha,x(0)} \, X_{\beta,x(-t)}
$$

$$
= \int \mathcal{D}x \, \mathcal{P}_{\widehat{x}} \, X_{\alpha,\widehat{x}(0)} \, X_{\beta,\widehat{x}(t)} = C_{\beta\alpha}(t), \tag{4.24}
$$

where we exploit time-translation invariance in the second equality, time-reversal invariance in the third, and finally the fact that $\mathcal{D}x = \mathcal{D}\widehat{x}$. Comparing with eq. (3.64), we obtain

$$
\int_{-\infty}^{t_0} \mathrm{d}t' \, K_{\beta\alpha}(t - t') = \int_{-\infty}^{t_0} \mathrm{d}t' \, K_{\alpha\beta}(t - t'). \tag{4.25}
$$

Taking the derivative with respect to t, we obtain the symmetry of the response functions:

$$
K_{\alpha\beta}(t) = K_{\beta\alpha}(t), \qquad \forall t. \tag{4.26}
$$

This relation can be seen as an extension of the Maxwell relations (2.30), which express the same symmetry for the derivatives of thermodynamic potentials.

As we are considering small perturbations, correlation functions must satisfy a linear system of equations:

$$
\frac{\mathrm{d}}{\mathrm{d}t} C_{\alpha\beta}(t) = -\sum_{\gamma} M_{\alpha\gamma} C_{\gamma\beta}(t), \qquad t > 0, \tag{4.27}
$$

where $(M_{\alpha\beta})$ is a positive-definite matrix since the equilibrium state must be stable. Therefore, correlations decay exponentially with time. Taking the derivative of eq. (4.24) with respect to t in the limit $t \to 0$, we obtain

$$
L_{\alpha\beta} = \sum_{\gamma} M_{\alpha\gamma} C_{\gamma\beta}(0) = \sum_{\gamma} M_{\beta\gamma} C_{\gamma\alpha}(0) = L_{\beta\alpha}. \tag{4.28}
$$

Thus the matrix $L_{\alpha\beta}$ is also symmetric. The relations (4.28) embodying this symmetry are known as the **Onsager reciprocity relations**. To interpret them physically, we introduce thermodynamic forces Y_α that describe deviations from equilibrium:

$$Y_\alpha = \frac{\partial S}{\partial X_\alpha}, \tag{4.29}$$

where $S(X) = S(X_1, X_2, \ldots)$ is the system entropy expressed in terms of the instantaneous values of the observables X_α. We assume that the macroscopic laws for the time rates of change of the X_α are linear in the thermodynamic forces Y_α, at least for moderate deviations from equilibrium:

$$\frac{dX_\alpha}{dt} = \frac{1}{k_B} \sum_\beta L_{\alpha\beta} Y_\beta, \tag{4.30}$$

where the Boltzmann constant is introduced for convenience. We now show that the matrix $(L_{\alpha\beta})$ in eq. (4.30) is the same matrix defined in eq. (4.28). Evaluating the correlation $C_{\alpha\gamma}(t)$ for short times, we obtain

$$\left. \frac{dC_{\alpha\gamma}}{dt} \right|_{t=0} = \sum_\beta L_{\alpha\beta} \left\langle X_\gamma Y_\beta \right\rangle^{eq}. \tag{4.31}$$

By the Boltzmann-Einstein principle (see eq. (2.60)), the equilibrium probability distribution of the observables $X = (X_\alpha)$ is proportional to $\exp(S(X)/k_B)$. We therefore obtain

$$\left\langle X_\gamma Y_\beta \right\rangle^{eq} = \mathcal{N} \int \prod_\alpha dX_\alpha \, X_\gamma \, \frac{\partial S}{\partial X_\beta} \, e^{S(X)/k_B} = \mathcal{N} k_B \int \prod_\alpha dX_\alpha \, X_\gamma \, \frac{\partial}{\partial X_\beta} \, e^{S(X)/k_B}, \tag{4.32}$$

where $\mathcal{N}$ is a normalization constant. Integrating by parts, we obtain

$$\left\langle X_\gamma Y_\beta \right\rangle^{eq} = -k_B \, \delta_{\gamma\beta}^K. \tag{4.33}$$

A comparison with eq. (4.27) confirms that the matrix $L_{\alpha\beta}$ introduced in eq. (4.30) is equal to that appearing in eq. (4.28). The coefficient $L_{\alpha\beta}$ appearing in eq. (4.30) relates the time derivative of the observable X_α with the force Y_β conjugate to the observable X_β. A similar relation connects the time derivative of X_β with the force Y_α conjugate to X_α via the coefficient $L_{\beta\alpha}$. The Onsager reciprocal relations impose that these coefficients are equal.

The Onsager reciprocity relations generalize to nonequilibrium steady states. In the steady state, the average entropy production rate $\dot{S}$ is expressed by

$$\dot{S} = \frac{1}{T} \sum_\alpha A_\alpha J_\alpha, \tag{4.34}$$

where the A_α are the affinities and the J_α the corresponding currents; see section 3.9. Since both the As and the Js vanish at equilibrium, we have close to equilibrium

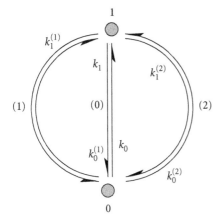

Figure 4.3. Jump network of the two-state, two-cycle model. Node 0 represents the inactive state, 1 the active state. Edge (0) represents spontaneous jumps due to thermal fluctuations. Edge (1) corresponds to substrate processing and edge (2) to hydrolysis of a high-energy molecule. Jumps along edges (1) and (2) involve interaction with particle reservoirs.

$$J_\alpha \approx \sum_\beta L_{\alpha\beta} A_\beta. \qquad (4.35)$$

In this case, the matrix $L = (L_{\alpha\beta})$ is called the **matrix of kinetic coefficients**. The Onsager reciprocity relations impose that the matrix L is symmetric as well, allowing us to relate kinetic coefficients appearing in different physical processes.

As an example, we consider an enzyme that transforms a substrate R into a product P. The jump network for this system is represented in fig. 4.3. The substrate is transformed into the product when jumping from the activated state 1 to the ground state 0 along pathway (1). The same transformation is assisted by hydrolysis of a high-energy molecule along pathway (2). Pathway (0) represents spontaneous activation/deactivation of the enzyme by interaction with the heat reservoir, without processing any molecule.

The jump (0) is not driven, while (1) and (2) entail a driving. Following the approach described in section 3.9, we choose edge (0) as the spanning tree and denote by $\mathcal{C}_\alpha$, $\alpha = (1, 2)$ the fundamental cycles obtained by joining edge (0) with edge (α). All cycles are considered positive in the counterclockwise direction. The master equation reads

$$\frac{dp_0}{dt} = \left(k_1 + k_1^{(1)} + k_1^{(2)}\right) p_1 - \left(k_0 + k_0^{(1)} + k_0^{(2)}\right) p_0;$$

$$\frac{dp_1}{dt} = \left(k_0 + k_0^{(1)} + k_0^{(2)}\right) p_0 - \left(k_1 + k_1^{(1)} + k_1^{(2)}\right) p_1. \qquad (4.36)$$

We express the jump rates by

$$k_1 = \omega_0\, e^{\epsilon/2k_BT};$$
$$k_1^{(1)} = \omega_1\, e^{(\epsilon+\delta_1)/2k_BT};$$
$$k_1^{(2)} = \omega_2\, e^{(\epsilon+\delta_2)/2k_BT};$$

$$k_0 = \omega_0\, e^{-\epsilon/2k_BT};$$
$$k_0^{(1)} = \omega_1\, e^{-(\epsilon+\delta_1)/2k_BT}; \qquad (4.37)$$
$$k_0^{(2)} = \omega_2\, e^{-(\epsilon+\delta_2)/2k_BT}.$$

Here ϵ is the energy difference between states 1 and 0, and $\delta_{1,2}$ are the drivings. The affinities are

$$A^{(1)} = k_B T \ln \frac{k_0 k_1^{(1)}}{k_1 k_0^{(1)}} = \delta_1;$$

$$A^{(2)} = k_B T \ln \frac{k_1 k_0^{(2)}}{k_0 k_1^{(2)}} = -\delta_2. \tag{4.38}$$

The steady-state probability distribution is given by

$$p_0^{st} = \frac{1}{\mathcal{N}} \left(k_1 + k_1^{(1)} + k_1^{(2)} \right);$$

$$p_1^{st} = \frac{1}{\mathcal{N}} \left(k_0 + k_0^{(1)} + k_0^{(2)} \right); \tag{4.39}$$

where

$$\mathcal{N} = k_0 + k_0^{(1)} + k_0^{(2)} + k_1 + k_1^{(1)} + k_1^{(2)}. \tag{4.40}$$

The average currents in the two fundamental cycles are given by

$$J^{(1)} = k_1^{(1)} p_1^{st} - k_1^{(0)} p_0^{st} = \frac{1}{\mathcal{N}} \left[2\omega_0 \omega_1 \sinh\left(\frac{\delta_1}{2 k_B T} \right) + 2\omega_1 \omega_2 \sinh\left(\frac{\delta_1 - \delta_2}{2 k_B T} \right) \right];$$

$$J^{(2)} = k_0^{(2)} p_0^{st} - k_1^{(2)} p_1^{st} = \frac{1}{\mathcal{N}} \left[2\omega_0 \omega_2 \sinh\left(\frac{\delta_2}{2 k_B T} \right) - 2\omega_1 \omega_2 \sinh\left(\frac{\delta_1 - \delta_2}{2 k_B T} \right) \right]. \tag{4.41}$$

Setting $\delta_{1,2} \ll k_B T$, we obtain $\mathcal{N} = 2 (\omega_0 + \omega_1 + \omega_2) \cosh(\epsilon / 2 k_B T)$ and the following expression of the matrix $L_{\alpha\beta}$:

$$L_{11} = \frac{\omega_1 (\omega_0 + \omega_2)}{\mathcal{N} k_B T};$$

$$L_{22} = \frac{\omega_2 (\omega_0 + \omega_1)}{\mathcal{N} k_B T}; \tag{4.42}$$

$$L_{12} = -\frac{\omega_1 \omega_2}{\mathcal{N} k_B T} = L_{21}.$$

The coefficients L_{12} and L_{21} are equal, as predicted by the Onsager reciprocity relations.

4.5 Detailed fluctuation relation

The detailed fluctuation relation is a stronger version of the integral fluctuation relation. It however requires the further assumption that the entropy production $s^{tot}(x)$ is odd under the mapping $x \longrightarrow \widehat{x}$:

$$s^{\text{tot}}(\widehat{x}) = -s^{\text{tot}}(x). \tag{4.43}$$

This property is called **involution**. Equation (4.12) seems to suggest that the entropy production is always an involution. There is however a subtle point. The entropy production of the backward trajectory is

$$s^{\text{tot}}(\widehat{x}) = k_{\text{B}} \ln \left[\frac{\mathcal{P}_{\widehat{x}}(\widehat{\lambda})}{\mathcal{P}_{\widehat{\widehat{x}}}(\widehat{\widehat{\lambda}})} \right]. \tag{4.44}$$

The time-reversal operation is an involution, therefore it is always true that $\widehat{\widehat{\lambda}} = \lambda$. However, the involution property also requires that $\mathcal{P}_{\widehat{\widehat{x}}}(\lambda) = \mathcal{P}_x(\lambda)$, i.e., that the backward-of-the-backward trajectory probability is equal to the forward trajectory probability. This requires that, taking the final distribution $P_{x_\text{f}}(t_\text{f})$ as the initial condition and solving the differential equation with the backward protocol, we recover the initial distribution $P_{x_0}(t_0)$ after a duration equal to $t_\text{f} - t_0$. This is not necessarily the case, since the distribution p_x at the end of the backward manipulation depends not only on the instantaneous value of the parameter $\lambda(t_0)$ but on the whole backward protocol $\widehat{\lambda}$. Examples in which the entropy production is and is not an involution are shown in fig. 4.4 for a two-state system.

Assuming that the entropy production is an involution, given an arbitrary function $f(s)$, we obtain

$$\left\langle f(s^{\text{tot}}(x)) \, e^{-s^{\text{tot}}(x)/k_{\text{B}}} \right\rangle_{\text{F}} = \left\langle f(s^{\text{tot}}(x)) \right\rangle_{\text{B}} = \left\langle f(-s^{\text{tot}}(\widehat{x})) \right\rangle_{\text{B}}. \tag{4.45}$$

We now choose the function $f(s^{\text{tot}}(x)) = \delta(s^{\text{tot}}(x) - s)$ that selects all trajectories x characterized by a given value s of s^{tot}. Substituting and rearranging terms, we obtain the **detailed fluctuation relation**

$$\frac{p(s^{\text{tot}}; \lambda)}{p(-s^{\text{tot}}; \widehat{\lambda})} = e^{s^{\text{tot}}/k_{\text{B}}}. \tag{4.46}$$

Nonequilibrium steady states are an important case where the involution property always holds. The reason is that, in this case, the distribution p_x does not depend on time and the dynamics for the backward and forward trajectories are identical. For nonequilibrium steady states, the detailed fluctuation relation becomes

$$\frac{p(s^{\text{tot}})}{p(-s^{\text{tot}})} = e^{s^{\text{tot}}/k_{\text{B}}}. \tag{4.47}$$

This expression helps clarify why the detailed fluctuation relation is a more powerful result than the integral fluctuation relation. The integral fluctuation relation provides a single global constraint on the distribution of the entropy production. However, the detailed fluctuation relation provides, *for any value of s^{tot}*, a relation between the probabilities of observing entropy productions $+s^{\text{tot}}$ and $-s^{\text{tot}}$. Thanks to the detailed fluctuation relation, in a steady-state system it is sufficient to know only $p(s^{\text{tot}})$ for positive (or negative) values of s^{tot} to reconstruct the entire distribution.

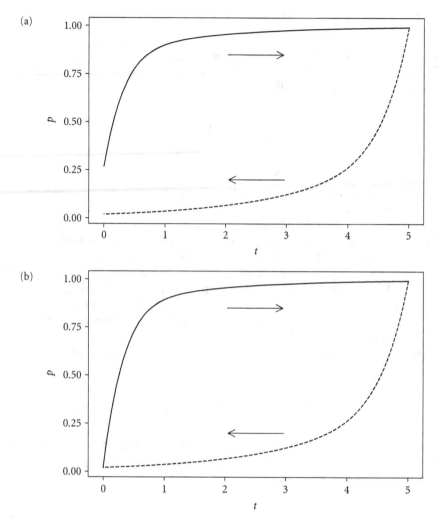

Figure 4.4. Involution property in a two-state system. The instantaneous probability $p(t)$ of the 0 state is plotted against t. (a) For a generic initial condition $p(t_0)$, performing the forward manipulation (solid line) and immediately following it with the backward manipulation (dashed line), the initial probability is not recovered. Involution does not hold. (b) With a carefully chosen initial condition, the involution property holds. In the model, the energy difference $\epsilon_1 - \epsilon_0$ between the two states is manipulated according to a linear protocol: $\epsilon_1 - \epsilon_0 = \lambda(t) = \lambda_0 + (\lambda_f - \lambda_0)(t/\mathcal{T})$. The jump rates are given by $k_{10} = \omega\, e^{-(\epsilon_1-\epsilon_0)/2k_BT}$, and analogously for k_{01}. We set $\lambda_0 = -k_BT$, $\lambda_1 = 2k_BT$, $\mathcal{T} = 5$, and $\omega = 1$.

4.6 The Jarzynski and Crooks relations

The Jarzynski and Crooks relations are fundamental results in stochastic thermo-dynamics. They can be seen as special cases of the integral and detailed fluctuation theorems when the initial and final states are equilibrium states. Specifically, we con-sider a system initially at thermodynamic equilibrium, bring it out of equilibrium by means of manipulations and/or drivings, and then let it relax to equilibrium again. The final equilibrium state does not need to be identical to the initial one since the initial and final values of the control parameter may be different. In this setting, the involution property always holds: once the reverse dynamics has brought back the energies to their initial values and the nonequilibrium drivings have been switched off, the initial equilibrium state is necessarily restored. Therefore, both the integral and the detailed fluctuation relations hold. Further, one has

$$p_X(t_0) = e^{(F(t_0) - \epsilon_x(t_0))/k_B T},$$
$$p_X(t_f) = e^{(F(t_f) - \epsilon_x(t_f))/k_B T},$$

(4.48)

where $F(t)$ is the equilibrium free energy corresponding to the value $\lambda(t)$ of the control parameter:

$$F(t) = -k_B T \ln \sum_x e^{-\epsilon_x(\lambda(t))/k_B T}.$$

(4.49)

Therefore, the entropy change of the system is

$$\Delta s^{\text{sys}} = k_B \ln \left[\frac{p_{x_0}(t_0)}{p_{x_f}(t_f)} \right] = \frac{1}{T} \left[\epsilon_{x_f}(t_f) - \epsilon_{x_0}(t_0) - \Delta F \right],$$

(4.50)

where $\Delta F = F(t_f) - F(t_0)$. We then express the total entropy production as

$$s^{\text{tot}} = \frac{q}{T} + \Delta s^{\text{sys}} = \frac{w - \Delta F}{T} = \frac{w^{\text{diss}}}{T},$$

(4.51)

where we use the stochastic first law, eq. (3.11), and define the **dissipated work** w^{diss} as the difference between the performed work and the free-energy change. The dissipated work represents the total amount of work that is released to the reservoir as heat. Substituting expression (4.51) into the integral fluctuation relation, and considering that the free-energy difference does not fluctuate, we obtain the **Jarzynski equality**:

$$\left\langle e^{-w/k_B T} \right\rangle_F = e^{-\Delta F/k_B T}.$$

(4.52)

The corresponding detailed fluctuation relation is the **Crooks relation**:

$$\frac{p(w; \lambda)}{p(-w; \widehat{\lambda})} = e^{(w - \Delta F)/k_B T}.$$

(4.53)

For both the Jarzynski and Crooks relations, the assumption that the final state is at equilibrium can be relaxed. We consider a process that starts at equilibrium at time t_0 but ends at t_f in a nonequilibrium state. We associate with it an auxiliary process of a longer time duration. The two processes are identical in the time interval $[t_0, t_f]$. The auxiliary process runs for longer times $t > t_f$ with all drivings set to zero, $\delta_{xx'}(t > t_f) = 0$, a fixed manipulation parameter $\lambda(t > t_f) = \lambda(t_f)$, and therefore constant energies $\epsilon_x(t > t_f) = \epsilon_x(t_f)$. The auxiliary process ends at a time $t_f' \gg t_f$, when equilibrium is reached. By construction, the auxiliary process starts and ends at equilibrium, so that its work distribution satisfies both the Jarzynski and Crooks relations. Moreover, in the time interval $[t_f, t'_f]$, no work is performed, so that the distribution of work is the same in the original and auxiliary processes. Therefore, the work distribution of the original process satisfies both the Jarzynski and Crooks relations.

The Jarzynski and Crooks relations are of paramount importance for experimental applications, as we discuss in chapter 7 in more detail.

4.7 Instantaneous quench

A simple example that helps to acquire familiarity with the Jarzynski and Crooks relations is a system undergoing an instantaneous quench. A **quench** is a manipulation protocol where the energies of the states are suddenly changed at a given time, and otherwise remain constant. Specifically, we consider a system at equilibrium at the initial time t_0. The energies $\epsilon_x(t_0)$ are constant up to the quench time $t_q > t_0$. The quench instantly alters the energies to the values $\epsilon_x(t_q)$. The system is then allowed to relax without further manipulation up to a time t_f, when we assume it has reached equilibrium. Since the manipulation is instantaneous, the work provided to the system is equal to the energy difference before and after the quench,

$$w = \epsilon_x(t_q) - \epsilon_x(t_0), \tag{4.54}$$

where x is the state of the system immediately before the quench. As the system is initially at equilibrium, the average work is

$$W = \sum_x [\epsilon_x(t_q) - \epsilon_x(t_0)]\, e^{(F(t_0) - \epsilon_x(t_0))/k_B T}. \tag{4.55}$$

During the quench, the heat reservoir does not play any role: heat is dissipated only during the free relaxation following the quench. The average total entropy production is given by

$$S^{\text{tot}} = W - \Delta F = \sum_x [\epsilon_x(t_q) - \epsilon_x(t_0)]\, e^{(F(t_0) - \epsilon_x(t_0))/k_B T} - \Delta F, \tag{4.56}$$

taking into account that $\epsilon_x(t_q) = \epsilon_x(t_f)$. The average entropy production satisfies $S^{\text{tot}} \geq 0$ for any choice of the initial and final energies, although that might be not obvious from eq. (4.56). In fact, it is easier to verify the Jarzynski equality directly:

$$\left\langle e^{-w/k_B T} \right\rangle_F = \sum_x e^{-(\epsilon_x(t_f) - \epsilon_x(t_0))/k_B T} e^{(F(t_0) - \epsilon_x(t_0))/k_B T}$$
$$= e^{F(t_0)/k_B T} \sum_x e^{-\epsilon_x(t_f)/k_B T} = e^{-\Delta F/k_B T}, \tag{4.57}$$

where $\Delta F = F(t_f) - F(t_0)$ and we exploit eq. (4.49). Positivity of the average entropy production follows from the Jensen inequality (A.11):

$$0 = \ln \left\langle e^{-(w - \Delta F)/k_B T} \right\rangle \geq -\frac{\langle w - \Delta F \rangle}{k_B T}. \tag{4.58}$$

The Crooks relation is easy to prove if, for each given value of w, there is never more than one state x, such as $\epsilon_x(t_f) - \epsilon_x(t_0) = w$. In this case, we find

$$\frac{p(w;\lambda)}{p(-w;\lambda)} = \frac{e^{(F(t_0)-\epsilon_x(t_0))/k_BT}}{e^{(F(t_f)-\epsilon_x(t_f))/k_BT}} = e^{(w-\Delta F)/k_BT}. \tag{4.59}$$

One can prove the Crooks relation also in the "degenerate" case, where there are multiple states characterized by the same energy difference w between initial and final states.

4.8 Fluctuation relations in practice

To apply fluctuation relations to real systems, we need to estimate the distribution of entropy production by replicating an experiment a certain number of times. It is useful to know the number $\mathcal{N}$ of experimental replicates necessary for this purpose. To be more specific, we consider a manipulation experiment with control parameter $\lambda(t)$ between two equilibrium states with $\lambda(t_0) = \lambda_0$ and $\lambda(t_f) = \lambda_f$. The experiment aims at estimating the free-energy difference $\Delta F = F(\lambda_f) - F(\lambda_0)$ by means of the Jarzynski equality (4.52). Solving the Jarzynski relation for ΔF, we obtain

$$\Delta F = -k_BT \ln\left\langle e^{-w/k_BT} \right\rangle_F = -k_BT \ln \int dw\, p(w;\lambda)\, e^{-w/k_BT}. \tag{4.60}$$

We wish to get a sense of the values of w that contribute the most to the integral in eq. (4.60). From the Crooks relation (4.53), we obtain

$$p(w;\lambda)\, e^{(\Delta F - w)/k_BT} = p(-w;\widehat{\lambda}), \tag{4.61}$$

where $p(w;\widehat{\lambda})$ is the distribution of w in the backward protocol $\widehat{\lambda}$. The left-hand side of eq. (4.61) is proportional to the argument of the integral in eq. (4.60). Therefore, the leading contribution to the integral comes from the most probable value of the work in the backward protocol, which we denote by $-w^*$. Its probability in the forward protocol is

$$p(-w^*;\widehat{\lambda})\, e^{-(\Delta F - w^*)/k_BT} = p(-w^*;\widehat{\lambda})\, e^{-w^{*\mathrm{diss}}/k_BT}, \tag{4.62}$$

where $w^{*\mathrm{diss}} = -w^* + \Delta F$ is the most probable value of the dissipated work in the backward protocol; see eq. (4.51). We expect $p(w;\widehat{\lambda})$ to be of order 1 at its maximum $-w^*$. We conclude that the probability r^* of obtaining w^* in the forward protocol is on the order of $e^{-w^{*\mathrm{diss}}/k_BT}$.

If a single event has the small probability r of taking place in an experiment, the number of times it takes place in $\mathcal{N}$ trials is approximately Poisson distributed with a parameter $\theta = r\mathcal{N}$, whose standard deviation is equal to $\sqrt{\theta}$, as discussed in appendix A.4. The relative standard error in $\mathcal{N}$ trials is therefore $\sigma = 1/\sqrt{\theta} = 1/\sqrt{r\mathcal{N}}$. If we wish to estimate r with relative standard error σ by the Jarzynski equality, the number of trials must be on the order of $\mathcal{N} = 1/(r^*\sigma^2)$. In conclusion, to reliably estimate the free-energy difference ΔF, we must perform a number of experiments on the order of

$$\mathcal{N} \approx \frac{e^{w^{*\mathrm{diss}}/k_BT}}{\sigma^2}. \tag{4.63}$$

The exponential dependence on $w^{*\text{diss}}$ means that the required number of experiments quickly becomes prohibitive as the protocol drives the system farther away from equilibrium.

A similar reasoning applies to the practical use of other fluctuation relations.

4.9 Adiabatic and nonadiabatic entropy production and the Hatano-Sasa relation

Throughout this book, we have discussed two ways to bring a mesoscopic system out of equilibrium: either by an external manipulation that introduces a time dependence in the energy of the different states or by drivings that dissipate energy every time a jump occurs. Either mechanism positively contributes to the total average entropy production S^{tot}; see section 3.8. In this section, we discuss a decomposition of the entropy production into two contributions, one originating from time dependence and one from drivings.

Given a trajectory x, we define the **adiabatic entropy production** by

$$s^{a}(x) = k_{B} \sum_{k=1}^{n} \ln \frac{k_{x_k x_{k-1}}(t_k) p^{\text{st}}_{x_{k-1}}(t_k)}{k_{x_{k-1} x_k}(t_k) p^{\text{st}}_{x_k}(t_k)}, \tag{4.64}$$

and the **nonadiabatic entropy production** by

$$s^{na}(x) = k_{B} \ln \frac{p_{x_0}(t_0)}{p_{x_f}(t_f)} + k_{B} \sum_{k=1}^{n} \ln \frac{p^{\text{st}}_{x_k}(t_k)}{p^{\text{st}}_{x_{k-1}}(t_k)}. \tag{4.65}$$

In both definitions, $p^{\text{st}}_{x_k}(t)$ is the **instantaneous stationary distribution**, i.e., the one that the system would reach if the control parameter were "frozen" at the value $\lambda(t)$. By summing eqs. (4.64) and (4.65), we find that the total entropy production is the sum of the adiabatic and nonadiabatic contributions:

$$s^{\text{tot}} = s^{a} + s^{na}. \tag{4.66}$$

To justify the names *adiabatic* and *nonadiabatic*, we first consider a case in which the jump rates are constant in time and the system is in a nonequilibrium steady state p^{st}. In such a case, $s^{na} = 0$, and therefore $s^{a} = s^{\text{tot}}$. This means that s^{a} is the entropy produced to maintain a nonequilibrium steady state in the absence of any manipulation. In the opposite case, where the system is manipulated but no drivings are present, one has $s^{a} = 0$ since p^{st} is, at every time, an equilibrium distribution satisfying detailed balance. Consequently, we have $s^{\text{tot}} = s^{na}$.

The heat associated with s^{a}, $q^{\text{hk}} = Ts^{a}$, is called the **housekeeping heat**, where the "house" is a metaphor for the nonequilibrium steady state. The remainder heat, $q^{\text{ex}} = q - q^{\text{hk}}$, is called the **excess heat**. The average nonadiabatic entropy production is

$$S^{na}(t) = \langle s^{na}(t) \rangle = \Delta S^{\text{sys}}(t) + \frac{Q^{\text{ex}}(t)}{T}, \tag{4.67}$$

where $\Delta S^{\text{sys}}(t)$ is the change in the average system entropy and

$$Q^{\text{ex}}(t) = \langle q^{\text{ex}} \rangle = k_{\text{B}}T \int_{t_0}^{t} dt \sum_x \left[\frac{dp_x(t)}{dt} \ln p_x^{\text{st}}(t) \right] \qquad (4.68)$$

is the average excess heat flowing into the reservoir, beyond the housekeeping heat.

We now consider a manipulation that brings a system from an initial steady state $p_{x_0}^{\text{st}}(t_0)$ to a final one $p_{x_f}^{\text{st}}(t_f)$. The associated nonadiabatic entropy production is expressed by

$$s^{\text{na}} = -k_{\text{B}} \sum_{k=0}^{n} \ln \frac{p_{x_j}^{\text{st}}(t_{k+1})}{p_{x_k}^{\text{st}}(t_j)} = \int_{t_0}^{t_f} dt \frac{d\lambda(t)}{dt} \left. \frac{\partial s_{x(t)}^{\text{st}}(\lambda)}{\partial \lambda} \right|_{\lambda=\lambda(t)}, \qquad (4.69)$$

where we introduce the function $s_x^{\text{st}}(t) = -k_{\text{B}} \ln[p_x^{\text{st}}(t)]$. In the absence of drivings, s^{st} is given by $s_x^{\text{st}} = [\epsilon_x(t) - F]/T$; see eq. (2.45).

If the final state is not a steady state, eq. (4.69) still holds. To show that, we follow the same logic we used to prove the Jarzynski equality with a nonequilibrium final state; see section 4.6. We consider an auxiliary process that is identical to the original process up to the end of the manipulation, and then runs for a sufficiently long time after t_f without any further manipulation, i.e., where $\epsilon_x(t) = \epsilon_x(t_f)$ for all x and $t > t_f$. The nonadiabatic entropy production for this auxiliary process is equal to that of the original process, since the partial derivative inside the integral in eq. (4.69) vanishes for $t > t_f$.

Although s^{na} vanishes when the system is in a time-independent steady state, it does not vanish for systems brought from an initial to a final steady state by an arbitrarily slow protocol. Indeed, from eq. (4.69), one has

$$s^{\text{na}} \xrightarrow{\text{slow manipulation}} \langle s^{\text{st}}(t_f) \rangle - \langle s^{\text{st}}(t_0) \rangle. \qquad (4.70)$$

Both the adiabatic and the nonadiabatic entropy production separately satisfy fluctuation relations. To show that this is true, it is useful to introduce the **conjugate jump rates**

$$k_{xx'}^{+}(t) = k_{x'x}(t) \frac{p_x^{\text{st}}(t)}{p_{x'}^{\text{st}}(t)}. \qquad (4.71)$$

The conjugate jump rates define a conjugate master equation whose solution is $p_x^{+}(t)$. A master equation and its conjugate share the same stationary distribution, $p_x^{\text{st+}}(t) = p_x^{\text{st}}(t)$, since

$$\sum_{x'} k_{xx'}^{+}(t) p_{x'}^{\text{st}}(t) = \sum_{x'} k_{x'x}(t) p_x^{\text{st}}(t) = \sum_{x'} k_{xx'}(t) p_{x'}^{\text{st}}(t) = \sum_{x'} k_{x'x}^{+}(t) p_x^{\text{st}}(t). \qquad (4.72)$$

The conjugate dynamics permits us to express both the adiabatic and the nonadiabatic entropy production as ratios of trajectory probabilities. Following similar steps as in section 4.1, we find

$$s^{\text{a}}(x) = k_{\text{B}} \ln \frac{\mathcal{P}_x(\lambda)}{\mathcal{P}_x^{+}(\lambda)}; \qquad (4.73)$$

$$s^{\text{na}}(x) = k_B \ln \frac{\mathcal{P}_x(\lambda)}{\mathcal{P}_{\widehat{x}}^+(\widehat{\lambda})}. \tag{4.74}$$

From these expressions, integral fluctuation relations for both s^{a} and s^{na} directly follow:

$$\left\langle e^{-s^{\text{a}}/k_B} \right\rangle_F = 1; \tag{4.75}$$

$$\left\langle e^{-s^{\text{na}}/k_B} \right\rangle_F = 1. \tag{4.76}$$

These relations imply that both the adiabatic and nonadiabatic contributions to the total average entropy production must be nonnegative. As we discussed before, for a system initially prepared in a steady state, eq. (4.76) can be written as

$$\left\langle \exp\left[-\frac{1}{k_B} \int_{t_0}^{t_f} dt \, \frac{d\lambda(t)}{dt} \, \partial_\lambda s_{x(t)}^{\text{st}}(\lambda(t)) \right] \right\rangle_F = 1. \tag{4.77}$$

Equation (4.77) is the **Hatano-Sasa relation**. This relation is the starting point for generalizations of fluctuation-dissipation relations to nonequilibrium steady states.

4.10 Systems with odd-parity variables

We have assumed so far that the mesoscopic degrees of freedom identified by the variable x are left invariant under time reversal. Examples of such mesostates are the position of a particle or the bound-unbound state of an enzyme. Some degrees of freedom of interest in physics behave differently under time reversal. For example, velocities, linear momenta, and magnetic momenta change sign under time reversal. In the case of magnetic momenta, this sign change is justified by the **Ampère principle**, i.e., the fundamental idea in electromagnetism stating that magnetic fields originate from electric currents. Since electric currents change sign under time reversal, so do their associated magnetic fields.

We call degrees of freedom such as position **even** under time reversal and degrees of freedom such as momenta **odd** under time reversal. In general, we may consider systems characterized by both even and odd degrees of freedom. We denote by $\tilde{x}$ the state obtained by time-reversing a degree of freedom x. With this notation, a degree of freedom that is even under time reversal satisfies $\tilde{x} = x$, whereas a degree of freedom that is odd under time reversal satisfies $\tilde{x} = -x$. In general, one can encounter both types of degrees of freedom in the same system. For example, a system could have mesostates x identified by two discrete variables: a variable y being even under time reversal and a variable z being odd. This means that if $x = (y, z)$, then $\tilde{x} = (y, -z)$. It is important to understand how to express the entropy production in these cases and to verify that fluctuation relations hold.

Systems with degrees of freedom of different parity under time reversal give us a chance to discuss a subtle aspect of the detailed balance condition

$$k_{xx'} p_{x'}^{\text{st}} = k_{x'x} p_x^{\text{st}}. \tag{4.78}$$

Equation (4.78) ensures that all probability currents vanish in the stationary state (see, e.g., section 2.6). The left- and right-hand sides of eq. (4.78) can be respectively interpreted as the rates by which jumps from x' to x or from x to x' take place. These rates are equal if detailed balance holds. If the states x and x' are even under time reversal, the two jumps are time-reversed images of each other. Equation (4.78) then embodies a fundamental condition for thermodynamic equilibrium: the lack of any statistical asymmetry under time reversal.

The situation is different if some degrees of freedom are not even under time reversal. In this case, the balance condition for jump rates under time reversal becomes

$$k_{xx'}p_{x'}^{\mathrm{st}} = k_{\tilde{x}'\tilde{x}}p_{\tilde{x}}^{\mathrm{st}}. \tag{4.79}$$

In these situations, we refer to eq. (4.78) as the mathematical detailed balance condition and to eq. (4.79) as the physical detailed balance condition. If states are even under time reversal, these two conditions are identical. They are also identical under the milder symmetry condition

$$k_{xx'} = k_{\tilde{x}\tilde{x}'}, \tag{4.80}$$

for all pairs x and x', since this condition also ensures that $p_x^{\mathrm{st}} = p_{\tilde{x}}^{\mathrm{st}}$ for all x. In systems with degrees of freedom of different symmetry under time reversal, the mathematical and physical detailed balance conditions are not equivalent. It follows from our discussion that thermodynamic equilibrium requires the physical detailed balance condition to be satisfied, whereas the mathematical detailed balance condition might not hold. This means that even in an equilibrium state, nonvanishing probability currents may exist. Such probability currents are called **dissipationless**.

We now consider the irreversibility of trajectories $k_B \ln \mathcal{P}_x(\lambda)/\mathcal{P}_{\hat{x}}(\hat{\lambda})$ for such a system. In this case, the expression of the backward trajectory must take into account the behavior of states under time reversal. This means that Eq. (4.4) must be generalized to

$$\widehat{x}(t) = \tilde{x}(\widehat{t}). \tag{4.81}$$

When evaluating the ratio of the probability of a trajectory and that of its time-reversed image, the contributions from dwells do not necessarily cancel out in the presence of odd-parity variables. Indeed, from the explicit expressions of the trajectory probabilities, we obtain

$$k_B \ln \frac{\mathcal{P}_x(\lambda)}{\mathcal{P}_{\hat{x}}(\hat{\lambda})} = \Delta s^{\mathrm{sys}} + k_B \sum_{j=1}^{n} \ln \frac{k_{x_j x_{j-1}}}{k_{\tilde{x}_{j-1}\tilde{x}_j}} - k_B \sum_{j=1}^{n+1} \int_{t_{j-1}}^{t_j} dt\, (k_{x_{j-1}}^{\mathrm{out}} - k_{\tilde{x}_{j-1}}^{\mathrm{out}}), \tag{4.82}$$

where the last term represents the contribution of the dwells.

If the symmetry condition expressed in eq. (4.80) is satisfied, then the irreversibility of trajectories can be interpreted as the total entropy production s^{tot}. Indeed, under these assumptions the second term on the right-hand side of eq. (4.82) is equal to s^{res}, whereas the third term vanishes.

In this case, entropy production satisfies an integral fluctuation theorem,

$$\left\langle e^{-s^{\mathrm{tot}}/k_B} \right\rangle = 1. \tag{4.83}$$

A corresponding detailed fluctuation theorem holds under additional assumptions, as discussed in sections 4.5 and 4.6.

4.11 Trajectory probability for Langevin equations (*)

In the following sections, we derive fluctuation relations for systems described by Langevin equations. Our first step is to express the probability density $\mathcal{P}(x)$ of a trajectory x generated by a Langevin equation,

$$\frac{dx}{dt} = \mu_P \mathcal{F}(x,t) + \sqrt{2D}\,\xi(t);\tag{4.84}$$

see also section 3.13. We consider as usual a time interval $[t_0, t_f]$ and denote by x_0 and x_f the initial and final states. We approximate the trajectory by discretizing the interval in $\mathcal{N}$ intervals as in section 2.7:

$$\mathcal{P}(x|x_0) \approx \prod_{\ell=1}^{\mathcal{N}} p(x_\ell; t_{\ell-1} + \Delta t | x_{\ell-1}; t_{\ell-1}),\tag{4.85}$$

where $x_{\mathcal{N}} = x_f$. At variance with eq. (2.87), here the conditional probabilities are densities, i.e., $p(x_\ell; t_{\ell-1} + \Delta t | x_{\ell-1}; t_{\ell-1})\,dx$ is the conditional probability that x at time $t_{\ell-1} + \Delta t$ falls between x_ℓ and $x_\ell + dx$, given that it is at $x_{\ell-1}$ at time $t_{\ell-1}$.

To evaluate $p(x_\ell, t_\ell | x_{\ell-1}, t_{\ell-1})$, we suppose that the time interval Δt is so short that we can approximate the increment $\Delta x_\ell = x_\ell - x_{\ell-1}$ by a term proportional to Δt and a random contribution due to the noise. The latter term is proportional to the increment ΔW_ℓ of the Wiener process during the time interval $[t_{\ell-1}, t_\ell]$. We therefore set

$$\Delta x_\ell \approx \mu_P \mathcal{F}(x_{\ell-1}, t_{\ell-1})\,\Delta t + \sqrt{2D}\,\Delta W_\ell.\tag{4.86}$$

Since ΔW_ℓ is a Gaussian random variable with vanishing average and variance equal to Δt, we obtain

$$p(x_\ell, t_\ell | x_{\ell-1}, t_{\ell-1}) \approx \frac{1}{\sqrt{4\pi D \Delta t}} \exp\left[-\frac{(x_\ell - x_{\ell-1} - \mu_P \mathcal{F}(x_{\ell-1}, t_{\ell-1})\,\Delta t)^2}{4 D \Delta t}\right].\tag{4.87}$$

Substituting this result into (4.85) leads to

$$\mathcal{P}(x|x_0) \approx \prod_{\ell=1}^{\mathcal{N}} \left\{ \frac{1}{\sqrt{4\pi D \Delta t}} \exp\left[-\frac{(x_\ell - x_{\ell-1} - \mu_P \mathcal{F}(x_{\ell-1}, t_{\ell-1})\,\Delta t)^2}{4 D \Delta t}\right]\right\}.\tag{4.88}$$

Formally, in the limit $\Delta t \to 0$, the expression of the probability density becomes

$$\mathcal{P}(x;\lambda) = \exp\left[-\mathcal{S}(x,\lambda)\right] p(x_0, t_0),\tag{4.89}$$

where the **action** $\mathcal{S}(x;\lambda)$ is defined by

$$\mathcal{S}(x;\lambda) = \int_{t_0}^{t_f} dt\, \frac{1}{4D} \left[\frac{dx(t)}{dt} - \mu_P \mathcal{F}(x(t), \lambda(t))\right]^2.\tag{4.90}$$

The expression of the action, eq. (4.90), implicitly contains stochastic quantities whose integral requires an interpretation, i.e., an explicit discretization rule. In our case, the discretization in eq. (4.88) implies that we evaluate $F(x,t)$ at the initial point of each interval. This means that the stochastic integrals appearing in the action must be interpreted according to the Ito calculus.

We formally define the trajectory measure by

$$\mathcal{D}x = \lim_{\Delta t \to 0} \prod_{\ell=1}^{\mathcal{N}} \frac{\mathrm{d}x_{\ell-1}}{\sqrt{4\pi D \Delta t}} \cdot \mathrm{d}x_\mathrm{f}. \qquad (4.91)$$

This expression allows us to explicitly express averages over trajectories.

4.12 Fluctuation relation for the Langevin equation (*)

In this section, we derive fluctuation relations for Langevin equations. Our first step is to verify that the relation (4.12) linking entropy production and the probability of forward and backward trajectories also holds for Langevin equations.

We consider a trajectory x generated by the Langevin equation (4.84). The probability density of x is given by eq. (4.89). We introduce the backward trajectory $\widehat{x}$ and the backward protocol $\widehat{\lambda}$ as in section 4.1. We now discretize the backward dynamics. The initial time of the ℓ-th interval of the backward trajectory, t_ℓ, corresponds to the final time of the same interval for the forward trajectory. Thus, following the same logic of section 4.11, we obtain

$$\Delta x_\ell \approx -\mu_\mathrm{P} \mathcal{F}(x_\ell, t_\ell)\, \Delta t + \sqrt{2D}\, \Delta W_\ell, \qquad (4.92)$$

where ΔW_ℓ is the increment of the Wiener process over a time interval of duration Δt. Then the probability of the backward trajectory is given by

$$\mathcal{P}(\widehat{x}; \widehat{\lambda}) = \exp\left[-\mathcal{S}(\widehat{x}, \widehat{\lambda})\right] p(x_\mathrm{f}, t_\mathrm{f}), \qquad (4.93)$$

where the action of the backward trajectory is

$$\begin{aligned}
\mathcal{S}(\widehat{x}; \widehat{\lambda}) &= \int_{t_0}^{t_\mathrm{f}} \mathrm{d}t\, \frac{1}{4D}\left[\frac{\mathrm{d}\widehat{x}(t)}{\mathrm{d}t} - \mu_\mathrm{P}\, \mathcal{F}(\widehat{x}(t), \widehat{\lambda}(t))\right]^2 \\
&= \int_{t_0}^{t_\mathrm{f}} \mathrm{d}t\, \frac{1}{4D}\left[-\frac{\mathrm{d}x}{\mathrm{d}t}\bigg|_{t=\widehat{t}} - \mu_\mathrm{P}\, \mathcal{F}(x(\widehat{t}), \lambda(\widehat{t}))\right]^2 \qquad (4.94)\\
&= \int_{t_0}^{t_\mathrm{f}} \mathrm{d}t\, \frac{1}{4D}\left[\frac{\mathrm{d}x(t)}{\mathrm{d}t} + \mu_\mathrm{P}\, \mathcal{F}(x(t), \lambda(t))\right]^2 .
\end{aligned}$$

Here the discretization must take into account eq. (4.92), i.e., $\mathcal{F}(x,t)$ is evaluated at the end of each time interval. Since this discretization is the opposite of the Ito discretization, we call it an **anti-Ito convention**. The measures of the backward and forward trajectories are equal, i.e., $\mathcal{D}\widehat{x} = \mathcal{D}x$; see eq. (4.91). We now evaluate the ratio between the probability of the forward and of the backward trajectories:

$$\frac{\mathcal{P}(x;\lambda)}{\mathcal{P}(\widehat{x};\widehat{\lambda})} = \frac{p(x_0,t_0)}{p(x_f,t_f)}\exp\left[-\mathcal{S}(x,\lambda)+\mathcal{S}(\widehat{x},\widehat{\lambda})\right] =$$

$$= \exp\left[\frac{\Delta s^{\mathrm{sys}}}{k_{\mathrm{B}}} - \mathcal{S}(x,\lambda)+\mathcal{S}(\widehat{x},\widehat{\lambda})\right]. \tag{4.95}$$

The difference between the actions of a forward trajectory and its backward counterpart is

$$\mathcal{S}(x;\lambda)-\mathcal{S}(\widehat{x};\widehat{\lambda}) = \int_{t_0}^{t_f} dt\,\frac{1}{4D}\left\{\left[\frac{dx(t)}{dt}-\mu_{\mathrm{P}}\,\mathcal{F}(x(t),\lambda(t))\right]^2\right.$$

$$\left.-\left[\frac{dx(t)}{dt}+\mu_{\mathrm{P}}\,\mathcal{F}(x(t),\lambda(t))\right]^2\right\} \tag{4.96}$$

$$= -\frac{1}{k_{\mathrm{B}}T}\int_{t_0}^{t_f} dt\,\frac{dx(t)}{dt}\circ\mathcal{F}(x(t),\lambda(t)),$$

where we use the Einstein relation (3.84). The integral appearing in eq. (4.96) is interpreted in the Stratonovich sense. The intuitive reason is that it is obtained as the average of one integral interpreted with the Ito convention and one interpreted with the anti-Ito convention. This result is more carefully derived in appendix A.10. Using the definition of heat for a Langevin equation, eq. (3.88), we obtain

$$-dt\,\frac{dx(t)}{dt}\circ\mathcal{F}(x(t),\lambda(t)) = -dx\circ\mathcal{F}(x(t),\lambda(t)) = -dq. \tag{4.97}$$

We therefore have

$$\mathcal{S}(x;\lambda)-\mathcal{S}(\widehat{x};\widehat{\lambda}) = -\frac{1}{k_{\mathrm{B}}T}\int_{t_0}^{t_f} dq = -\frac{s^{\mathrm{res}}}{k_{\mathrm{B}}}, \tag{4.98}$$

where s^{res} is the entropy change of the reservoir. Substituting this result into eq. (4.95), we obtain the irreversibility relation

$$\frac{\mathcal{P}(x;\lambda)}{\mathcal{P}(\widehat{x};\widehat{\lambda})} = e^{s^{\mathrm{tot}}(x;\lambda)/k_{\mathrm{B}}}. \tag{4.99}$$

All fluctuation relations we derived from master equations can be derived for Langevin equations from the irreversibility relation (4.99) following the same steps. In this case too, when the initial distribution is the equilibrium one, we have for any trajectory

$$T\,s^{\mathrm{tot}}(x) = w(x) - \Delta F, \tag{4.100}$$

where $\Delta F = F(\lambda_f) - F(\lambda_0)$. Using this relation, we can directly obtain the Jarzynski and Crooks equalities for Langevin equations.

4.13 Brownian particle in a time-dependent harmonic potential (*)

As an illustration, we evaluate the work distribution for a particle undergoing Brownian motion in a manipulated harmonic potential. The position of the particle evolves according to a Langevin equation (3.82), where the driving is absent and the potential is

$$U(x, \lambda) = \frac{1}{2} \lambda \, x^2. \tag{4.101}$$

We assume that the Einstein relation (3.84) holds. Since the Langevin equation with this potential is linear in x, its solution at time t, with a given initial condition, is a linear functional of the Wiener process, which is Gaussian. Therefore, if the initial distribution of x is Gaussian, it will remain Gaussian at all later times; see also appendix A.4. Moreover, if $\langle x \rangle$ vanishes at the initial time, it will vanish by symmetry at any later time. We look therefore for a time-dependent distribution of the form

$$p(x; t, \lambda) = \frac{1}{\sqrt{2\pi \gamma(t)}} \exp\left(-\frac{x^2}{2\gamma(t)}\right). \tag{4.102}$$

The distribution satisfies the Fokker-Planck equation

$$\frac{\partial p}{\partial t} = \mu_{\mathrm{P}} \frac{\partial}{\partial x} \left[\lambda(t) x \, p + k_{\mathrm{B}} T \frac{\partial p}{\partial x} \right]. \tag{4.103}$$

We obtain a simpler equation by introducing the **generating function**:

$$\phi(q, t) = \int \mathrm{d}x \, e^{qx} \, p(x, t). \tag{4.104}$$

Properties of generating functions are discussed in appendix A.4. Given the generating function $\phi(q, t)$ of the variable x, the average $\langle x \rangle_t$ and variance $\sigma^2(t) = \langle x^2 \rangle_t - \langle x \rangle_t^2$ of x are respectively given by

$$\langle x \rangle_t = \left. \frac{\partial \phi(q, t)}{\partial q} \right|_{q=0} ; \qquad \sigma^2(t) = \left. \frac{\partial^2 \ln \phi(q, t)}{\partial q^2} \right|_{q=0}. \tag{4.105}$$

Multiplying both sizes of eq. (4.103) by e^{qx} and integrating, we obtain an evolution equation for the generating function:

$$\frac{\partial \phi}{\partial t} = \mu_{\mathrm{P}} \left[-\lambda(t) q \frac{\partial \phi}{\partial q} + k_{\mathrm{B}} T q^2 \, \phi \right]. \tag{4.106}$$

If $p(x; t)$ is given by (4.102), then $\phi(q, t)$ is given by

$$\phi(q, t) = e^{\gamma(t) q^2 / 2}. \tag{4.107}$$

Substituting in (4.106), we obtain an equation for $\gamma(t)$:

$$\frac{\mathrm{d}\gamma}{\mathrm{d}t} = 2\mu_{\mathrm{P}} \left[-\lambda(t) \gamma(t) + k_{\mathrm{B}} T \right]. \tag{4.108}$$

This is a linear equation that can be analytically solved. We do not however need its solution. To obtain the fluctuation relation, we consider the work w accumulated up to time t:

$$w(t) = \int_{t_0}^{t} dt' \frac{d\lambda(t')}{dt'} \left. \frac{\partial U}{\partial \lambda} \right|_{\lambda=\lambda(t'), x=x(t')}. \tag{4.109}$$

The joint probability distribution $p(x, w; t, \lambda)$ of x and w satisfies the evolution equation

$$\frac{\partial}{\partial t} p(x, w; t, \lambda) = \mu_P \frac{\partial}{\partial x} \left[\lambda(t)x\, p + k_B T \frac{\partial p}{\partial x} \right] - \dot{\lambda}(t) \left. \frac{\partial U}{\partial \lambda} \right|_{\lambda=\lambda(t), x=x(t)} \frac{\partial p}{\partial w}$$

$$= \mu_P \frac{\partial}{\partial x} \left[\lambda(t)x\, p + k_B T \frac{\partial p}{\partial x} \right] - \frac{1}{2} \frac{d\lambda(t)}{dt} x^2 \frac{\partial p}{\partial w}. \tag{4.110}$$

The corresponding generating function is

$$\phi^{(x,w)}(q_1, q_2, t) = \int dx\, dw\, e^{q_1 x + q_2 w} p(x, w; t, \lambda) \tag{4.111}$$

and satisfies the equation

$$\frac{\partial \phi^{(x,w)}}{\partial t} = \mu_P \left[-\lambda(t)\, q_1 \frac{\partial \phi^{(x,w)}}{\partial q_1} + k_B T q_1^2\, \phi^{(x,w)} \right] + \frac{1}{2} \frac{d\lambda(t)}{dt}\, q_2 \frac{\partial^2 \phi^{(x,w)}}{\partial q_1^2}. \tag{4.112}$$

This equation also admits a Gaussian solution of the form

$$\phi^{(x,w)}(q_1, q_2, t) = e^{\alpha(q_2, t) + \gamma(q_2, t) q_1^2/2}, \tag{4.113}$$

where α and γ evolve according to

$$\frac{\partial \alpha}{\partial t} = \frac{q_2}{2} \frac{d\lambda(t)}{dt} \gamma(q_2, t); \tag{4.114a}$$

$$\frac{\partial \gamma}{\partial t} = 2\mu_P \left[-\lambda(t)\, \gamma(q_2, t) + k_B T \right] + q_2 \frac{d\lambda(t)}{dt} \gamma^2(q_2, t). \tag{4.114b}$$

Equation (4.114b) is a Riccati equation. It can be analytically solved in some simple cases, but is easy to solve numerically.

The quantity $\alpha(q_2, t)$ is the cumulant generating function of the accumulated work w (see eq. (A.52)):

$$\alpha(q_2, t) = \ln \left\langle e^{q_2\, w(t)} \right\rangle. \tag{4.115}$$

The Jarzynski equality (4.52) implies that

$$\alpha\left(-\frac{1}{k_B T} \right) = \ln \left\langle e^{-w(t)/k_B T} \right\rangle = -\frac{F(\lambda(t)) - F(\lambda(t_0))}{k_B T}. \tag{4.116}$$

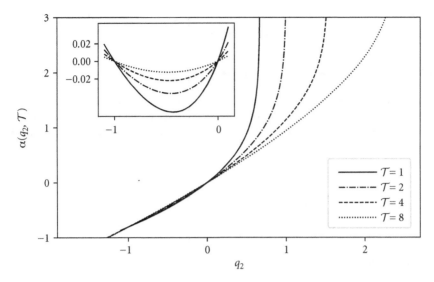

Figure 4.5. Cumulant generating function $\alpha(q_2, \mathcal{T})$ of the work performed on the system for the linear protocol (4.120) starting from equilibrium, plotted as a function of q_2 (in units of $(k_BT)^{-1}$). We set $\lambda_0 = 1$, $\lambda_1 = 5$ (in units of k_BT), and different values of $\mathcal{T}$ (in units of μ_P^{-1}). The cumulant generating function diverges at positive finite values of q_2, which depends on $\mathcal{T}$. The curves cross at $q_2 = 0$ by normalization, and at $q_2 = -1/k_BT$ due to the Jarzynski equality. This is highlighted in the inset, where $\alpha(q_2, \mathcal{T}) - q_2 \Delta F$ is plotted against q_2. See Speck [158].

Indeed, one can verify that if $q_2 = -1/k_BT$, eqs. (4.114a and b) admit the solution

$$\gamma(t) = \frac{k_BT}{\lambda(t)}, \qquad t_0 \leq t \leq t_f, \tag{4.117}$$

with

$$\alpha(t) = -\frac{1}{2}\ln\frac{\lambda(t)}{\lambda(t_0)} = -\frac{\Delta F(t)}{k_BT}. \tag{4.118}$$

Here $\Delta F = F(\lambda(t)) - F(\lambda(t_0))$, where the free energy is

$$F(\lambda) = -k_BT \ln \int dx\, e^{-\lambda x^2/2k_BT} = -\frac{k_BT}{2}\ln\frac{2\pi k_BT}{\lambda}. \tag{4.119}$$

Numerical solution of eqs. (4.114a and b) yields the cumulant generating function $\alpha(q_2, t_f)$ of the work performed on the system during the manipulation. Figure 4.5 shows the cumulant generating function of the work for a linear protocol,

$$\lambda(t) = \lambda_0 + (\lambda_f - \lambda_0)\frac{t}{\mathcal{T}}, \qquad 0 \leq t \leq \mathcal{T}, \tag{4.120}$$

with $\lambda_0 = 1$ and $\lambda_f = 5$ (in units of k_BT), and for different values of $\mathcal{T}$ (in units of μ_P^{-1}). The curves are visibly nonparabolic, which means that the distribution of w is not Gaussian. All curves cross at $q_2 = -1$ (in units of $(k_BT)^{-1}$) in agreement with the Jarzynski equality. They also exhibit a divergence for positive values of q_2, which is the signature of the exponential decay of $p(w, t_f)$ for large values of w.

4.14 Brownian motion with inertia (*)

If the inertia of a Brownian particle is not negligible, the stochastic equations describing its motion take a slightly different form. A particle of mass m in one dimension subject to a potential $U(r, \lambda)$ and to random noise evolves according to a stochastic version of the Newton equation:

$$m\frac{\mathrm{d}^2 r}{\mathrm{d}t^2} = -\partial_r U(r, \lambda) - \mu_P \frac{\mathrm{d}r}{\mathrm{d}t} + \sqrt{2D}\,\xi(t), \qquad (4.121)$$

where r is the particle position and $\mu_P\, \mathrm{d}r/\mathrm{d}t$ represents the effects of friction. We recast this equation in terms of the conjugate variables (p_r, r), where $p_r = m\, \mathrm{d}r/\mathrm{d}t$ is the momentum:

$$\frac{\mathrm{d}r}{\mathrm{d}t} = \frac{p_r}{m};$$

$$\frac{\mathrm{d}p_r}{\mathrm{d}t} = -\partial_r U(r, \lambda) - \frac{\mu_P}{m} p_r + \sqrt{2D}\,\xi(t). \qquad (4.122)$$

By going through the steps leading to the Fokker-Planck equation, we obtain the evolution equation satisfied by the probability density $p(p_r, r; t)$:

$$\frac{\partial p(p_r, r; t)}{\partial t} = \frac{\partial}{\partial r}\left[-\frac{p_r}{m} p(p_r, r; t) \right]$$
$$+ \frac{\partial}{\partial p_r}\left[\left(\frac{\partial U}{\partial r} + \frac{\mu_P}{m} p_r \right) p(p_r, r; t) + D\frac{\partial p(p_r, r; t)}{\partial p_r} \right]. \qquad (4.123)$$

This equation is known as the **Kramers equation**. Provided that the Einstein relation (3.84) holds, the Kramers equation has the equilibrium solution

$$p^{\mathrm{eq}}(p, r) = \exp\left[\frac{F(\lambda) - \mathcal{H}(p_r, r; \lambda)}{k_B T} \right], \qquad (4.124)$$

where

$$\mathcal{H}(p_r, r; \lambda) = \frac{p_r^2}{2m} + U(r, \lambda) \qquad (4.125)$$

is the **Hamiltonian** and

$$F(\lambda) = -k_B T \ln \int \mathrm{d}p_r\, \mathrm{d}r\, e^{-\mathcal{H}(p_r, r, \lambda)/k_B T} \qquad (4.126)$$

is the free energy.

The trajectory probability of the Kramers equation can be derived following the same reasoning as in section 4.11. In this case too, the trajectory probability can be expressed in the form of eq. (4.89). For the Kramers equation, the action reads

$$S(p_r, r; \lambda) = \int_{t_0}^{t_f} dt \, \frac{1}{4D} \left[\frac{dp_r(t)}{dt} + \frac{\partial}{\partial r} U(x(t), \lambda(t)) + \frac{\mu_P}{m} p_r(t) \right]^2. \qquad (4.127)$$

We now compare the probability of a trajectory $x = (p_r, r)$ of the Kramers equation to that of its time-reversed image $\widehat{x}$. In this case, we have to take into account that the momenta change sign under time reversal. This is therefore another example in which the state $\widehat{x}(t)$ in the time-reversed trajectory is different from $x(\widehat{t})$:

$$\widehat{x} = (\widehat{p}_r(t), \widehat{r}(t)) = (-p_r(\widehat{t}), r(\widehat{t})); \qquad (4.128)$$

see also section 4.10. Taking into account this transformation, the action of the backward trajectory is expressed by

$$S(\widehat{p}_r, \widehat{r}; \widehat{\lambda}) = \int_{t_0}^{t_f} dt \, \frac{1}{4D} \left[\frac{dp_r(t)}{dt} + \partial_r U(x(t), \lambda(t)) - \frac{\mu_P}{m} p_r(t) \right]^2. \qquad (4.129)$$

The ratio between the probability density $\mathcal{P}(x, \lambda)$ and that of its time reverse $\mathcal{P}(\widehat{x}, \widehat{\lambda})$ can be evaluated as in section 4.12, leading to the expression

$$\frac{\mathcal{P}(x|x_0; \lambda)}{\mathcal{P}(\widehat{x}|x_f; \widehat{\lambda})} = \exp\left[-\frac{1}{k_B T} \left(S(x; \lambda) - S(\widehat{x}; \widehat{\lambda}) \right) \right], \qquad (4.130)$$

where

$$S(x; \lambda) - S(\widehat{x}; \widehat{\lambda}) = \frac{1}{k_B T} \int_{t_0}^{t_f} dt \, \frac{p_r(t)}{m} \left(\frac{dp_r(t)}{dt} + \partial_r U(r(t), \lambda(t)) \right). \qquad (4.131)$$

In this case, we have

$$dt \, \frac{p_r(t)}{m} \left(\frac{dp_r(t)}{dt} + \partial_r U(r(t), \lambda(t)) \right) = d\left(\frac{p_r^2}{2m} \right) + dU - d\lambda \, \partial_\lambda U$$
$$= d\mathcal{H} - dw = -dq, \qquad (4.132)$$

from which the irreversibility relation (4.12) follows.

Since the states of a system described by the Kramers equation are not even under time reversal, it is possible to have dissipationless probability currents in the equilibrium state; see section 4.10. As a simple example, consider a charged Brownian particle in two dimensions subject to a magnetic field $\vec{B}$ directed normally to the plane. The Langevin equation for this system reads

$$\frac{d\vec{v}}{dt} = -\mu_P \vec{v} + \frac{q}{m} \vec{v} \times \vec{B} + \sqrt{2D} \vec{\xi}, \qquad (4.133)$$

where $\vec{v} = (v_1, v_2)$ is the velocity of the particle, m its mass, and q its charge. The corresponding Kramers equation reads

$$\frac{\partial p(\vec{v}; t)}{\partial t} = \frac{\partial J_1}{\partial v_1} + \frac{\partial J_2}{\partial v_2} \tag{4.134}$$

$$= \frac{\partial}{\partial v_1} \left[\left(\mu_P v_1 + \frac{q}{m} B v_2 \right) p + D \frac{\partial p}{\partial v_1} \right] + \frac{\partial}{\partial v_2} \left[\left(\mu_P v_2 - \frac{q}{m} B v_1 \right) p + D \frac{\partial p}{\partial v_2} \right].$$

The equilibrium distribution is the Maxwell-Boltzmann one, $p^{eq}(\vec{v}) \propto e^{-v^2/2mk_B T}$. Substituting in the above equation, and taking into account the Einstein relation (3.84), we obtain

$$J_1 = -\frac{qB}{m} v_2 \, p^{eq}(\vec{v}); \qquad J_2 = \frac{qB}{m} v_1 \, p^{eq}(\vec{v}). \tag{4.135}$$

One can check that the right-hand side of eq. (4.134) indeed vanishes, although the current $\vec{J}$ does not vanish, as anticipated. In this case, eq. (4.13), which states that the probability densities of a trajectory and its reverse are equal, does not hold. However, according to the Ampère principle, the Langevin equation satisfied by the backward velocity $\widehat{\vec{v}}$ in the backward dynamics is given by eq. (4.133), since the sign of $\vec{B}$ must be changed. We have therefore, for any trajectory $\mathbf{v} = (\vec{v}(t))$,

$$\dot{s}^{tot}(\mathbf{v}) = k_B \ln \frac{\mathcal{P}_\mathbf{v}(\vec{B})}{\mathcal{P}_{\widehat{\mathbf{v}}}(-\vec{B})} = 0. \tag{4.136}$$

4.15 Hamiltonian systems (*)

For Hamiltonian systems, fluctuation relations like the Jarzynski equality and the Crooks relation take a very simple form. We identify the state ξ of a Hamiltonian system by n pairs of momentum p_r and coordinate r variables:

$$\xi = ((p_{r1}, r_1), \ldots, (p_{rn}, r_n)) = (p_r, r). \tag{4.137}$$

The Hamiltonian $\mathcal{H}(\xi; \lambda)$ depends on an external parameter λ. The system is manipulated during the time interval $t_0 \leq t \leq t_f$ by changing the parameter λ according to a protocol $\lambda = (\lambda(t))$, such that $\lambda(t_0) = \lambda_0$ and $\lambda(t_f) = \lambda_f$. During this time interval, the system is thermally isolated. Therefore, its state changes according to the canonical equations of motion:

$$\frac{dp_{ri}}{dt} = -\frac{\partial \mathcal{H}}{\partial r_i},$$

$$\frac{dr_i}{dt} = \frac{\partial \mathcal{H}}{\partial p_{ri}}, \tag{4.138}$$

for $i = 1, \ldots, n$. Prior to the instant t_0, the value of λ is constant and the system is in equilibrium with a reservoir at temperature T. The initial distribution of ξ is therefore equal to the equilibrium distribution

$$p^{eq}(\xi; \lambda_0) = e^{(F(\lambda_0) - \mathcal{H}(\xi; \lambda_0))/k_B T}, \tag{4.139}$$

where

$$F(\lambda) = -k_B T \ln \int d\xi \; e^{-\mathcal{H}(\xi; \lambda)/k_B T} \tag{4.140}$$

is the free energy corresponding to a given value of λ.

Since the equations of motion (4.138) are deterministic, the trajectories $\boldsymbol{\xi} = (\xi(t))$ in the time interval $t_0 \le t \le t_f$ are uniquely identified by the initial condition $\xi_0 = (\xi(t_0))$. We have, on the other hand,

$$
\begin{aligned}
\frac{d}{dt} \mathcal{H}(\xi(t); \lambda(t)) &= \sum_{i=1}^{n} \left[\frac{dp_{ri}}{dt} \frac{\partial \mathcal{H}}{\partial p_{ri}} + \frac{dr_i}{dt} \frac{\partial \mathcal{H}}{\partial r_i} \right] + \frac{\partial \mathcal{H}}{\partial t} \\
&= \sum_{i=1}^{n} \left[-\frac{\partial \mathcal{H}}{\partial r_i} \frac{\partial \mathcal{H}}{\partial p_{ri}} + \frac{\partial \mathcal{H}}{\partial p_{ri}} \frac{\partial \mathcal{H}}{\partial r_i} \right] + \frac{\partial \mathcal{H}}{\partial t} = \frac{d\lambda}{dt} \frac{\partial \mathcal{H}}{\partial \lambda}.
\end{aligned}
\tag{4.141}
$$

We obtain therefore, for $t_0 \le t \le t_f$,

$$\mathcal{H}(\xi(t); \lambda(t)) = \mathcal{H}(\xi_0; \lambda_0) + \int_{t_0}^{t} dt' \frac{d\lambda(t')}{dt'} \frac{\partial}{\partial \lambda} \mathcal{H}(\xi(t'); \lambda(t')). \tag{4.142}$$

Identifying the value of the Hamiltonian with the system energy, the integral appearing on the right-hand side is none other than the manipulated work $w(\boldsymbol{\xi}; \lambda)$ defined in eq. (3.9). Since $\xi(t)$, once the protocol λ is given, depends only on ξ_0, w can be considered as a function $w(t, \xi_0)$ of the instant t and the initial state ξ_0.

We now evaluate $\langle e^{-w(t,\xi_0)/k_B T} \rangle$, where the average is taken over the distribution (4.139) of the initial condition ξ_0. We obtain

$$
\begin{aligned}
\left\langle e^{-w(t,\xi_0)/k_B T} \right\rangle &= \int d\xi_0 \; e^{-w(t,\xi_0)/k_B T} \; p^{eq}(\xi_0) \\
&= \int d\xi_0 \; e^{-(\mathcal{H}(\xi(t,\xi_0); \lambda(t)) - \mathcal{H}(\xi_0; \lambda_0))/k_B T} \\
&\quad \times e^{(F(\lambda_0) - \mathcal{H}(\xi_0; \lambda_0))/k_B T} \\
&= \int d\xi_0 \; e^{(-\mathcal{H}(\xi(t,\xi_0); \lambda(t)) + F(\lambda_0))/k_B T}.
\end{aligned}
\tag{4.143}
$$

We change the variable of integration from the initial state ξ_0 to the evolved state $\xi(t, \xi_0)$. According to the Liouville theorem, the volume of the region $d\xi(t, \xi_0)$ evolved at time t from a region $d\xi_0$ at time $t = t_0$ is equal to the volume of the original region. Therefore, the determinant of the Jacobian associated with this change of variable is equal to one, and we obtain

$$\left\langle e^{-w(t,\xi_0)/k_B T} \right\rangle = \int d\xi \; e^{(-\mathcal{H}(\xi; \lambda(t)) + F(\lambda_0))/k_B T} = e^{(F(\lambda_0) - F(\lambda(t)))/k_B T}, \tag{4.144}$$

where we use the definition (4.140) of the free energy. Setting $t = t_f$, we obtain the Jarzynski equality (4.52). This implies that the average dissipated work

$$W^{\text{diss}} = \langle w(\boldsymbol{\xi}; \boldsymbol{\lambda}) \rangle - \Delta F \tag{4.145}$$

is nonnegative.

In this derivation, we neither considered the reverse process nor made an explicit connection with entropy production. In fact, the dynamics in the time interval from t_0 to t_f satisfies the Liouville theorem and therefore the system entropy does not change. We now consider a case in which, at the end of the manipulation, the system is again put in contact with the reservoir. The heat $q(\boldsymbol{\xi}; \boldsymbol{\lambda})$ released to this reservoir is equal to

$$\begin{aligned} q(\boldsymbol{\xi}; \boldsymbol{\lambda}) &= \mathcal{H}(\xi_f; \lambda_f) - \langle \mathcal{H}(\xi; \lambda_f) \rangle^{\text{eq}} \\ &= \mathcal{H}(\xi_0; \lambda_0) + w(t_f, \xi_0; \boldsymbol{\lambda}) - (F(\lambda_f) + TS(\lambda_f)) \\ &= (w(t_f, \xi_0; \boldsymbol{\lambda}) - \Delta F) + T (s(\xi_0) - S(\lambda_f)), \end{aligned} \tag{4.146}$$

where $\Delta F = F(\lambda_f) - F(\lambda_0)$, $\langle \mathcal{H}(\xi; \lambda_f) \rangle^{\text{eq}}$ is evaluated with the equilibrium distribution associated with λ_f, and S_f is the corresponding equilibrium entropy. We have indeed

$$\langle \mathcal{H}(\xi; \lambda_f) \rangle^{\text{eq}} = F(\lambda_f) + T S_f. \tag{4.147}$$

Thus the total entropy produced along the trajectory $\boldsymbol{\xi}$ is given by

$$s^{\text{tot}}(\boldsymbol{\xi}; \boldsymbol{\lambda}) = \frac{q(\boldsymbol{\xi}; \boldsymbol{\lambda})}{T} + S_f - s(\xi_0) = \frac{w(t_f, \xi_0) - \Delta F}{T}. \tag{4.148}$$

As we discussed, this entropy is entirely produced at the end of the manipulation protocol.

We now consider the reverse process. We define the backward protocol $\widehat{\boldsymbol{\lambda}}$ as being made up of a long initial time interval in which the system is allowed to equilibrate with the reservoir, with the value λ_f of the parameter, followed by a manipulation time interval for $t_0 \leq t \leq t_f$ and followed again by a long time interval in contact with the reservoir, with the value λ_0 of the parameter. The value $\widehat{\lambda}(t)$ of the parameter λ in the interval $t_0 \leq t \leq t_f$ is given by

$$\widehat{\lambda}(t) = \lambda(\widehat{t}), \tag{4.149}$$

where $\widehat{t} = t_f + t - t_0$. We focus on the dynamics during the interval $t_0 \leq t \leq t_f$. Given the trajectory $\boldsymbol{\xi}$, the backward trajectory $\widehat{\boldsymbol{\xi}}$ is defined by $\widehat{\xi}(t) = \widetilde{\xi}(\widehat{t})$, where $\widetilde{\xi}$ is the time-reversed image of the phase-space point ξ. If $\xi = (p_r, r)$, we have

$$\widetilde{\xi} = (-p_r, r) = ((-p_{r1}, r_1), \ldots, (-p_{rn}, r_n)). \tag{4.150}$$

We now compare the probabilities $\mathcal{P}(\boldsymbol{\xi}; \boldsymbol{\lambda})$ and $\mathcal{P}(\widehat{\boldsymbol{\xi}}; \widehat{\boldsymbol{\lambda}})$ of the forward and backward trajectories in the forward and backward protocols, respectively. As we stressed before, the probability of a trajectory $\boldsymbol{\xi}$ is equal to the probability of its initial condition ξ_0. It follows that

$$\frac{\mathcal{P}(\boldsymbol{\xi}; \boldsymbol{\lambda})}{\mathcal{P}(\widehat{\boldsymbol{\xi}}; \widehat{\boldsymbol{\lambda}})} = \frac{p^{\text{eq}}(\xi_0; \lambda_0)}{p^{\text{eq}}(\widetilde{\xi}_f; \lambda_f)}. \tag{4.151}$$

Assuming that the Hamiltonian $\mathcal{H}(\xi;\lambda)$ is time-reversal invariant, i.e, that $\mathcal{H}(\tilde{\xi};\lambda) = \mathcal{H}(\xi;\lambda)$, $\forall\lambda$, we obtain

$$\frac{\mathcal{P}(\xi;\lambda)}{\mathcal{P}(\widehat{\xi};\widehat{\lambda})} = e^{(F(\lambda_0)-F(\lambda_f)-(\mathcal{H}(\xi_0;\lambda_0)-\mathcal{H}(\xi_f;\lambda_f)))/k_BT}$$

$$= e^{-(w(\xi;\lambda)-\Delta F)/k_BT} = e^{-s^{\mathrm{tot}}(\xi;\lambda)/k_B}. \tag{4.152}$$

Since the initial and final states are equilibrium states, we have

$$w(\widehat{\xi};\widehat{\lambda}) = -w(\xi;\lambda). \tag{4.153}$$

Thus, summing over all trajectories ξ with a given value w of the work, we obtain the Crooks relation (4.53) in the form

$$\frac{p(w;\lambda)}{p(-w;\widehat{\lambda})} = e^{-(w-\Delta F)/k_BT}. \tag{4.154}$$

The Jarzynski equality follows by integrating over w.

We use this result to relate the average dissipated work defined in eq. (4.145) and the Kullback-Leibler divergence between the phase-space distributions in the forward and in the reverse process. To this aim, we express both the forward and the reverse process in terms of the forward time, so that corresponding points in the time axis have the same value of λ (fig. 4.6). Using the Liouville theorem, eq. (4.151) assumes the form

$$\frac{p(\xi_f,t_f;\lambda)}{p(\tilde{\xi}_f,t_f;\widehat{\lambda})} = e^{(w(\xi;\lambda)-\Delta F)/k_BT}. \tag{4.155}$$

Taking the logarithm, we obtain

$$w(\xi;\lambda) = \Delta F + k_BT \ln \frac{p(\xi_f,t_f;\lambda)}{p(\tilde{\xi}_f,t_f;\widehat{\lambda})}. \tag{4.156}$$

Upon averaging, we express the average work $W = \langle w(\xi;\lambda) \rangle$ performed on the system as

$$W = \Delta F + k_BT \int d\xi_f \, p(\xi_f;\lambda) \ln \frac{p(\xi_f,t_f;\lambda)}{p(\tilde{\xi}_f,t_f;\widehat{\lambda})}$$

$$= \Delta F + k_BT \, D_{\mathrm{KL}} \left(p(\xi_f,t_f;\lambda) \| p(\tilde{\xi}_f,t_f;\widehat{\lambda}) \right). \tag{4.157}$$

For any value of t satisfying $t_0 \le t \le t_f$ and for any point ξ in phase space, we have

$$\frac{p(\xi,t;\lambda)}{p(\tilde{\xi},t;\widehat{\lambda})} = \frac{p(\xi_f,t_f;\lambda)}{p(\tilde{\xi}_f,t_f;\widehat{\lambda})}, \qquad t_0 \le t \le t_f, \tag{4.158}$$

because $\widehat{\xi}$ is the unique trajectory going through $\tilde{\xi}$ at time t under the backward protocol $\widehat{\lambda}$ if the trajectory ξ goes through ξ at time t by the protocol λ. This implies that we can

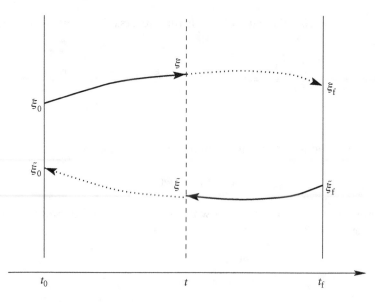

Figure 4.6. Scheme of the correspondence between the forward trajectory ξ under the forward protocol λ and its time-reversed image $\widehat{\xi}$ under the backward protocol $\widehat{\lambda}$. The trajectory ξ starts from x_0 at time t_0, goes through ξ at time t, and ends at ξ_f at time t_f. The backward trajectory is parameterized in terms of $\widehat{t}$ rather than t. It starts at $\tilde{\xi}_f$ at t_f, goes through $\tilde{\xi}$ at time t, and ends at $\tilde{\xi}_0$ at time t_0. See Kawai et al. [86].

evaluate the Kullback-Leibler divergence at any instant in time in the interval $[t_0, t_f]$, obtaining the same result. This property is peculiar to Hamiltonian systems.

We now consider a mesoscopic system coupled to a thermal reservoir and assume that the system plus the reservoir are thermally isolated and evolve according to Hamiltonian dynamics. In these settings, the stochastic dynamics of the mesoscopic system originates from coarse graining, i.e., from integrating out the degrees of freedom of the thermal reservoir. It is interesting to study how eq. (4.157) is affected by such a coarse-graining operation. Specifically, we partition the phase space of a Hamiltonian system into regions Ω_x and we assign each such region Ω_x to a mesostate x. We then use eq. (4.155) to evaluate the average of $e^{-w(\xi;\lambda)/k_BT}$ over each of these regions:

$$\left\langle e^{-w(\xi;\lambda)/k_BT} \right\rangle_x = \frac{1}{p_x(t;\lambda)} \int_{\Omega_x} d\xi \, p(\xi,t;\lambda) \, e^{-w(\xi;\lambda)/k_BT}, \qquad (4.159)$$

where $p_x(t;\lambda)$ is the probability that, at time t, the microstate $\xi(t)$ of the system belongs to the region Ω_x:

$$p_x(t;\lambda) = \int_{\Omega_x} d\xi \, p(\xi,t;\lambda). \qquad (4.160)$$

We then have, by eq. (4.155),

$$\left\langle e^{-w(\xi;\lambda)/k_BT} \right\rangle_x = \frac{p_x(t;\widehat{\lambda})}{p_x(t;\lambda)} \, e^{-\Delta F/k_BT}. \qquad (4.161)$$

By the Jensen inequality, this relation implies

$$\langle w(\boldsymbol{\xi}; \lambda) \rangle_x \geq \Delta F + k_{\mathrm{B}} T \ln \frac{p_x(t; \lambda)}{p_x(t; \widehat{\lambda})}. \tag{4.162}$$

Applying the same reasoning to $w(\widehat{\boldsymbol{\xi}}; \widehat{\lambda})$, we obtain

$$\left\langle w(\widehat{\boldsymbol{\xi}}; \widehat{\lambda}) \right\rangle_x \geq -\Delta F - k_{\mathrm{B}} T \ln \frac{p_x(t; \lambda)}{p_x(t; \widehat{\lambda})}. \tag{4.163}$$

Taking the averages, we obtain the following inequalities for the average work W_{F} and W_{B} respectively performed in the forward and reverse processes, for $t_0 \leq t \leq t_{\mathrm{f}}$:

$$W_{\mathrm{F}} \geq \Delta F + k_{\mathrm{B}} T \, D_{\mathrm{KL}}(p_x(t; \lambda) \| p_x(t; \widehat{\lambda})); \tag{4.164}$$

$$W_{\mathrm{B}} \geq -\Delta F + k_{\mathrm{B}} T \, D_{\mathrm{KL}}(p_x(t; \widehat{\lambda}) \| p_x(t; \lambda)). \tag{4.165}$$

For stochastic systems, the Kullback-Leibler divergence in general depends on t. These results can also be directly obtained from the equality (4.157) by observing that the Kullback-Leibler divergence cannot increase upon coarse graining; see section 2.10.

We illustrate these results by evaluating the dissipated work and the probability distributions for the overdamped harmonic oscillator discussed in section 4.13. The distribution $p(x, t; \lambda)$ satisfies the Fokker-Planck equation

$$\frac{\partial p}{\partial t} = \mu_{\mathrm{P}} \frac{\partial}{\partial x} \left[\lambda(t) x p + k_{\mathrm{B}} T \frac{\partial p}{\partial x} \right], \tag{4.166}$$

where μ_{P} is the friction coefficient. In our case, if the initial distribution is Gaussian, it remains Gaussian at all later times. We have found in section 4.13 that the mean of the Gaussian vanishes at all times if it does so initially, and that its variance $\gamma(t)$ satisfies the equation

$$\frac{d\gamma}{dt} = 2\mu_{\mathrm{P}} \left[-\lambda(t) \gamma(t) + k_{\mathrm{B}} T \right]. \tag{4.167}$$

The free energy difference ΔF is given by

$$\Delta F = \frac{k_{\mathrm{B}} T}{2} \ln \frac{\lambda_{\mathrm{f}}}{\lambda_0}, \tag{4.168}$$

which is obtained by direct integration.

If $p(x, \gamma)$ is a Gaussian distribution with vanishing mean and variance equal to γ, we have

$$D_{\mathrm{KL}}(p(x, \gamma) \| p(x, \widehat{\gamma})) = \frac{1}{2} \left(\frac{\gamma}{\widehat{\gamma}} - 1 + \ln \frac{\widehat{\gamma}}{\gamma} \right). \tag{4.169}$$

The average accumulated work up to time t satisfies the equation

$$\frac{dW(t)}{dt} = \frac{1}{2} \frac{d\lambda(t)}{dt} \langle x^2 \rangle_t = \frac{1}{2} \frac{d\lambda(t)}{dt} \gamma(t), \tag{4.170}$$

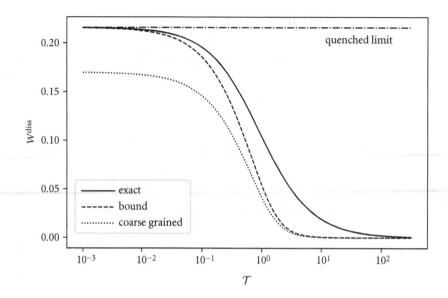

Figure 4.7. Dissipated work $W^{\mathrm{diss}} = \langle w \rangle - \Delta F$ as a function of the duration $\mathcal{T} = t_f - t_0$ of the manipulation for the overdamped harmonic oscillator linearly manipulated from $\lambda_0 = 3$ to $\lambda_f = 1$. Units are such that $\mu_P = 1$ and $k_B T = 1$. The quenched limit is given by eq. (4.171). We also show the bound obtained from eq. (4.164) by evaluating $p(x, t; \lambda)$ and $p(x, t; \widehat{\lambda})$ for $t = \mathcal{T}/2$. The weaker bound is obtained by convoluting the distribution with a Gaussian of width $\sigma^2 = 1/12$, which is equal to the variance of a uniform distribution over an interval of length $\Delta x = 1$. See Kawai et al. [86].

where $\gamma(t)$ is the solution of eq. (4.167). We can therefore solve the system of equations (4.167, 4.170) and obtain the dissipated work W^{diss} and the variances $\gamma(t)$, $\widehat{\gamma}(t)$, respectively associated with the forward and backward protocols. In this way, we obtain W^{diss} as a function of the manipulation duration $\mathcal{T}$, as shown in fig. 4.7. For $\mathcal{T} \to 0^+$, the dissipated work tends to the quenched limit, given by

$$
W^{\mathrm{quench}} = \int dx\, (H(x; \lambda_f) - H(x; \lambda_0))\, p^{\mathrm{eq}}(x; \lambda_0) = \frac{1}{2}\, (\lambda_f - \lambda_0)\, \langle x^2 \rangle^{\mathrm{eq}}_{\lambda_0}
$$
$$
= \frac{k_B T}{2} \left(\frac{\lambda_f}{\lambda_0} - 1 \right).
$$
(4.171)

For all values of $\mathcal{T}$, the dissipated work is larger than the bound obtained from eq. (4.164) by evaluating $p(x, t; \lambda)$ and $p(x, t; \widehat{\lambda})$ for $t = \mathcal{T}/2$, and also larger than a weaker bound obtained by coarse-graining the distributions $p(x, t; \lambda)$ and $p(x, t; \widehat{\lambda})$ with a procedure that approximates averaging over an interval of length $\Delta x = 1$; see fig. 4.7.

4.16 Further reading

There exists a vast literature on fluctuation relations. Indeed, different forms of these relations hold when the observables of interest, the dynamics obeyed by the system, or the underlying hypotheses are different. Evans, Cohen, and Morriss [51] proposed an early fluctuation relation for the shear stress in a nonequilibrium system in contact with a thermostat. Gallavotti and Cohen [56, 57] derived a detailed fluctuation relation for the entropy production rate in deterministic chaotic systems. Their work has been so influential that the symmetry implied by the detailed fluctuation relation (see eq. (4.46)) is often referred to as the Gallavotti-Cohen symmetry. Onsager [121, 122] derived the reciprocity relations.

Jarzynski [81] and Crooks [33, 34] developed fluctuation relations for systems manipulated from an equilibrium state to another equilibrium state. Their work paved the way for the use of nonequilibrium measurements to estimate equilibrium free-energy differences and therefore had an enormous impact. Their success led to the rediscovery of works by Bochkov and Kuzovlev [19, 20, 21, 22] that pioneered similar fluctuation relations, although with a different definition of work, as discussed by Jarzynski [82].

In parallel, Kurchan [92] showed that, for long times and in the steady state, the entropy production of a general system of Langevin equation satisfies the Gallavotti-Cohen symmetry. Lebowitz and Spohn [98] extended this result to a general Markovian dynamics. Seifert [148] clarified that, for Langevin dynamics, the integral fluctuation relation also holds for finite times and time-dependent drivings. Oono and Paniconi [123] introduced the concept of housekeeping heat in non-equilibrium steady states.

We have seen that fluctuation relations emerge rather naturally when relating entropy production to the ratio of probabilities of forward and backward trajectories. Although this idea was implicitly present in some of the early derivations, Maes [106] points out this aspect more explicitly. This very powerful concept extends to many other fluctuation relations, as reviewed by Harris and Schütz [73] and Gawedzki [61]. The discussion in section 4.9 follows the lines of Esposito and Van den Broeck [50], and the relation (4.77) is derived in Hatano and Sasa [75]. Decomposition of the entropy production for systems with odd-parity variables is discussed in Spinney and Ford [159] and Lee et al. [100]. Verley and Lacoste [173, 174] review its application to fluctuation-dissipation relations out of equilibrium. Cuetara et al. [36] derive a quite general version of the fluctuation relation. Rao and Esposito [137] provide a unified perspective on many fluctuation theorems derived in the literature. A pedagogical introduction to ensemble and trajectory thermodynamics and to fluctuation relations is due to Van den Broeck and Esposito [170].

The example of the Brownian particle in a time-dependent harmonic potential is discussed by Speck [158]. The relation between dissipated work and the divergence of phase-space distributions is pointed out by Kawai et al. [86].

4.17 Exercises

4.1   Assuming that the distribution of s^{tot} is Gaussian, prove eq. (4.17).

4.2   Consider the dragged particle introduced in section 4.3. Show that the probability distribution of the total entropy s^{tot} produced in a time interval (t_0, t_f) satisfies the detailed fluctuation theorem.

4.3   A particle with mobility μ_P is immersed in a fluid at temperature T, and it is subject to a constant force f along the x-axis for a time interval of duration $\mathcal{T}$. Show that the Jarzynski equality implies that the Einstein relation between the mobility and the diffusion coefficient D is satisfied.

4.4   Consider a gas made of a single molecule contained in a cylinder with adiabatic walls with a movable piston. The molecule is initially at equilibrium at temperature T in a volume V_0. The piston is then moved with constant speed v until the volume of the gas is V. Show explicitly the validity of the Jarzynski equality in the case $V > V_0$ and very large speed v. Assume that $V - V_0 \ll V_0$.

4.5   Imagine that you are shown a movie of a mesoscopic system undergoing manipulation while in contact with a reservoir at temperature T. The movie can be either the record of the forward (F) process, in which the system is manipulated out of an initial equilibrium state, or the record of the backward (B) process, projected in reverse order. Show that, to assess the probability that the movie shows the F process, it is sufficient to evaluate the dissipated work $w - \Delta F$, where the work w is computed from the trajectory exhibited in the movie. Evaluate the probability $p_F | w$ as a function of w.

4.6   By exploiting eq. (A.20), show that the average nonadiabatic entropy production rate is nonnegative at all times.

4.7   Consider a system obeying a master equation satisfying the generalized detailed balance condition (3.7). By exploiting eq. (A.20), show that the minimal rate of entropy production needed to keep the system in a nonequilibrium steady state p_x^{st} is given by

$$\dot{S}^{min} = -k_B T \left. \frac{d}{dt} D_{KL}(p(t) \| p^{eq}) \right|_{p=p^{st}},$$

where the derivative is evaluated with the master equation that is obtained by setting all the drivings $\delta_{xx'}$ to zero, and where p_x^{eq} is the corresponding equilibrium distribution.

4.8   The position of a Brownian particle evolves according to the Langevin equation (3.82) with the potential (4.101). Evaluate the probability density $p_F(w)$ of the work when the potential strength λ is suddenly changed from λ_0 to λ_f, assuming that the particle is initially in thermal equilibrium. Evaluate also the corresponding distribution $p_R(w)$ for the reverse process, assuming that the particle is initially at equilibrium with the strength λ_f. Verify the detailed fluctuation relation

$$\frac{p_F(w)}{p_B(-w)} = e^{(\Delta F - w)/k_B T},$$

where $\Delta F = F(\lambda_f) - F(\lambda_0)$. Show that the distributions exhibit exponential tails.

4.9   Consider eq. (4.110) in the slow manipulation regime described by equation (4.120), with $\mathcal{T} \gg \Gamma^{-1}$. Obtain the equation satisfied by $\gamma_1(p, t)$ and $\sigma_w^2(t)$ and verify the relation (4.118).

4.10   Consider a weakly interacting, dilute gas contained in a cylinder with a piston, initially at equilibrium at temperature T. The system undergoes an adiabatic quasi-static transformation in which its volume is changed from V_0 to V_f. Assuming that the gas is able to equilibrate with itself during the transformation, verify the Jarzynski equality for this system. Optionally, evaluate the probability distribution of the work performed during the transformation and show that it satisfies the Jarzynski equality.

CHAPTER 5

Thermodynamics of Information

Maxwell, then Boltzmann and Gibbs, already hinted at a connection between thermodynamics and information. They emphasized the statistical nature of the second law long before the birth of information theory. Despite this long history, thermodynamics of information still puzzles physicists. Stochastic thermodynamics places information in the context of concrete physical models, dramatically easing the study of its role in thermodynamics.

5.1 A brief history

In his *Theory of Heat* (1871), Maxwell illustrated the statistical nature of the second principle by a very famous argument:

> We now suppose that such a vessel is divided into two portions, A and B, by a division in which there is a small hole, and that a being, who can see the individual molecules, opens and closes this hole, so as to allow only the swifter molecules to pass from A to B, and only the slower molecules to pass from B to A. He will thus, without expenditure of work, raise the temperature of B and lower that of A, in contradiction to the second law of thermodynamics.

Thus a "demon" able to determine the energy of a single molecule could use this information to reduce the entropy of the vessel, driving it away from equilibrium. It would then in principle be possible to extract work from this disequilibrium.

Szilard proposed an intriguing conceptual experiment that made the connection between the second law and information even sharper (fig. 5.1). He considered a closed cylindrical container of volume V, in contact with a heat reservoir at temperature T, which contains an ideal gas made of a single molecule. The molecule is initially free to wander in the whole cylinder. An agent divides the cylinder into two equal chambers of volumes $V/2$ by inserting a piston in its middle. The agent then observes if the molecule is in the left or right chamber. If the particle is in the left chamber, the agent slowly moves the piston to the right, until it reaches the end of the cylinder. During this process, the molecule keeps bouncing on the piston, yielding work given by the expression for an isothermal expansion, $-W = k_B T \ln 2$. If the particle is in the right chamber, the agent moves the piston to the left until it reaches the end of the cylinder, extracting the same amount of work. After the expansion, the piston is removed, the cylinder returns to the initial state, and the cycle is repeated.

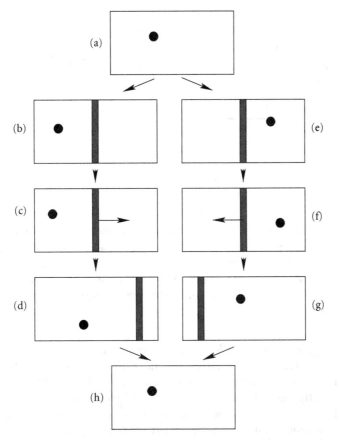

Figure 5.1. The Szilard thought experiment. A cylindrical container containing a single molecule (a) is divided in two parts by a movable wall. An agent identifies whether the particle is on the left or right of the wall. If it is on the left (b) the agent slowly moves the wall to the right (c) until it reaches the end of the container (d). In like manner, if the particle is on the right (e), the wall is slowly moved to the left (f, g). This procedure allows the agent to extract work equal to $k_BT \ln 2$ from the heat reservoir. At the end (h), the position of the particle is unknown and the system is back in its initial state.

This procedure can be generalized to a case where the piston divides the cylinder into two unequal chambers. In this case, the agent extracts from the reservoir an average work per cycle given by

$$-W = k_BT\, H(p). \tag{5.1}$$

Here $H(p) = -(V_L/V) \ln(V_L/V) - (1 - V_L/V) \ln(1 - V_L/V)$ is the Shannon entropy of the particle distribution in the two chambers, where V_L is the volume of the left chamber. This expression already suggests a correspondence between the information about the location of the particle and the maximum possible work extraction. This idea was generalized by Bennett and popularized by Feynman. They considered machines extracting work from a reservoir by exploiting very long sequences of microscopic containers, each containing a single molecule in a *known* left or right location (fig. 5.2).

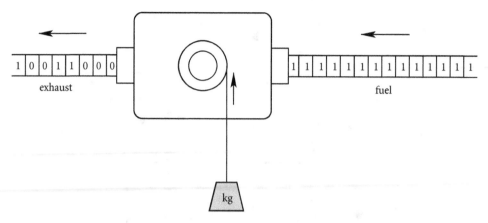

Figure 5.2. The Bennett-Feynman information-fueled engine. A tape (fuel) containing a large number of Szilard cylinders, each with the molecule in state 1 (right), is fed into the machine. Once inside, each cylinder undergoes the Szilard manipulation, and an average amount of work $-W = k_BT \ln 2$ is extracted. At the end of the manipulation, the location of the molecule in the cylinder is randomized (exhaust). See the discussion in Feynman [53, pp. 146–147].

How can we reconcile this result with the second law of thermodynamics? The Kelvin statement of the second law of thermodynamics forbids cyclic work extraction from a single heat reservoir, as pointed out in section 2.1. Yet the Szilard engine appears to accomplish exactly this feat. The solution to this dilemma followed a rather wandering path. Szilard himself suggested that the demon could be exorcised if the measurement of the particle's position required dissipation. Brillouin made the argument more explicit by considering detailed measurement protocols. However, Bennett argued that a dissipationless measurement could be performed provided the measuring apparatus was in a "standard state" before the measurement.

The discussion can be made more concrete by replacing the demon with a mechanical device taking actions depending on the measurement outcome. Importantly, this device should necessarily store the measurement outcome in a physical memory, at least for a short time. In this framework, Landauer pointed out the necessity of spending work to cyclically erase this memory. In the context of the thermodynamics of computation, Bennett generalized it in the following way:

> Any *logically irreversible* manipulation of information, such as the erasure of a bit or the merging of two computation paths, must be accompanied by a corresponding *entropy increase* in noninformation bearing degrees of freedom of the information-processing apparatus or its environment.

This statement is referred to as the **Landauer principle.** Quantitatively, the Landauer principle requires the expenditure of at least $k_BT \ln 2$ of work for erasing 1 bit, a quantity known as the **Landauer bound.**

5.2   Back to nonequilibrium free energy

To analyze the Szilard engine with the tools of stochastic thermodynamics, we first need to extend the concept of nonequilibrium free energy introduced in section 2.3. Our starting point is the Gibbs relation (2.132), which links the thermodynamic entropy and information entropy of an ensemble. In classical statistical mechanics, this link is valid only for macroscopic systems at equilibrium. We do not set this limitation in stochastic thermodynamics, where the system entropy is defined by eq. (3.25) for mesoscopic systems that are not necessarily at equilibrium. The price to pay for this generality is to accept an ensemble perspective: the system entropy in stochastic thermodynamics is defined at the level of a single system, but only as a member of an ensemble prepared according to a certain probability distribution.

In the ensemble perspective, the probability distribution p^{eq} of a system in thermal equilibrium satisfies a variational principle. For instance, if $\langle \epsilon \rangle_p = \sum_x p_x \epsilon_x$ is fixed, then $H(p^{eq}) \geq H(p)$, $\forall p$, where p^{eq} is the canonical distribution. Similarly, the Shannon entropy of a grand canonical distribution cannot be smaller than that of any distribution with the same values of average energy and average number of particles.

We derive a more general result by considering a system in contact with a reservoir at temperature T. We assume that the system is initially in a state described by a probability distribution p over the states, with Shannon entropy $H(p)$. We generalize the **nonequilibrium free energy** to this case by setting

$$F^{neq} = \langle \epsilon \rangle - k_B T H(p). \tag{5.2}$$

Also this broader definition reduces to the conventional free energy at equilibrium. For all probability distributions p, one has

$$F^{neq}(p) \geq F^{neq}(p^{eq}) = F. \tag{5.3}$$

To prove it, we note that

$$
\begin{aligned}
\Delta F^{neq} &= F^{neq}(p) - F^{neq}(p^{eq}) \\
&= \sum_x \left[\left(p_x \epsilon_x - p_x^{eq} \epsilon_x \right) + k_B T \left(p_x \ln p_x - p_x^{eq} \ln p_x^{eq} \right) \right] \\
&= \sum_x p_x \left(\epsilon_x + k_B T \ln p_x - F \right) = k_B T \sum_x p_x \ln \frac{p_x}{p_x^{eq}} \\
&= k_B T D_{KL}(p \| p^{eq}) \geq 0,
\end{aligned}
\tag{5.4}
$$

where $D_{KL}(p \| p^{eq})$ is the Kullback-Leibler divergence introduced in eq. (2.133). This result generalizes the analogous one obtained in section 2.3 (cf. eq. (2.35)), which applies to equilibrium states for which the thermodynamic entropy S is defined.

5.3 Information in stochastic thermodynamics

A common ingredient in the Maxwell demon paradox, the Szilard engine, and Landauer's idea is the manipulation of thermodynamic systems at the level of single molecules. Stochastic thermodynamics provides a natural framework to analyze these systems in a concrete way, besides the general ideas discussed in section 5.1. In particular, it proves useful to also consider the measuring device as a concrete physical system.

For example, we revisit the Szilard engine by considering a system (sys) as being made up of an **object** (obj) and a **measuring device** (dev). The object represents the part of the system that is being measured: in the case of the Szilard engine, the object is the particle in the cylinder. We describe the object with a binary mesoscopic variable

$x \in \{L, R\}$, depending on whether the particle is in the left or right chamber, respectively. We similarly describe the state of the measuring device by a binary mesoscopic variable $y \in \{L, R\}$. After performing an error-free measurement, the state of the measuring device matches the state of the particle.

We assume that initially the location of the particle and the state of the measuring device are independent and each uniformly distributed over the two states. The entropy of the system is therefore

$$S^{\text{sys}} = k_B \ln 4 = 2 k_B \ln 2. \tag{5.5}$$

After an error-free measurement, the system can only be found in the states $(x, y) = (L, L)$ or (R, R), with equal probabilities. This means that the total entropy is reduced to $k_B \ln 2$ by the measurement. This can be achieved either by providing work, equal at least to $-T \Delta S^{\text{sys}} = k_B T \ln 2$, or by increasing the entropy of another system, for example, a memory attached to the measuring device. When the expansion is completed and the piston is replaced in the middle of the cylinder, the states of the particle and the measuring device are again uncorrelated, and the system can be found in any of its four states.

We now consider a more general case where the object can be found in mesostates x, characterized by energies ϵ_x, initially at equilibrium at temperature T. To perform a measurement, the object is connected to a measuring device whose states are described by the variable y. We assume for simplicity that all these states have equal energy $\epsilon_y = 0$ and that the device is initially at equilibrium. The interdependence between x and y is quantified by the mutual information

$$I(\text{obj} : \text{dev}) = \sum_{x,y} p_{x,y} \ln \frac{p_{x,y}}{p_x p_y}. \tag{5.6}$$

At a time t_0 before the measurement, the equilibrium distribution of the system factorizes, $p_x^{\text{eq}}(t_0) = p_x^{\text{eq}} p_y^0$, where p_y^0 is the initial reference state of the measuring device. This implies that the mutual information vanishes. The measurement alters $p_{x,y}$ by introducing dependences between x and y. We assume that the measurement does not perturb the state x of the object, so that the marginal distribution $p_x = \sum_y p_{x,y}$ (see section A.3) remains unaltered. The change of system entropy therefore reads

$$\Delta S^{\text{sys}} = S(t_m) - S(t_0)$$
$$= -k_B \sum_{x,y} \left[p_{x,y}(t_m) \ln p_{x,y}(t_m) - p_{x,y}(t_0) \ln p_{x,y}(t_0) \right], \tag{5.7}$$

where t_m is a time immediately after the measurement. We write the final distribution as $p_{x,y}(t_m) = p_{y|x}(t_m) p_x(t_m)$, obtaining

$$\Delta S^{\text{sys}} = -k_B \sum_{x,y} \left[p_{x,y}(t_m) \ln \frac{p_{y,x}(t_m)}{p_x(t_m)} - p_{x,y}(t_0) \ln p_y(t_0) \right], \tag{5.8}$$

taking into account that the distribution of x is unaltered by the measurement and that $p_{x,y}(t_0)$ factorizes. Adding and subtracting $S^{\text{dev}}(t_m) = -k_B \sum_y p_y(t_m) \ln p_y(t_m)$, we obtain

$$\Delta S^{\text{sys}} = -k_B I(\text{obj} : \text{dev}) - \Delta S^{\text{dev}}. \tag{5.9}$$

Since the system interacts only with the heat reservoir, $\Delta S^{\text{sys}} < 0$ implies that an amount of heat at least equal to $Q = T \Delta S^{\text{res}} = -T \Delta S^{\text{sys}}$ must have been released to the heat reservoir in order to comply with the second law. It is instructive to consider two limiting cases:

- The measurement does not alter the marginal distribution of the device, so that $\Delta S^{\text{dev}} = 0$. In this case, eq. (5.9) predicts that an amount of work at least equal to $k_B T \, I(\text{obj} : \text{dev})$ has to be dissipated into the reservoir during the measurement.
- The measurement is adiabatic, $\Delta S^{\text{sys}} = 0$. Equation (5.9) then implies that $k_B I(\text{obj} : \text{dev}) = -\Delta S^{\text{dev}}$. This means that the mutual information acquired by the measurements has been "dumped" into the memory of the measuring device, which necessarily increases its entropy.

In light of this discussion, the operation of the Szilard engine can be restated as follows:

1. Initially, a partition is placed in the middle of the cylinder, but no measurement has yet been done. The system entropy is equal to $2 \, k_B \ln 2$.
2. An error-free measurement is performed, so that $y = x \in \{L, R\}$. The system entropy is now equal to $k_B \ln 2$. Therefore, either an amount of work equal to at least $k_B T \ln 2$ has been performed on the system and passed as heat to the reservoir, or the entropy of the measuring device has increased by $k_B \ln 2$, or a combination of the two has occurred.
3. By performing a free expansion, an amount of work up to $k_B T \ln 2$ is gleaned from the reservoir.
4. The partition is placed again in the middle of the cylinder, and the correlation between x and y is now broken.

If the measurement in point (2) is dissipative, a nonnegative amount of work, on average, is performed on the system, in agreement with Kelvin's formulation of the second law. Otherwise, work has been extracted but the operation of the engine is not cyclic, as the entropy of the device has been increased. To complete the cycle, we have to restore the initial state of the device and include the cost of that operation. This cost can be estimated by the change in nonequilibrium free energy $F^{\text{neq,dev}}(t_m) - F^{\text{neq,dev}}(t_0)$, which is equal to $T \Delta S^{\text{dev}}$ since the states of the device are isoenergetic. In our case, the minimum work that has to be performed is $W \geq k_B T \ln 2$, in agreement with the Landauer bound.

5.4 The Sagawa-Ueda relation

The link between information and stochastic thermodynamics can also be studied with fluctuation relations. We consider a mesoscopic object that is manipulated according to an experimental protocol. As in section 5.3, we suppose that an instantaneous measurement is performed at time t_m, and we denote its outcome by y. **Feedback control** is a modification of the manipulation protocol according to the outcome of the measurement. Using the Szilard engine as an example, feedback control represents the decision of placing the piston on either the right or left of the wall. Because of feedback control, the control parameter specifying the manipulation is a function not only of time, but also of the measurement outcome, $\lambda = (\lambda(t, y))$.

As in the previous section, we express the joint distribution of the state of the object and the measuring device at a time t_m immediately after the measurement by

$$p_{x,y}(t_m) = p_{y|x}(t_m)p_x(t_m). \tag{5.10}$$

The mutual information quantifies the average amount of information gained with the measurement. In parallel with the entropy, we define the **stochastic mutual information**

$$i_{x:y} = \ln \frac{p_{x,y}(t_m)}{p_x(t_m)p_y(t_m)} = \ln \frac{p_{y|x}(t_m)}{p_y(t_m)}, \tag{5.11}$$

so that $I = \langle i \rangle$. The joint probability of a trajectory x of the object and a measurement outcome y is

$$\mathcal{P}_{y,x}(\lambda) = p_{y|x}(\lambda)\mathcal{P}_x(\lambda). \tag{5.12}$$

These definitions allow us to generalize the integral fluctuation relation in the presence of feedback control:

$$\left\langle e^{-s^{tot}/k_B - i} \right\rangle_F = \int \mathcal{D}x \, dy \, p_{y|x}(t_m) \mathcal{P}_x(\lambda) \frac{\mathcal{P}_{\widehat{x}}(\widehat{\lambda})}{\mathcal{P}_x(\lambda)} \frac{p_y(t_m)}{p_{y|x}(t_m)}$$

$$= \int \mathcal{D}x \, dy \, \mathcal{P}_{\widehat{x}}(\widehat{\lambda}) \, p_y(t_m) = 1. \tag{5.13}$$

Here we make use of $p_{y|x}(t_m) = p_{y|x}(t_m)$: since the measurement is instantaneous, its outcome depends only on the state of the object at time m. Equation (5.13) is the **Sagawa-Ueda** relation. As in the Jarzynski equality (eq. (4.52)), if the protocol starts and ends at equilibrium, we have $s^{tot} = w - \Delta F$. There is however a subtle difference: in this case, ΔF is a stochastic quantity, since the final free energy depends in principle on the outcome of the measurement. Equation (5.13) then implies

$$W - \langle \Delta F \rangle \geq -k_B T I. \tag{5.14}$$

Equation (5.14) quantifies the apparent violations of the second law due to feedback control. In a single, instantaneous measurement, the average total entropy production can be negative, up to a minimum given by $-k_B$ times the mutual information between the object and the measuring device immediately after the measurement.

5.5 The Mandal-Jarzynski machine

The Mandal-Jarzynski machine is an explicit, analytically solvable model of a mesoscopic physical system that is capable of trading information for work. Its dynamics shows at play many of the concepts discussed so far.

The Mandal-Jarzynski machine is characterized by three mesostates A, B, and C, all of equal energy (fig. 5.3a). The machine interacts with a tape whose states (bits) can be 0 or 1. Bits 0 and 1 also have the same energies. At time intervals equal to τ, the tape shifts to the right, so that the machine interacts with a new bit. We denote by r the fraction of 1 bits in the incoming tape and by $d = 1 - 2r$ the excess of bits 0. The machine is

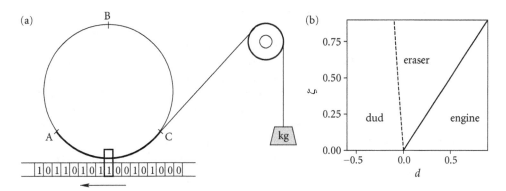

Figure 5.3. (a) Sketch of the Mandal-Jarzynski machine. (b) Phase diagram. Regime boundaries indicate (solid line) vanishing rotation speed and (dotted line) vanishing Shannon entropy change of the tape (the machine swaps 0s and 1s at equal average rates). The time interval τ is set to 0.8. See Mandal and Jarzynski [108].

also connected to a mass m. Depending on the operating regime, it can either lift it (performing work) or lower it (extracting work).

Jumps between states $A \longleftrightarrow B$ and $B \longleftrightarrow C$ occur at rates $k_{AB} = k_{BA} = k_{BC} = k_{CB} = 1$, regardless of the state of the tape and the mass. In contrast, jumps between states A and C are coupled to both the tape and the mass:

- A jump $A \longrightarrow C$ can occur only if the bit read by the machine is 1. During such a jump, the bit is "flipped" to 0. Similarly, a jump $C \longrightarrow A$ can occur only if the bit is 0 and is flipped it to 1 when the jump occurs.
- When the machine jumps from C to A, the mass is lifted by a height Δh. Conversely, when $A \longrightarrow C$, the mass is lowered by Δh.

The jump rates between states A and C satisfy the generalized detailed balance relation

$$\frac{k_{AC}}{k_{CA}} = e^{-mg\,\Delta h/k_{\mathrm{B}}T}, \tag{5.15}$$

as the work by the "external agent" (the mass in this case) along the jump $A \rightarrow C$ is equal to $mg\,\Delta h$; see eq. (3.7). A choice compatible with eq. (5.15) is $k_{AC} = 1 - \zeta$ and $k_{CA} = 1 + \zeta$, with $\zeta = \tanh(mg\,\Delta h/2k_{\mathrm{B}}T)$. The system is at equilibrium if the fraction r of 1s in the tape is equal to $p_1^{\mathrm{eq}} = e^{-mg\,\Delta h/k_{\mathrm{B}}T}/(1 + e^{-mg\,\Delta h/k_{\mathrm{B}}T}) = (1 - \zeta)/2$.

The steady state of the model can be evaluated analytically. It turns out that the machine can operate in three distinct regimes, depending on the choice of parameters (fig. 5.3b):

- In the first regime, the machine is effectively an **information engine**, as it extracts work from the heat reservoir, at the cost of dumping entropy into the bit sequence. The machine rotates clockwise on average, i.e., $A \longrightarrow B \longrightarrow C \longrightarrow A$. This occurs when $d \geq \zeta$, i.e., when the entropy of the bit sequence is sufficiently low and the mass sufficiently light. A machine operating in this regime is a concrete example of a Maxwell demon.

- In the second regime, the machine acts as an **eraser**, consuming mechanical work to reduce the Shannon entropy of the tape. The rotation is counterclockwise, i.e., the mass is lowered on average, and the Shannon entropy of the tape is reduced. In the particular case of full erasure, i.e., when the output tape is made up of all 0s or 1s, the machine spends at least $W \geq k_B T H(p)$, where $H(p) = -p_0 \ln p_0 - p_1 \ln p_1$ is the Shannon entropy of the incoming sequence of bits. This operating regime is therefore consistent with the Landauer principle.
- In the last regime (**dud**), the machine both dissipates work and increases entropy, without therefore performing any useful operation. The boundary between the eraser and the dud regime is given by the condition $H(p) = H(p^{eq})$, as for the engine-eraser boundary, but with $p_0 < \frac{1}{2}$, i.e., $d < 0$.

The Mandal-Jarzynski machine provides us with an opportunity to clarify one important aspect of the Landauer principle: the work spent to reduce the entropy of a memory may not be irreversibly dissipated into the heat reservoir. This work (or at least a part of it) can be retrieved by another machine working as an information engine. From this perspective, the Maxwell demon and the Landauer principle are two sides of the same coin: information can be converted into work and back. In Landauer's words, "Information is physical."

5.6 Copying information

Living systems rely on their capacity to replicate information at the molecular level with high accuracy and speed. Important examples of information copying in biology are DNA replication and DNA-to-RNA transcription. At variance with erasure, information copying is not necessarily irreversible. For example, we consider a system with two states $x = 0, 1$, initially prepared in the state $x = 0$. It is possible to copy the state y of another binary system into x without losing information. For example, we can apply the reversible transformation

$$(x, y) \longrightarrow (x', y), \tag{5.16}$$

where

$$x' = x \, \text{XOR} \, y = \begin{cases} y, & \text{if } x = 0; \\ 1 - y, & \text{if } x = 1. \end{cases} \tag{5.17}$$

Thus, provided that x is initially in a reference state (0 in this case), copying can be performed without dissipation, i.e., there is no equivalent of the Landauer bound for copying. However, information can be copied without dissipation only in the limit of vanishing speed, whereas copying at finite speed involves a thermodynamic cost. Further, copying at finite temperature necessarily implies errors, and correcting these errors can significantly increase dissipation. Understanding the tradeoffs between error rate, speed, and dissipation in copying is crucial to characterize the performance of biological copying machines. Besides biology, these trade-offs are also becoming relevant for artificial computing, as microprocessors become smaller and faster.

To be more concrete, we consider a mesoscopic machine that sequentially replicates a long, preexisting polymer, made up of two different kinds of monomers (fig. 5.4a). At each time, the monomer at the tip of the copied polymer can be removed, or a new monomer can be added. We denote by k_r^+, k_r^- the rates of incorporation/removal of right monomers and by k_w^+, k_w^- the corresponding rates for wrong monomers. We assume for simplicity that the rates do not depend on the specific pairing of monomers

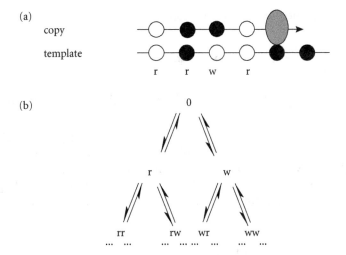

Figure 5.4. (a) Scheme of a molecular machine (gray oval) that copies a long existing polymer, called the template. Due to thermal fluctuations, the machine sometimes incorporates w monomers that do not match the template. (b) Sketch of the corresponding jump network. The process is iteratively repeated, so that the whole jump network is an infinite tree.

but only on whether the match is correct. The jump network for the complete system (the machine plus the copied polymer) is an infinite tree, sketched in fig. 5.4b. Each state in this tree is a string rwwr..., identifying the sequence of right and wrong monomers that have been incorporated in the copy. The generalized detailed balance condition (3.7) reads

$$k_r^+/k_r^- = e^{(-\Delta\epsilon_r+\delta)/k_BT}; \qquad k_w^+/k_w^- = e^{(-\Delta\epsilon_w+\delta)/k_BT}. \tag{5.18}$$

Here $\Delta\epsilon_r$ and $\Delta\epsilon_w$ are the energies of a right and wrong monomer incorporation, respectively. The parameter δ is an external chemical driving that promotes monomer incorporation. We assume the chemical driving to be equal for right and wrong monomers. Increasing δ drives the system away from equilibrium. We remark that this system can be driven away from equilibrium although its jump network, depicted in fig. 5.4b, does not contain any loop. This fact does not contradict the results in section 2.6, since the number of states here is infinite.

We write the four jump rates as

$$\begin{aligned} k_r^+ &= \omega_r\, e^{\delta/k_BT}; & k_r^- &= \omega_r\, e^{\Delta\epsilon_r/k_BT}; \\ k_w^+ &= \omega_w\, e^{\delta/k_BT}; & k_w^- &= \omega_w\, e^{\Delta\epsilon_w/k_BT}. \end{aligned} \tag{5.19}$$

We choose this expression of the rates on the physical assumption that the chemical driving affects the incorporation rates, whereas specific binding energies affect monomer stability in the copy and therefore their removal rates. In general, right and wrong monomers can be discriminated either due to a difference in their binding energies $\Delta\epsilon_i, i \in \{r, w\}$ (energetic discrimination), or due to a difference in the intrinsic rates ω_i (kinetic discrimination), or both.

The master equation for this system reads

$$\frac{d}{dt} p_{...r} = k_r^+ p_{...} + k_r^- p_{...rr} + k_w^- p_{...rw} - (k_r^+ + k_w^+ + k_r^-)p_{...r},$$

$$\frac{d}{dt} p_{...w} = k_w^+ p_{...} + k_r^- p_{...wr} + k_w^- p_{...ww} - (k_r^+ + k_w^+ + k_w^-)p_{...w}, \tag{5.20}$$

where ...r is a copy with an r monomer at the tip, and similarly for ...w. To solve the master equation, we assume that the probability of producing a given string rwrr... with N_r right and N_w wrong monomers in the steady state is $p_{rwrr...} \propto (1 - \eta)^{N_r} \eta^{N_w}$, where η is the error probability to be determined a posteriori. Substituting this guess into the master equation, we find that it is a solution, provided that η satisfies the condition

$$\frac{\eta}{1 - \eta} = \frac{k_w^+ - \eta k_w^-}{k_r^+ - (1 - \eta)k_r^-}. \tag{5.21}$$

The corresponding elongation speed is

$$v = k_r^+ - (1 - \eta)k_r^- + k_w^+ - \eta k_w^-. \tag{5.22}$$

We discuss two interesting limiting cases:

Equilibrium limit. In the equilibrium limit, the elongation speed v is vanishingly small. We solve the equation for η in this limit and evaluate the corresponding rates using eq. (5.21). We find that the error rate tends to $\eta^{eq} = e^{-\Delta \epsilon_w / k_B T}/(e^{-\Delta \epsilon_r / k_B T} + e^{-\Delta \epsilon_w / k_B T})$, as expected from detailed balance. The corresponding driving δ is smaller than the average work needed for the incorporation of a right monomer: $\delta^{stall} = \epsilon_R + k_B T \ln(1 - \eta_{eq})$. The second term in the expression of the stall driving is due to the compositional disorder of the copy: in the absence of driving, entropy increase would naturally lead to a nonequilibrium elongation of the chain.

Fully irreversible limit. In the limit of strong driving, $\delta \gg 1$, which implies a large velocity, the error eventually tends to $\eta = k_w^+/(k_r^+ + k_w^+) = \omega_w/(\omega_r + \omega_w)$. Physically, this ratio is related to the difference in the activation barrier between the reactions leading to incorporations of right and wrong monomers; see the discussion in section 3.12.

As the driving δ increases, the copying speed also increases, while the value of the error moves from the equilibrium to the irreversible limit. Since the error in the fully irreversible limit can be either larger or smaller than that in the equilibrium limit, depending on the other parameters, the trade-off between speed and error can be either positive or negative. This fact is illustrated in fig. 5.5.

Out of equilibrium, the elongation of the copy polymer is a dissipative process that produces entropy. The average entropy production rate can be expressed as

$$T\dot{S}^{tot} = k_B T \left[J^r \ln \left(\frac{k_r^+}{k_r^-} (1 - \eta) \right) + J^w \ln \left(\frac{k_w^+}{k_w^-} \eta \right) \right]$$

$$= v(1 - \eta)(\delta - \epsilon_r + k_B T \ln(1 - \eta)) + v\eta (\delta - \epsilon_w + k_B T \ln \eta) \tag{5.23}$$

$$= v \left[\Delta W - \Delta F - k_B T D_{KL}(\eta \| \eta^{eq}) \right] \geq 0,$$

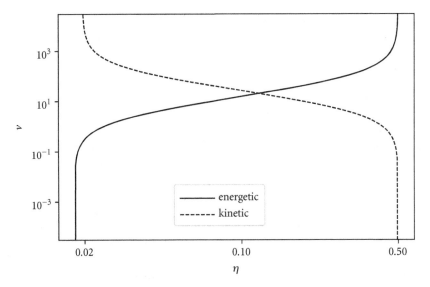

Figure 5.5. Kinetic and energetic discrimination. The two curves show polymerization speed v versus error η as the driving δ increases from 0 to 10 k_BT for different choices of the other parameters. In both curves, higher speed corresponds to larger values of δ, but the effect of increasing the driving on the error is opposite for the two kinds of discrimination. Parameters: (energetic) $\omega_r = \omega_w = 1$, $\epsilon_r = 1\,k_BT$, $\epsilon_w = 5\,k_BT$; (kinetic) $\epsilon_r = \epsilon_w = 1\,k_BT$, $\omega_r = 50$, $\omega_w = 1$. Both axes are logarithmic.

where $v = J^r + J^w$ is the average net rate at which monomers are incorporated into the copy, $\Delta W = \delta$ is the work per incorporated monomer, and $\Delta F = k_BT\,\ln(e^{-\Delta\epsilon_r/k_BT} + e^{-\Delta\epsilon_w/k_BT})$ is the equilibrium free energy change after incorporating each monomer. A derivation of eq. (5.23) using the Schnakenberg formula (3.39) is left as exercise 5.2. Equation (5.23) also holds for more general models where incorporation of monomers does not occur in a single jump, but is a more complex process characterized by several intermediate states, potentially correcting errors.

To interpret eq. (5.23), we consider the limiting case in which right and wrong monomers are isoenergetic, so that $\eta^{eq} = 1/2$. We further assume that it is possible to transform a right monomer into a wrong one and vice versa. For example, the different monomers can be distinguished by a molecular tag (such as methylation) that can be reversibly added and removed to the monomers by another enzyme. In this case, the joint system of the template plus the copy is conceptually equivalent to the tape of the Mandal-Jarzynski machine discussed in section 5.5. In particular, a precise match of the template and the copy can be fed to a Mandal-Jarzynski machine operating as an information engine. Such a machine would extract mechanical work by increasing the error in the copy. This thought experiment shows that the last term in eq. (5.23) represents free energy reversibly incorporated into the copy as information, which can in principle be transformed again into mechanical work.

5.7 Information cost in sensing

Living organisms need to constantly keep track of the state of relevant variables in their environment. They perform this task by means of electrical, chemical, and mechanical signaling pathways, called sensory systems. Important properties of sensory systems are a fast response to environmental changes, complemented by a slower adaptation, i.e., a long-term storage of the average state of the environment into a memory. Systems possessing these properties are aptly called **sensory adaptation systems**. Stochastic thermodynamics allows us to investigate the fundamental minimum energy cost for the functioning of a sensory adaptation system.

We focus on a minimal model of a sensory adaptation system. This system monitors a binary variable, the **signal** $e \in \{0, 1\}$, representing a relevant property of the environment. The state x of the sensory adaptation system is specified by two internal binary variables: the **activity** $a \in \{0, 1\}$ and the **memory** $m \in \{0, 1\}$. At constant signal, the state $m \in \{0, 1\}$ must be closely correlated with e, while a is left free to adapt to sudden changes in the environment. This means that the dynamics of m must be much slower than that of a.

For simplicity, we consider a sensory adaptation system that is not driven by chemical energy. This means that, at constant signal, its dynamics satisfies the detailed balance condition (2.83), and the system eventually settles down to an equilibrium state.

We write the energy function $\epsilon = \epsilon_{x,e} = \epsilon_{a,m,e}$ of the sensory adaptation system based on physical considerations. The lowest-energy states must be those where the memory matches the signal, $m = e$, ensuring stability of the memory. The energies of these states are independent of a, since the activity should not play any role if the signal matches the memory. Instead, when $m \neq e$, states in which the activity matches the environment ($a = e$) should be energetically more stable than those with $a \neq e$. These requirements lead to the expression

$$\epsilon_{a,m,e} = |e - m| \left(\Delta_m + |a - e| \, \Delta_a \right). \tag{5.24}$$

The parameters $\Delta_m > 0$ and $\Delta_a > 0$ are penalties due to memory and activity mistracking, respectively. We express them by

$$\Delta_{a,m} = -k_B T \ln \delta_{a,m}, \tag{5.25}$$

where $\delta_{a,m} \ll 1$ are tolerances for activity (a) and memory (m) tracking. We assume that individual jumps change the state of either a or m, but not both simultaneously:

$$k_{(1-a,m),(a,m)} = \omega_a \, e^{\epsilon_{a,m,e}/k_B T};$$

$$k_{(a,1-m),(a,m)} = \omega_m \, e^{\epsilon_{a,m,e}/k_B T}. \tag{5.26}$$

Since the activity is fast compared to the memory, we take $\omega_a \gg \omega_m$. The probability distribution $p_x(t)$ follows a master equation with the rates (5.26). The environment-dependent equilibrium state is given by

$$p_{x|e}^{\text{eq}} \propto e^{-\epsilon_{a,m,e}/k_B T}. \tag{5.27}$$

To illustrate the dynamics, we consider an environment that has been in state $e = 0$ for some time and suddenly switches to $e = 1$ at time $t = 0$. The initial distribution is $p_{x|e=0}^{\text{eq}}$. In a time on the order of ω_a^{-1}, the activity variable tends to the new environmental state. At longer times, of order ω_m^{-1}, m starts aligning with the new environmental state while a slowly resets toward a state where 0 and 1 are equiprobable. The dynamics is shown in fig. 5.6.

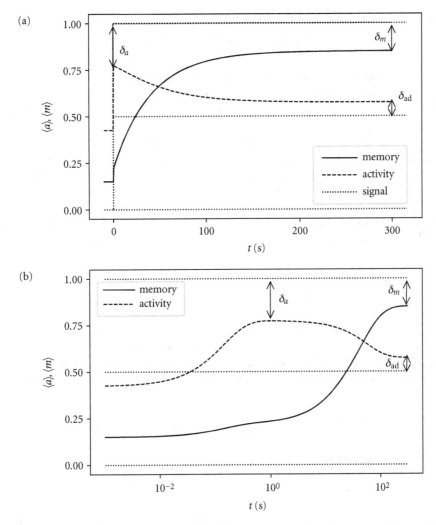

Figure 5.6. Average activity $\langle a \rangle$ and average memory $\langle m \rangle$ as a function of time t for a signal switching from $e=0$ to $e=1$ at $t=0$. (a) Linear time axis. (b) Logarithmic time axis. Parameters: $\delta_m = 0.35$, $\delta_a = 0.01$, $\omega_a = 1/50\,\mathrm{s}^{-1}$, $\omega_m = 1/250\,\mathrm{s}^{-1}$, $k_\mathrm{B}T = 1$. See Sartori et al. [142].

The information carried by the sensory adaptation system about the environment is quantified by the mutual information

$$I(\mathrm{sys} : \mathrm{env}) = \sum_{x,e} p_{x|e} p_e \ln \frac{p_{x|e}}{p_x}, \tag{5.28}$$

where $p_x = \sum_e p_{x|e} p_e$. We assume that the environment is unbiased, so that $p_e = 1/2$. By using the chain rule for the mutual information (2.145) and the fact that the distribution of the system (sys) is the joint distribution of the activity (act) and the memory (mem), we obtain

$$I(\mathrm{sys} : \mathrm{env}) = I(\mathrm{mem} : \mathrm{env}) + I(\mathrm{act} : \mathrm{env} | \mathrm{mem}), \tag{5.29}$$

where

$$I(\text{act} : \text{env}|\text{mem}) = \sum_{a,m,e} p_{a,m|e} p_e \ln \frac{p_{a|m,e}}{p_{a|m}} \tag{5.30}$$

is the conditional mutual information between the activity and the environment, given the memory. This quantity is positive but quite small in the steady state. The reason is that most of the information on the environment is carried by the memory, $I(\text{sys} : \text{env}) \approx I(\text{mem} : \text{env}) \approx 1$ bit.

We now study a case in which the environmental signal is initially either 0 or 1 with equal probability. At time t, the signal can switch with probability $1/2$. The information borne by the sensory adaptation system can be split into information about the initial environment $\text{env}(t_0)$, which must be erased, and information about the final environment $\text{env}(t_f)$, which must be acquired. Given the instantaneous state $p_{x|e(t_0),e(t_f)}(t)$ conditioned to the initial and final environments, we define the acquired information $\Delta I^{\text{meas}}(t)$ on the final environment and the erased information $\Delta I^{\text{eras}}(t)$ by

$$\Delta I^{\text{meas}}(t) = I(\text{sys}(t) : \text{env}(t_f)) - I(\text{sys}(t_0) : \text{env}(t_f));$$
$$\Delta I^{\text{eras}}(t) = I(\text{sys}(t_0) : \text{env}(t_0)) - I(\text{sys}(t) : \text{env}(t_0)|\text{env}(t_f)). \tag{5.31}$$

Here $I(\text{sys}(t) : \text{env}(t_f))$ is the mutual information between the state x of the sensory adaptation system at time t and the final environment. Similarly, $I(\text{sys}(t) : \text{env}(t_0)|\text{env}(t_f))$ is the conditional mutual information between the system at time t and the initial environment, given the final environment. They are respectively defined by

$$I(\text{sys}(t) : \text{env}(t_f)) = \sum_{x,e_i,e_f} p_{x|e_i,e_f}(t) p_{e_i,e_f} \ln \frac{p_{x|e_i,e_f} p_{e_i,e_f}(t)}{p_x(t) p_{e_f}},$$

$$I(\text{sys}(t) : \text{env}(t_0)|\text{env}(t_f)) = \sum_{x,e_i,e_f} p_{x|e_i,e_f}(t) p_{e_i,e_f} \ln \frac{p_{x|e_i,e_f}(t)}{p_{x|e_f}(t)}, \tag{5.32}$$

where $p_x(t) = \sum_{e_i,e_f} p_{x|e(t_0),e(t_f)}(t) p_{e(t_0),e(t_f)}$ and $p_{e(t_f)} = \sum_{e(t_0)} p_{e(t_0),e(t_f)}$. Analogous definitions hold for $I(\text{sys} : \text{env}(t_0))$ and $p_{e(t_0)}$. One can verify that $I(\text{sys}(t_0) : \text{env}(t_f)) = \lim_{t\to\infty} I(\text{sys}(t) : \text{env}(t_0)|\text{env}(t_f)) = 0$. These quantities can be further split into contributions from the activity and the memory via the chain rule; see section 2.10. The behavior of ΔI^{meas} and ΔI^{eras} as a function of time are shown in fig. 5.7, along with their decomposition into information carried by m and a, respectively. In particular, the activity carries significant information about the measurement at intermediate times, whereas it carries quite small erasure information.

Upon switching from e_i to e_f, the environment performs work on the sensory adaptation system. The average work is given by $W = \sum_{e_i,e_f} \langle \epsilon_{x,e_f} \rangle^{\text{eq}}_{p_{x,e_i}} p_{e_i,e_f} \geq 0$. This work is in part dissipated in the reservoir during relaxation to the new equilibrium state p^{eq}_{x,e_f}, thereby producing entropy. The average total entropy produced up to time t given (e_i, e_f) is expressed by

$$S^{\text{tot}}_{e_i,e_f}(t) = \Delta S^{\text{sys}}_{e_i,e_f}(t) + S^{\text{res}}_{e_i,e_f}(t), \tag{5.33}$$

where

$$\Delta S^{\text{sys}}_{e_i,e_f}(t) = k_B \left[H(p_{x|e_i,e_f}(t)) - H(p_{x|e_i,e_f}(0)) \right], \tag{5.34}$$

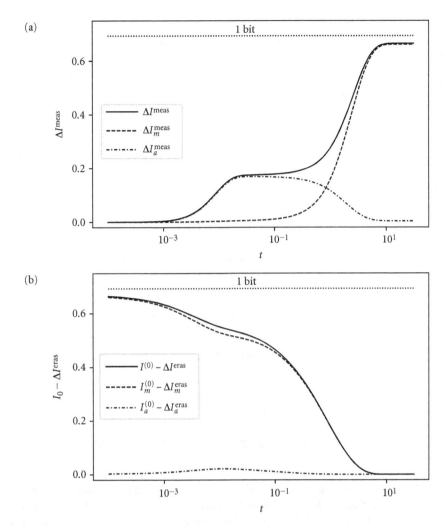

Figure 5.7. (a) Information acquired about the new state of the environment as a function of time. The information is split into information carried by the memory m and information carried by the activity a, given env(t_f) and m. (b) Information lost about the old state of the environment as a function of time. The difference $I^{(0)} - \Delta I^{\text{eras}}(t)$, where $I^{(0)} = I(\text{sys}(t_0) : \text{env}(t_0))$, is split into information carried by the memory m and information carried by the activity a, given env(t_0) and m. In both plots, t denotes the time elapsed since the switch; the scale is logarithmic. Parameters: $\delta_m = 0.01$, $\delta_a = 0.01$, $\omega_a = 1/50 \, \text{s}^{-1}$, $\omega_m = 1/250 \, \text{s}^{-1}$, $k_B T = 1$. See Sartori et al. [142].

in which $H(p_x)$ is the Shannon entropy of the distribution p_x, and

$$S^{\text{res}}_{e_i,e_f}(t) = \frac{1}{T} Q^{\text{res}}_{e_i,e_f}(t) = \frac{1}{T} \left(\langle \epsilon_{x,e_f} \rangle_{p^{\text{eq}}_{x|e_i}} - \langle \epsilon_{x,e_f} \rangle_{p_{x|e_i,e_f}(t)} \right). \tag{5.35}$$

Here $Q^{\text{res}}_{e_i,e_f}(t)$ is the average heat released to the reservoir up to time t, equal to the difference between the initial average energy of the sensory adaptation system (immediately after the switch) and its current average. By the second law of stochastic thermodynamics, $S^{tot} \geq 0$ for any finite time interval. Therefore, $S^{tot}(t)$ is a monotonically nondecreasing function of t. Since this is true for any choice of e_i and e_f, it is also true on average.

We split the total entropy production into two nonnegative terms, one pertaining to the measuring process and the other to the erasure. Averaging out $H(p_{x|e_i,e_f}(t))$, we obtain the conditional entropy

$$H(\mathrm{sys}(t)|\mathrm{env}(t_0),\mathrm{env}(t_f)) = H(\mathrm{sys}(t)|\mathrm{env}(t_f)) - I(\mathrm{sys}(t):\mathrm{env}(t_0)|\mathrm{env}(t_f))$$
$$= H(\mathrm{sys}(t)) - I(\mathrm{sys}(t):\mathrm{env}(t_f)) - I(\mathrm{sys}(t):\mathrm{env}(t_0)|\mathrm{env}(t_f)), \tag{5.36}$$

where we apply the chain rule twice. Averaging (5.33) and substituting this relation, we obtain

$$\Delta S^{tot}(t) = \underbrace{k_B \Delta H(\mathrm{sys}(t)) + \frac{1}{T}Q^{res}(t) - k_B I(\mathrm{sys}(t):\mathrm{env}(t_f))}_{\Delta S^{meas}(t)}$$
$$\underbrace{- k_B I(\mathrm{sys}(t):\mathrm{env}(t_0)|\mathrm{env}(t_f))}_{\Delta S^{eras}(t)} . \tag{5.37}$$

The quantity $\Delta S^{meas}(t)$ cannot decrease in time:

$$\frac{d\Delta S^{meas}}{dt} = -k_B \sum_{x,e_f} p_{e_f} \frac{dp_{x|e_f}(t)}{dt} \ln \frac{p_{x|e_f}(t)}{p_{x|e_f}^{eq}}$$
$$= -k_B \sum_{e_f} p_{e_f} \frac{d}{dt} D_{KL}(p_{x|e_f}(t)\|p_{x|e_f}^{eq}) \geq 0. \tag{5.38}$$

The last equality holds because the Kullback-Leibler divergence between the instantaneous and the stationary distribution decreases monotonically with time; see eq. (A.20). The nonnegativity of $d\Delta S^{eras}/dt$ is proved in a similar way. Integrating with respect to t, we find that both ΔS^{meas} and ΔS^{eras} are nonnegative.

We thus obtain, taking into account that $\lim_{t\to\infty} Q = W - \Delta \langle\epsilon\rangle$, and recalling the definition of the nonequilibrium free energy $F^{neq} = \langle\epsilon\rangle_p - k_B T H(p)$ in section 2.3,

$$W - \langle\Delta F\rangle \geq k_B T \left[I(\mathrm{sys}(+\infty):\mathrm{env}(t_f)) - I(\mathrm{sys}(0):\mathrm{env}(t_f))\right]$$
$$= k_B T \, \Delta I^{meas}; \tag{5.39a}$$

$$T \Delta S^{eras} = k_B T \left(I(\mathrm{sys}(0):\mathrm{env}(t_0)|\mathrm{env}(t_f)) - I(\mathrm{sys}(+\infty):\mathrm{env}(t_0)|\mathrm{env}(t_f))\right)$$
$$= k_B T \, \Delta I^{eras} \geq 0. \tag{5.39b}$$

The first relation corresponds to the Sagawa-Ueda relation (5.14), and shows that, in the present system, measurement is energetically costly and that the work is provided by the switch in the signal. The second relation shows that memory erasure takes place without energy costs but is necessarily dissipative. The entropy production as a function of time is shown in fig. 5.8.

Most biologically relevant sensory adaptation systems are driven in a nonequilibrium steady state by dissipation of chemical energy and therefore do not satisfy

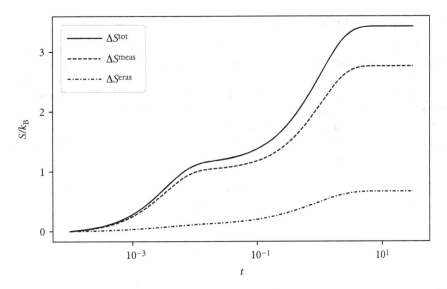

Figure 5.8. Total entropy production of the sensory adaptation process in the model and its decomposition in measurement ΔS^{meas} and erasure ΔS^{eras} contributions, expressed as a function of the time elapsed since the switch. Parameters: $\delta_m = 0.01$, $\delta_a = 0.01$, $\omega_a = 1/50\,\text{s}^{-1}$, $\omega_m = 1/250\,\text{s}^{-1}$, $k_B T = 1$. Timescale is logarithmic. See Sartori et al. [142].

detailed balance. In spite of this fundamental difference, inequalities closely related to (5.39) also hold for such systems. They are based on the separate fluctuation relations for adiabatic and nonadiabatic entropy production discussed in section 4.9. Indeed, in a nonequilibrium steady state, the entropy production is entirely captured by the adiabatic contribution defined in (4.64). A change in the signal is accompanied by nona-diabatic entropy production (see eq. (4.67)), which is also associated with excess heat released to the reservoir. In our case, the average excess heat is given by

$$Q^{\text{ex}}(t) = k_B T \sum_{e_i, e_f} p_{e_i, e_f} \int_{t_0}^{t} dt \sum_{x} \frac{dp_{x|e_i, e_f}(t)}{dt} \ln p_{x|e_f}^{\text{st}}. \tag{5.40}$$

As a consequence of the fluctuation relation (4.74), averaged over e_i, e_f, one has

$$S^{\text{na}}(t) = k_B \, \Delta H(\text{sys}(t)|\text{env}(t_0), \text{env}(t_f)) + \frac{Q^{\text{ex}}(t)}{T} \geq 0. \tag{5.41}$$

The nonadiabatic entropy production $S^{\text{na}}(t)$ can be further split, as above, in the contributions of measure and erasure, yielding

$$k_B \, \Delta H(\text{sys}(t)) + \frac{Q^{\text{ex}}(t)}{T} \geq k_B \, \Delta I^{\text{meas}}(t) \geq 0;$$

$$S^{\text{na}}(t) \geq k_B \, \Delta I^{\text{eras}}(t) \geq 0. \tag{5.42}$$

Here $\Delta I^{\text{meas}}(t)$ and $\Delta I^{\text{eras}}(t)$ are defined as in (5.31). In this case, the system usually receives most of the energy necessary for the process by the inner working of the cell, rather than from the signal, as in the equilibrium case.

5.8 Information reservoirs

We have discussed two ways of handling information in stochastic thermodynamics: via a measurement-and-feedback scheme, as in the Szilard thought experiment, and via interaction with a tape, as done by the Mandal-Jarzynski machine. In this section, we introduce a unified framework to describe these two setups, where the measuring device and the tape are seen as particular cases of a general **information reservoir**. This framework also permits us to compare the total entropy production in the two cases. We further compare the entropy production by the information reservoir with that of a comparable heat reservoir.

We consider a system with two states, d (down) and u (up), with $\epsilon_d = 0, \epsilon_u = \epsilon > 0$, in contact with a heat reservoir at temperature T. The jump rates induced by interactions with this reservoir satisfy the detailed balance relation

$$\frac{k^\uparrow}{k^\downarrow} = e^{-\epsilon/k_B T}, \tag{5.43}$$

where $k^\uparrow = k_{ud}$ and $k^\downarrow = k_{du}$. In the absence of other interactions, the equilibrium distribution is

$$p_d^{eq} = \frac{1}{1 + e^{-\epsilon/k_B T}}; \qquad p_u^{eq} = \frac{e^{-\epsilon/k_B T}}{1 + e^{-\epsilon/k_B T}}. \tag{5.44}$$

Besides the heat reservoir, the system is also in touch with an information reservoir. Interactions with the information reservoir take place at random times with rate γ and can swap the two states. If the system is in state u at the time of swapping, the system releases an amount of work ϵ, e.g., by lifting a weight attached to it. Otherwise, if the system is in state d, the weight is lowered, granting the system an amount of work ϵ. After the interaction, we swap the labels of the states, so that the lowest-energy state always has the label d. The information-mediated jump rates are

$$\bar{k}^\uparrow = \gamma\, r; \qquad \bar{k}^\downarrow = \gamma\, (1-r). \tag{5.45}$$

Here r is a parameter between 0 and 1. We also take into account interactions with the reservoir that do not swap the states. The corresponding rates are

$$\bar{k}_d = \gamma\, (1-r); \qquad \bar{k}_u = \gamma\, r. \tag{5.46}$$

The evolution of the system coupled to both reservoirs is described by a master equation:

$$\begin{aligned}
\frac{dp_u}{dt} &= \left(k^\uparrow + \bar{k}^\uparrow\right) p_d - \left(k^\downarrow + \bar{k}^\downarrow\right) p_u; \\
\frac{dp_u}{dt} &= \left(k^\downarrow + \bar{k}^\downarrow\right) p_u - \left(k^\uparrow + \bar{k}^\uparrow\right) p_d.
\end{aligned} \tag{5.47}$$

The system eventually reaches a steady state, characterized by a probability distribution,

$$p_u^{st} = \frac{k^\uparrow + \bar{k}^\uparrow}{k+\gamma}; \qquad p_d^{st} = \frac{k^\downarrow + \bar{k}^\downarrow}{k+\gamma}, \tag{5.48}$$

where $k = k^\uparrow + k^\downarrow$ and we use the relation $\gamma = \bar{k}^\uparrow + \bar{k}^\downarrow$. Physically, the information reservoir can be interpreted in different ways:

Measurement-and-feedback (MF). In this interpretation, the information reservoir is a device that measures the state of the system at rate γ, obtaining a result $y \in \{0, 1\}$, where 1 corresponds to u and 0 to d. The parameter r represents the error probability of the measurement. This means that $p_{u|1} = p_{d|0} = 1 - r$ and $p_{u|0} = p_{d|1} = r$. In this interpretation, we assume $0 < r < 1/2$. If the outcome of the measurement is $y = 1$, the states of the system are swapped. Therefore, if the system is in u, it jumps to d with probability $(1 - r)$, while if it is in d, it jumps to u with probability r. We thus recover the information-mediated rates (5.45, 5.46). The probability p_1^{dev} that the measurement yields 1, which is also the probability the system is in state u immediately after an interaction, is given by

$$p_1^{dev} = (1 - r)\, p_u^{st} + r\, p_d^{st} = p_u^{st} + r - 2\, r p_u^{st}. \tag{5.49}$$

The average work performed on the system per unit time is

$$\dot{W} = \epsilon\, \gamma \left[r p_d^{st} - (1 - r)\, p_u^{st} \right] = -\gamma \epsilon \left(p_u^{st} - r \right). \tag{5.50}$$

Since the system is in a steady state, $\dot{W}$ is equal to the average heat released to the heat reservoir per unit time, $\dot{W} = T \dot{S}^{res}$. The total entropy production is the sum of S^{res} plus the change in the joint entropy of the system and the measuring device. The mutual information between the system and the device for a single interaction is given by

$$I = H(\text{dev}) - H(\text{dev}|\text{sys}) = H(\text{dev}) - H(r), \tag{5.51}$$

where $H(\text{dev})$ is the Shannon entropy of the device and $H(r)$ is the Shannon entropy associated with the error rate r. The Sagawa-Ueda relation (5.14) then implies

$$\dot{W} + \gamma\, k_B T\, I \geq 0, \tag{5.52}$$

with I given by eq. (5.51).

TAPE. The information reservoir is interpreted as a tape containing an infinite sequence of bits. With rate γ, the tape shifts by 1 bit and is connected to the system. If the incoming bit on the tape is 1 (0), the state of the system is labeled u (d). If the system changes state by interacting with the heat reservoir, the bit on the tape is also switched, so that only the jumps $(0, d) \longleftrightarrow (1, u)$ are allowed. The switching rates are again given by eq. (5.45). In this interpretation, the probabilities that incoming and outgoing bits are equal to 1 are given by r and p_u^{st}, respectively. Therefore, the entropy of the outgoing tape changes at a rate $\gamma\, k_B \left[H(p^{st}) - H(r) \right]$. We then have

$$T \dot{S}^{tot}_{TAPE} = \dot{W} + \gamma\, k_B T \left[H(p^{st}) - H(r) \right] \geq 0, \tag{5.53}$$

where $\dot{W}$ is given by eq. (5.50). Using the relation

$$\epsilon = k_B T \left[\ln \left(1 - p_u^{eq} \right) - \ln p_u^{eq} \right], \tag{5.54}$$

we obtain

$$T \dot{S}^{\text{tot}}_{\text{TAPE}} = \gamma \, k_B T \left[D_{\text{KL}}(r \| p^{\text{eq}}) - D_{\text{KL}}(p^{\text{st}} \| p^{\text{eq}}) \right] \geq 0. \tag{5.55}$$

Since

$$p_u^{\text{st}} - p_u^{\text{eq}} = \frac{\gamma}{k + \gamma} \left(r - p_u^{\text{eq}} \right), \tag{5.56}$$

the difference of the Kullback-Leibler divergences is nonnegative. On the other hand, since $H(\text{dev}) \geq H(p^{\text{st}})$, the bound in eq. (5.53) is tighter than the Sagawa-Ueda inequality (5.52). In particular, the difference $H(p^{\text{st}}) - H(r)$ can be negative, while the mutual information I is always nonnegative. When $H(p^{\text{st}}) - H(r)$ is negative, the system works as an eraser, spending work to reduce the tape entropy.

Generalized detailed balance (GDB). A third alternative is to treat the information reservoir in the same way as a heat reservoir and compute the average entropy production rate using the Schnakenberg formula (3.39). In this case, the average entropy production rate is given by

$$\dot{S}^{\text{tot}}_{\text{GDB}} = k_B \left(k^{\downarrow} p_u^{\text{st}} - k^{\uparrow} p_d^{\text{st}} \right) \ln \frac{k^{\downarrow}}{k^{\uparrow}} + k_B \left(\bar{k}^{\downarrow} p_u^{\text{st}} - \bar{k}^{\uparrow} p_d^{\text{st}} \right) \ln \frac{\bar{k}^{\downarrow}}{\bar{k}^{\uparrow}}, \tag{5.57}$$

where we separate the contributions coming from interactions with the two different reservoirs. This expression can be rearranged in the form

$$T \dot{S}^{\text{tot}}_{\text{GDB}} = \dot{W} + \gamma \, k_B T \left(p_u^{\text{st}} - r \right) \ln \frac{1 - r}{r} \geq 0. \tag{5.58}$$

This expression of the entropy production rate diverges for $r \to 0$, in contrast with the previous cases.

Summarizing, the same dynamics yields three different expressions of the entropy production rate, depending on how we interpret the information reservoir. The reason is that some thermodynamic costs related to information handling are not taken into account in the MF and TAPE cases: in MF, the costs associated with the acquisition and erasure of the information in the measuring device and, in TAPE, the cost of producing the initial state of the tape. In the latter case, it is instructive to consider a regime in which $p_u^{\text{st}} > r$, so that $\dot{W} < 0$, and the system lifts the weight on average. Equation (5.53) sets the minimal work needed to restore the tape to its previous condition:

$$\dot{W} \geq k_B T \gamma \left(H(p^{\text{st}}) - H(r) \right). \tag{5.59}$$

In the regime $\gamma \ll k$, we have $p^{\text{st}} \approx p^{\text{eq}}$. Therefore, to reset the fraction of 1 on the tape to r in this regime, we need an amount of work $\epsilon = k_B T \ln[(1 - r)/r]$. The corresponding minimal power applied to the system is given by

$$\dot{W} = k_B T \gamma \left(p_u^{\text{st}} - r \right) \ln \frac{1 - r}{r} \geq k_B T \gamma \left[H(p^{\text{st}}) - H(r) \right]. \tag{5.60}$$

If we first extract work by increasing the tape entropy according to eq. (5.53), and then spend some work to reset the tape, the total entropy production is bounded by (5.58). Therefore, the divergence of the entropy production for $r \to 0$ represents the thermodynamic cost of producing an error-free tape.

5.9 Fluctuation relations with information reservoirs (*)

We have seen that different interpretations of an information reservoir lead to different expressions of the average entropy production, that satisfy different inequalities. In turn, these inequalities descend from different fluctuation relations, which we derive in this section.

Our first step is to find how to account for interactions with the information reservoir that do not lead to a jump (cf. eq. (5.46)). To this aim, we formally consider two copies of our two-state system, A and B. We swap copies every time an information-mediated interaction occurs, even if the state x of the system does not change. The jump network therefore has four states $x\alpha$, where $x \in \{d, u\}$ represents the physical state of the system and $\alpha \in \{A, B\}$ denotes the copy. The jump rates $k^{\uparrow,\downarrow}$ connect states of the same copy and $\bar{k}^{\uparrow,\downarrow}$, $\bar{k}_{d,u}$ connect different copies; see eqs. (5.43, 5.45, 5.46). Since the dynamics is symmetric between the two copies, we have $p_{x\alpha}^{st} = p_x^{st}/2$, $\forall x, \alpha$.

We compare the probability density of a trajectory x with that of its reverse $\widehat{x}$. The backward trajectory satisfies a master equation with the same jump rates k within each copy, but different ones $\widehat{k}$ between the copies. In particular, we choose the reverse rates $\widehat{k}$ to satisfy

$$\widehat{k}^{\uparrow} + \widehat{k}_d = \widehat{k}^{\downarrow} + \widehat{k}_u = \gamma. \tag{5.61}$$

With this choice, the escape rates of forward and backward trajectories are equal. The ratio of the probability densities of forward and backward trajectories is

$$\frac{\mathcal{P}_x(k, \bar{k})}{\mathcal{P}_{\widehat{x}}(\widehat{k}, \bar{k})} = \frac{p_{x_0}(t_0)}{p_{x_n}(t_f)} \, e^{s^{res}(x)/k_B} = e^{s^{tot}(x)/k_B}, \tag{5.62}$$

where

$$s^{res}(x) = \frac{k_B}{2} \sum_{x \neq x'} \mathcal{J}_{x,x'}(x) \ln \frac{k_{xx'}}{k_{x'x}} + k_B \sum_{x,x'} \mathcal{R}_{xx'}(x) \ln \frac{\bar{k}_{xx'}}{\bar{k}_{x'x}} \tag{5.63}$$

is the entropy change in the reservoirs associated with the trajectory x. Here $\mathcal{J}_{xx'}(x)$ is the net integrated current between x' and x,

$$\mathcal{J}_{xx'}(x) = \sum_{\ell=0}^{n} \sum_{\alpha} \left(\delta_{x\alpha,x_{n+1}\alpha_{n+1}}^K \delta_{x'\alpha,x_n\alpha_n}^K - \delta_{x'\alpha,x_{n+1}\alpha_{n+1}}^K \delta_{x\alpha,x_n\alpha_n}^K \right), \tag{5.64}$$

and $\mathcal{R}_{xx'}(x)$ is the total number of switches between the two copies connecting the state x' with the state x:

$$\mathcal{R}_{xx'}(x) = \sum_{\ell=0}^{n} \left(\delta_{xA,x_{n+1}\alpha_{n+1}}^K \delta_{x'B,x_n\alpha_n}^K + \delta_{xB,x_{n+1}\alpha_{n+1}}^K \delta_{x'A,x_n\alpha_n}^K \right). \tag{5.65}$$

The term in $\mathcal{R}_{xx'}$ has this form because the switch between A and B has the same weight as the switch between B and A. From eq. (5.62), we obtain the integral fluctuation relation

$$\left\langle e^{-s^{tot}/k_B} \right\rangle = 1, \tag{5.66}$$

which implies

$$S^{\text{tot}} = \langle s^{\text{tot}} \rangle \geq 0. \tag{5.67}$$

In the steady state, since the entropy of the system does not change, this leads to

$$\dot{S}^{\text{tot}} = \lim_{\mathcal{T} \to \infty} \frac{\langle s^{\text{res}} \rangle}{\mathcal{T}} = \frac{k_B}{2} \sum_{x \neq x'} J_{x,x'} \ln \frac{k_{xx'}}{k_{x'x}} + k_B \sum_{x,x'} R_{xx'} \ln \frac{\bar{k}_{xx'}}{k_{x'x}} \geq 0, \tag{5.68}$$

where $\mathcal{T}$ is the duration of the time interval and

$$J_{xx'} = \lim_{\mathcal{T} \to \infty} \frac{\langle \mathcal{J}_{xx'} \rangle}{\mathcal{T}} = k_{xx'}\, p_{x'}^{\text{st}} - k_{x'x}\, p_{x}^{\text{st}};$$

$$R_{xx'} = \lim_{\mathcal{T} \to \infty} \frac{\langle \mathcal{R}_{xx'} \rangle}{\mathcal{T}} = \bar{k}_{xx'}\, p_{x'}^{\text{st}}. \tag{5.69}$$

The different expressions of the reservoir entropy change in the different schemes are obtained by suitably choosing the reverse rates $\widehat{k}$:

- In the measurement-and-feedback interpretation, we choose the jump rates according to the outcome of measurements along the forward trajectory:

$$\widehat{k}_{ud} = \widehat{k}_{du} = \gamma\, p_1^{\text{dev}}; \qquad \widehat{k}_{dd} = \widehat{k}_{uu} = \gamma\, (1 - p_1^{\text{dev}}). \tag{5.70}$$

We then obtain

$$\dot{S}_{\text{MF}}^{\text{tot}} = \frac{k_B}{2} \sum_{x \neq x'} J_{x,x'} \ln \frac{k_{xx'}}{k_{x'x}} + k_B\, \gamma\, [H(\text{dev}) - H(r)] \geq 0. \tag{5.71}$$

The last term is proportional to the mutual information between the system and the measuring device and is therefore nonnegative.

- In the tape interpretation, we determine the rates $\widehat{k}$ characterizing the backward trajectory by swapping the fraction of 1 in the incoming and outgoing tape with respect to the forward trajectory:

$$\widehat{k}_{uu} = \widehat{k}_{ud} = \gamma\, p_u^{\text{st}}; \qquad \widehat{k}_{du} = \widehat{k}_{dd} = \gamma\, p_d^{\text{st}}. \tag{5.72}$$

This choice leads to

$$\dot{S}_{\text{TAPE}}^{\text{tot}} = \frac{k_B}{2} \sum_{x \neq x'} J_{x,x'} \ln \frac{k_{xx'}}{k_{x'x}} + k_B\, \gamma\, [H(p^{\text{st}}) - H(r)] \geq 0. \tag{5.73}$$

The term associated with information handling is a product of the rate by which the tape is processed and the difference between the entropy of outgoing and incoming bits. This difference can be either positive or negative, and, since it is not antisymmetric, it does not satisfy the conservation laws of probability currents.

- In the generalized detailed balance interpretation, we choose $\widehat{k}_{xx'} = \bar{k}_{xx'}$. The "jumps" with $x' = x$ do not contribute and we obtain

$$\dot{S}^{\text{tot}}_{\text{GDB}} = \frac{k_B}{2} \sum_{x \neq x'} J_{x,x'} \ln \frac{k_{xx'}}{k_{x'x}} + k_B J' \ln \frac{1-r}{r} \geq 0, \qquad (5.74)$$

where

$$J' = \bar{k}^{\downarrow} p_u^{\text{st}} - \bar{k}^{\uparrow} p_d^{\text{st}} = \gamma (p_u^{\text{st}} - r) \qquad (5.75)$$

is the net current between u and d due to interactions with the information reservoir. In this interpretation, we recover the familiar expression (5.58) of the entropy production.

The different entropy production rates satisfy the relations

$$\dot{S}^{\text{tot}}_{\text{GDB}} - \dot{S}^{\text{tot}}_{\text{TAPE}} = k_B \gamma\, D_{\text{KL}}(p^{\text{st}} \| r) \geq 0;$$
$$\dot{S}^{\text{tot}}_{\text{MF}} - \dot{S}^{\text{tot}}_{\text{TAPE}} = k_B \gamma\, [H(\text{dev}) - H(p^{\text{st}})] \geq 0. \qquad (5.76)$$

The tightest bound on $\dot{W}$ is provided by $\dot{S}^{\text{tot}}_{\text{TAPE}}$. There is no general inequality between $\dot{S}^{\text{tot}}_{\text{GDB}}$ and $\dot{S}^{\text{tot}}_{\text{MF}}$.

This formalism can be generalized to arbitrary jump networks. If only some states interact with information reservoirs, the entropy production rates take slightly different forms. For example, if only states d and u undergo information-mediated jumps, we obtain

$$\dot{S}^{\text{tot}}_{\text{TAPE}} = \frac{k_B}{2} \sum_{x \neq x'} J_{x,x'} \ln \frac{k_{xx'}}{k_{x'x}} + k_B \gamma\, \bar{p}\, [H(\bar{p}_u) - H(r)] \geq 0,$$

$$\dot{S}^{\text{tot}}_{\text{MF}} = \frac{k_B}{2} \sum_{x \neq x'} J_{x,x'} \ln \frac{k_{xx'}}{k_{x'x}} + k_B \gamma\, \bar{p}\, [H(\text{dev}) - H(\bar{p}_u)] \geq 0, \qquad (5.77)$$

where $\bar{p} = p_d^{\text{st}} + p_u^{\text{st}}$ and $\bar{p}_u = p_u^{\text{st}}/\bar{p}$. Further, the second equality in eq. (5.75) is replaced by

$$J' = \gamma\, \bar{p}\, (\bar{p}_u - r). \qquad (5.78)$$

5.10 Further reading

The formulation of the demon paradox, in Maxwell's words, can be found in [111, p. 338]. The original paper by Szilard [162] is translated in [163]. Leff and Rex [101] collect and discuss a number of historically important papers, including Szilard's paper. Brillouin [25] contributed to the solution of Maxwell's paradox; see also Bennett [13, sec. 5]. Landauer established his principle in [96] and further elaborated on it in [97]; see also Bennett [14]. Sagawa and Ueda [140] originally derived the theory presented in section 5.4.

The Mandal-Jarzynski machine was introduced in [108]. Bennett [12] introduced the model of information copying discussed in this chapter, and noted that information can be copied without dissipation [13, 14]. Sartori and Pigolotti [143] made a distinction between kinetic and energetic discrimination in copying. Bennett [12], Andrieux and Gaspard [4], and Sartori and Pigolotti [144] further discussed the dissipation cost of copying. Sartori et al. [142] proposed the model of sensory adaptation. Barato and Seifert [7, 8] developed the theory of stochastic thermodynamics with information reservoirs. The theory was extended by Shiraishi et al. [155]. A parallel development is due to Horowitz and Esposito [77].

Parrondo et al. [124] review general aspects of information thermodynamics. A recent review, which also addresses aspects of computation theory, is due to Wolpert [176].

5.11 Exercises

5.1   Show that the work W performed on a system in contact with a reservoir at temperature T, initially in an arbitrary distribution p_x, satisfies the inequality

$$-W \leq F^{\text{neq}}(p) - F,$$

where $F^{\text{neq}}(p)$ is the nonequilibrium free energy given by eq. (5.2) and F is the corresponding equilibrium free energy. Also show that

$$F^{\text{neq}}(p) - F = D_{\text{KL}}(p\|p^{\text{eq}}),$$

where p_x^{eq} is the equilibrium probability distribution.

5.2   Compute the average entropy production rate of a template-assisted polymerization process, eq. (5.23), using the Schnakenberg formula, eq. (3.39).

5.3   Consider a Markovian version of the Mandal-Jarzynski machine described in section 5.5. The system can be in one of three states $x \in \{0, 1, 2\}$. The energies of these states are, respectively, $\epsilon_0 = 0$, $\epsilon_1 = -\epsilon/2$, $\epsilon_2 = +\epsilon/2$, with $\epsilon > 0$. The jump rates between 0 and the other two states satisfy detailed balance, i.e.,

$$\frac{k_{01}}{k_{10}} = \frac{k_{20}}{k_{02}} = e^{-\epsilon/2k_B T}.$$

The intrinsic frequency of these jumps is equal to 1. The system is in contact with a tape, which controls the jumps between 1 and 2. The jump $2 \longrightarrow 1$ can only take place if the bit on the tape is 0, and if it takes place, the bit is set to 1. In the same way, the reverse jump can only take place if the bit on the tape is 1, and if it takes place, the bit is set to 0. These jumps are connected to an external weight, so that, if the state jumps from 2 to 1, an amount of work equal to ϵ is provided to the environment, and if the reverse jump takes place, the same amount of work is released to the system. The interactions with the tape take place at random times with frequency γ, and after each interaction a new case of the tape is advanced. The fraction of 1s in the incoming tape is r. Thus the system is described by a master equation with suitable rates, as discussed in section 5.8. Evaluate the phase diagram of the system in the regime $\gamma \ll 1$. Give the expressions of the entropy production rate in the three interpretations discussed in section 5.9. Generalize to larger values of γ.

5.4   Consider a system with three states: $x \in \{0, 1, 2\}$, with $\epsilon_1 = \epsilon_2 = \epsilon = 0$, $\epsilon_0 = \epsilon > 0$. Jumps between 0 and 1 take place in contact with a heat reservoir at temperature T_h, and those between 0 and 2 with a reservoir at temperature $T_c < T_h$. Jumps between 1 and 2 are assisted by a tape: If the system is in 1 and the bit on the tape is 0, the system switches to 2 and the bit on the tape is replaced by 1. If the system is in 2 and the bit on the tape is 1, the system switches to 1 and the bit is replaced by 0. Interactions between the system and the tape take place at rate γ. The fraction of 1s on the tape is r. Assume that γ is much smaller than the jump rates between 0 and 1, 2.

Evaluate the entropy production rate and identify the working regimes. Show that the system can operate as a refrigerator, moving energy from the lower- to the higher-temperature reservoir while increasing the entropy of the tape. Identify other possible regimes. Generalize to larger values of γ.

5.5   Consider a two-state system with two edges between the states, labeled by I and II. Both edges are associated with information-mediated jumps, driven by interaction with tapes. The fraction of ones in the two incoming tapes are $r_I \leq \frac{1}{2}$ and $r_{II} \leq \frac{1}{2}$. The jump rates from 0 to 1 are $\gamma_I r_I$ for edge I and $\gamma_{II} r_{II}$ for edge II. The reverse jump rates are $\gamma_I(1 - r_I)$ and $\gamma_{II}(1 - r_{II})$.

Evaluate the entropy production rate. Assuming $r_I < r_{II} \leq \frac{1}{2}$, information is written on tape I and erased from tape II. Express the efficiency of the process and evaluate it in the regimes $\gamma_{II} \gg \gamma_I$ and $\gamma_{II} \ll \gamma_I$, respectively.

5.6   A 1-bit memory is represented by a system with three states: $x \in \{-1, 0, +1\}$, where $\epsilon_{\pm 1} = 0$ and $\epsilon_0 \gg k_B T$, and only the jumps $0 \rightleftharpoons \pm 1$ are allowed, all with equal attempt frequency. The state 0 therefore acts as a barrier between the ± 1 states. The dynamics satisfies detailed balance. One can erase the contents of the memory by manipulating the energies according to the following protocol:

1. The energy barrier ϵ_0 is continuously lowered to 0.
2. Then the energies ϵ_0 and ϵ_{+1} are tilted by setting $\epsilon_0(t) = -\epsilon(t)/2$, $\epsilon_{+1}(t) = -\epsilon(t)$, where $\epsilon(t)$ grows from 0 to a positive value ϵ_f. This biases the occupation toward the state $+1$.
3. The energy barrier ϵ_0 is continuously reset to its initial value.
4. Finally, the energy ϵ_{+1} is continuously reset to its initial value.

The outcome of the process is that, independent of the initial state, the final state of the system is equal to $+1$ with high probability.

By taking advantage of the relation (4.164), show that, for such a system, the Landauer bound holds; namely, that the average work W performed on the system satisfies $W \geq k_B T \ln 2$.

Give an interpretation of the inverse process and show that it is related to the Szilard engine.

Evaluate W numerically for different values of the duration $\mathcal{T}$ of the manipulation and check that the Landauer bound is approached as $\mathcal{T} \to \infty$. Also show numerically that the average work W in the reverse process, with the initial condition $p_x(t_0) = \delta_{x,+1}^K$, satisfies the inequality $W \geq -k_B T \ln 2$, and interpret this result.

CHAPTER 6

Large Deviations: Theory and Practice

The law of large numbers governs the behavior of empirical averages of random quantities when the number of realizations is large. It predicts that values of an empirical mean far from the corresponding theoretical average are extremely unlikely. Large deviation theory describes how the probability of these improbable events exponentially declines with the number of realizations. It has a tremendous impact on statistical physics in general and stochastic thermodynamics in particular. In this chapter, we introduce the main concepts and results of large deviation theory, focusing on those that are more useful for stochastic thermodynamics.

6.1 Large deviations in a nutshell

We consider a city in which it rains with probability r on any given day. Rain events on different days are uncorrelated. In a period of $\mathcal{T} \gg 1$ day, what is the distribution of the number τ of rainy days? Alternatively, what is the distribution of the fraction $f = \tau/\mathcal{T}$ of rainy days?

With a slight abuse of language, we call observables such as τ **extensive**, since their average scales linearly with the observation time $\mathcal{T}$. This definition is analogous to extensive quantities in classical thermodynamics, whose value is proportional to the system size. Similarly, we call variables such as f **intensive**, since their average does not depend on $\mathcal{T}$.

To compute the distributions of f (or equivalently of τ), we introduce the variables $x(t)$, which assume the value 1 if it rains on day t and 0 otherwise. The variable $\tau = \sum_{t=1}^{\mathcal{T}} x(t)$ is a sum of independent dichotomous variables and therefore follows the binomial distribution

$$p_\tau(\mathcal{T}) = \binom{\mathcal{T}}{\tau} r^\tau (1-r)^{\mathcal{T}-\tau}. \tag{6.1}$$

We wish to describe the behavior of f for large values of $\mathcal{T}$. Since f is a normalized sum of independent, identically distributed random variables with mean r and variance $r(1-r)$, we can invoke the central limit theorem and obtain that its distribution is approximately Gaussian:

$$p(f;\mathcal{T}) \approx \sqrt{\frac{\mathcal{T}}{2\pi\, r(1-r)}}\, \exp\left[-\mathcal{T}\frac{(f-r)^2}{2r(1-r)}\right]. \tag{6.2}$$

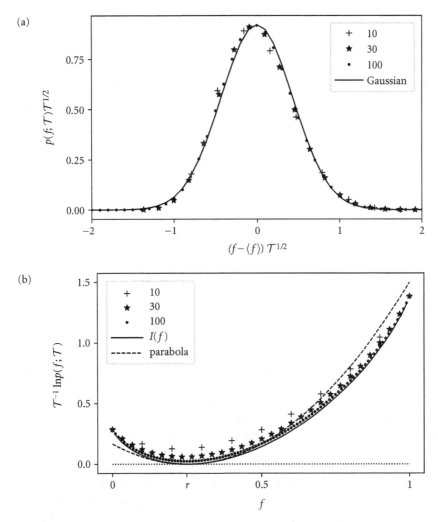

Figure 6.1. Binomial distribution, eq. (6.1), for $r = 0.25$ and different values of $\mathcal{T}$, plotted as a function of $f = \tau/\mathcal{T}$. (a) Rescaling with the average and the variance: $y = (f - \langle f \rangle) \mathcal{T}^{1/2}$. Already at moderate values of $\mathcal{T}$, the distribution of y is very close to a Gaussian distribution, in agreement with the central limit theorem. (b) Logarithmic rescaling: $I(f) = -\mathcal{T}^{-1} \ln p(f; \mathcal{T})$. The logarithm of the distribution of f does not approach a parabola, but rather the rate function $I(f) = f \ln(f/r) + (1-f) \ln[(1-f)/(1-r)]$.

It is instructive to quantitatively compare eq. (6.1) and eq. (6.2) in a concrete case. We take a time interval of one year, $\mathcal{T} = 365$, and a probability of rain, $r = 0.25$, compatible with a city like Kansas City where it rains about 91 days per year on average. We first evaluate the probability that it rains for 100 days during a year, slightly more than the average. The binomial distribution assigns a probability $p \approx 0.0270$ to this event, compared to $p \approx 0.0267$ for the Gaussian. Thus the Gaussian approximation is rather good in this case. The approximation of $p(f; \mathcal{T})$ by a Gaussian for the case of a binomial distribution is shown in fig. 6.1a. However, the agreement is significantly worse if we consider years where the probability significantly departs from the average. For example, the binomial distribution assigns a probability $p \approx 9.94 \cdot 10^{-24}$ to the event that it rains for almost half a year in Kansas City, $\tau = 180$. The corresponding probability according to the Gaussian distribution is $p \approx 4.9 \cdot 10^{-27}$, about three orders of magnitude smaller.

This example shows that the Gaussian approximation implied by the central limit theorem can be rather inaccurate on the tails of a distribution, at least in a relative sense. We wish to circumvent this problem and characterize the distribution of f for large $\mathcal{T}$, including arbitrarily extreme events. This asymptotic behavior is described by the function

$$\lim_{\mathcal{T} \to \infty} \frac{1}{\mathcal{T}} \ln p(f; \mathcal{T}) = -I(f), \tag{6.3}$$

where

$$I(f) = f \ln \frac{f}{r} + (1 - f) \ln \frac{1 - f}{1 - r}. \tag{6.4}$$

We compactly express this asymptotic behavior by the notation

$$p(f; \mathcal{T}) \asymp e^{-\mathcal{T} I(f)}. \tag{6.5}$$

When a relation of the form (6.5) holds for a probability distribution $p(f)$, we say that $p(f)$ satisfies a **large deviation principle**. The function $I(f)$ is called the **rate function** (or also the **Cramér function**) of the variable f. The rate function $I(f)$ for the binomial distribution is shown in fig. 6.1b. The rate function is always nonnegative, vanishes for $f = \langle f \rangle$, and in most cases is approximated by a parabola when f is close to its average, in agreement with the central limit theorem. However, the rate function might significantly deviate from a parabola when the absolute difference $|f - \langle f \rangle|$ is large, e.g., on the order of $\langle f \rangle$ itself, as shown in fig. 6.1b for the case of the binomial distribution. These values of f are called **large deviations**, to contrast them with the small deviations on the order of $\mathcal{T}^{-1/2}$ that are ruled by the central limit theorem. **Large deviation theory** is the branch of probability theory that studies asymptotic forms of probability distributions described by eq. (6.5) when a parameter like $\mathcal{T}$ becomes large.

It might be useful to further stress the difference between the central limit theorem and large deviation theory. In our example, the probability of values of f that are substantially different from the mean decreases exponentially with $\mathcal{T}$. The central limit theorem essentially ignores these values, as their probability is very small anyway. This is the reason why the Gaussian approximation to the distribution of f improves as $\mathcal{T}$ grows, regardless of the details of the original distribution (see also the discussion at the end of appendix A.4). Instead, taking the logarithm of the distribution, as in eq. (6.3), acts as a magnifying glass on the probability of rare events far from the mean, and therefore reveals the problem-specific shape of the tails of $p(f)$. For this reason, large deviation theory is sometimes referred to in a pictorial way as "the central limit theorem in logarithmic scale." The Gaussian behavior is recovered if we perform a Taylor expansion of the rate function to second order around its minimum:

$$I(f) \approx \frac{(f - \langle f \rangle)^2}{2 \tilde{\sigma}_f^2}, \tag{6.6}$$

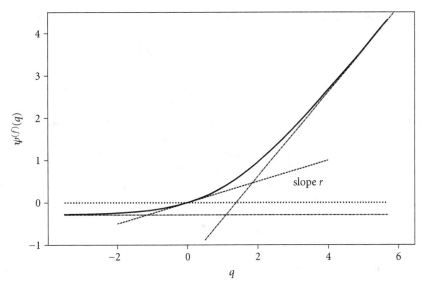

Figure 6.2. Scaled cumulant generating function $\psi^{(f)}(q)$ associated with a binomially distributed variable f with $r = 0.25$; see (6.1). The rate function $I(f)$ is shown in fig. 6.1b. The function $\psi^{(f)}(q)$ is obtained from $I(f)$ by means of the inverse Gärtner-Ellis theorem; see eq. (6.10). The slope of $\psi^{(f)}(q)$ at $q=0$ is equal to $\langle f \rangle$, i.e., in our case, to r. The scaled cumulant generating function is convex and lies between the two asymptotes $y = \lim_{f \to 0} I(f) = \ln r$ (for $q \to -\infty$) and $y = \lim_{f \to 1} I(f) + q = \ln(1 - r) + q$ (for $q \to +\infty$).

where $\tilde{\sigma}_f^2$ is the **scaled variance** of f, defined by

$$\tilde{\sigma}_f^2 = \lim_{T \to \infty} T \sigma_f^2. \tag{6.7}$$

The approximation in eq. (6.6) is valid in the range of small deviations governed by the central limit theorem.

Fluctuations of intensive and extensive quantities can also be studied via the cumulant generating function $\Phi^{(f)}(q; T) = \ln \langle e^{T q f} \rangle = \ln \langle e^{q \tau} \rangle$, discussed in appendix A.4. For large T, the cumulant generating function grows linearly with T. We extract its asymptotic behavior by introducing the **scaled cumulant generating function**

$$\psi^{(f)}(q) = \lim_{T \to \infty} \frac{1}{T} \ln \langle e^{q T f} \rangle = \lim_{T \to \infty} \frac{1}{T} \Phi^{(f)}(q; T). \tag{6.8}$$

As the name suggests, the scaled cumulant generating function "rescales" cumulants with the appropriate power of T, so that they approach a finite limit as $T \to \infty$. The first scaled cumulant $\partial_q \psi^{(f)}(q; T)|_{q=0}$ is equal to $\langle f \rangle$, and the second scaled cumulant $\partial_q^2 \psi^{(f)}(q; T)|_{q=0}$ is equal to the scaled variance $\tilde{\sigma}_f^2$ introduced in eq. (6.7). As an example, we show in fig. 6.2 the scaled cumulant generating function corresponding to the binomial distribution.

The scaled cumulant generating function contains precious information about large deviations. Indeed, the **Gärtner-Ellis theorem** directly links the scaled cumulant generating function with the rate function of the same random variable. According to this theorem, if the rate function of f is convex, it can be obtained as the **Legendre-Fenchel transform** of $\psi^{(f)}(q)$:

$$I(f) = \sup_q \left[qf - \psi^{(f)}(q) \right]. \tag{6.9}$$

The Legendre-Fenchel transform differs from the Legendre transform by the appearance of the supremum rather than the maximum in its definition. This generalization permits us to treat nonconvex or nondifferentiable functions and functions achieving their supremum for $q \to \infty$, as discussed in appendix A.2. The Legendre-Fenchel transform of a function is convex, and therefore if the rate function $I(f)$ satisfies eq. (6.9), it is also convex. This is the case of the binomial distribution. For convex functions, the Legendre-Fenchel transformation is an involution. This means that the scaled cumulant generating function is also the Legendre-Fenchel transform of the rate function:

$$\psi^{(f)}(q) = \sup_f \left[qf - I(f) \right]. \tag{6.10}$$

It turns out that it is easier to heuristically prove eq. (6.10):

$$
\begin{aligned}
\psi^{(f)}(q) &= \lim_{\mathcal{T} \to \infty} \frac{1}{\mathcal{T}} \ln \int_{-\infty}^{\infty} df\, P(f; \mathcal{T})\, e^{\mathcal{T} qf} \\
&= \lim_{\mathcal{T} \to \infty} \frac{1}{\mathcal{T}} \ln \int_{-\infty}^{\infty} df\, e^{-\mathcal{T}[I(f) - qf]} = \sup_f \left[qf - I(f) \right],
\end{aligned}
\tag{6.11}
$$

where the last equality follows from evaluating the integral using the method of steepest descent. The Gärtner-Ellis theorem (6.9) then descends from the fact that the Legendre-Fenchel transformation is an involution on convex functions.

6.2 Currents, traffic, and other observables

The basic observables in stochastic thermodynamics are the stochastic work $w(\boldsymbol{x})$, the stochastic heat $q(\boldsymbol{x})$, and the stochastic entropy production $s^{\text{tot}}(\boldsymbol{x})$, introduced in chapter 3. These quantities are the stochastic counterparts of the conventional macroscopic thermodynamic observables; see section 2.1. Practical applications of stochastic thermodynamics often involve a broader range of stochastic observables, such as the net distance traveled by a molecular motor or the number of ATP molecules consumed in a chemical reaction.

Large deviation theory is useful in characterizing these observables when measured in the steady state over a long, but finite, stretch of time $\mathcal{T}$. As in the previous section, we call *extensive observables* those whose average grows proportionally to $\mathcal{T}$. For example, the observables w, q, and s^{tot} are extensive. By dividing an extensive observable by the duration $\mathcal{T}$, we define its intensive counterpart, whose average does not depend on $\mathcal{T}$.

We consider, as usual, a mesoscopic system described by a master equation. A first class of observables are the **static observables**. Extensive static observables are functionals of the trajectory that are expressed by

$$\mathcal{A}(\boldsymbol{x}) = \sum_y \alpha_y \tau_y, \tag{6.12}$$

where the αs are given coefficients and τ_y is the **empirical dwell time** in state y, defined as the total amount of time spent in state y along the trajectory:

$$\tau_y(x) = \int_{t_0}^{t_f} dt\, \delta^K_{y,x(t)}. \tag{6.13}$$

A second class of observables are the **dynamic observables**. Dynamic observables are functionals of the trajectory associated with the empirical number of jumps between pairs of states. A generic extensive dynamic observable is expressed by

$$\mathcal{J}(x) = \sum_{x \neq x'} n_{xx'}(x)\, d_{xx'}. \tag{6.14}$$

Here the **empirical jump numbers** $n_{xx'}(x)$ are defined by

$$n_{xx'}(x) = \sum_{k=1}^{n} \delta^K_{xx_k} \delta^K_{x'x_{k-1}}, \tag{6.15}$$

where the x_ℓs are the sequence of states visited by the trajectory. The matrix $d_{xx'}$ of **distances** fully specifies the functional. Dynamic observables defined by an antisymmetric distance matrix $d_{xx'}$ are called **integrated empirical currents**. Integrated empirical currents are odd under time reversal ($x \longrightarrow \widehat{x}$). Relevant examples of empirical currents are

- The entropy change of the heat reservoir $s^{\mathrm{res}}(x)$, corresponding to the choice $d_{xx'} = k_B \ln(k_{xx'}/k_{x'x})$.
- The total entropy production s^{tot}, corresponding to $d_{xx'} = k_B \ln[(k_{xx'} p_{x'})/(k_{x'x} p_x)]$.

The intensive counterparts of empirical dwell times are the **empirical frequencies**

$$f_y = \frac{\tau_y}{\mathcal{T}} = \frac{1}{\mathcal{T}} \int_{t_0}^{t_f} dt\, \delta^K_{y,x(t)}, \tag{6.16}$$

whereas the intensive version of integrated empirical currents are the **empirical currents**

$$j = \frac{\mathcal{J}(x)}{\mathcal{T}} = \frac{1}{\mathcal{T}} \sum_{x \neq x'} n_{xx'}(x)\, d_{xx'}. \tag{6.17}$$

We use the symbol $j_{xx'}$ to denote the particular empirical current between the pair of states x and x':

$$j_{xx'} = \frac{\mathcal{J}_{xx'}(x)}{\mathcal{T}} = \frac{n_{xx'} - n_{x'x}}{\mathcal{T}}. \tag{6.18}$$

We sometimes consider dynamic observables whose distance matrix $d_{xx'}$ is symmetric, and that are therefore even under time reversal. An important example is the **traffic** between a given pair of states x and x':

$$t_{xx'}(x) = \frac{n_{xx'}(x) + n_{x'x}(x)}{\mathcal{T}}. \tag{6.19}$$

In the steady state, intensive observables associated with static and dynamic observables of master equations satisfy large deviation principles. The intuitive reason is that we can then partition a long interval $\mathcal{T}$ into smaller time intervals and consider the contribution to each observable from each of these intervals. Since the dynamics is Markovian, in the steady state these contributions are weakly correlated random variables. This means that their sum scales asymptotically with $\mathcal{T}$, like the sum of independent random variables discussed in section 6.1.

Close to equilibrium, i.e., in the linear response regime, the large deviation functions of arbitrary currents are approximately parabolic, corresponding to Gaussian distributions. To prove it, we consider a system in the steady state driven out of equilibrium by a small driving δ. In the absence of the driving, the system is at equilibrium and $\psi^{(j)}(q) = \psi^{(j)}(-q)$ because of invariance under time reversal. The driving δ breaks this symmetry. Thus, for small δ, the scaled cumulant generating function has the form

$$\psi(q) = \psi^+(q) + \delta\,\psi^-(q), \tag{6.20}$$

where $\psi^+(q)$ is an even function and $\psi^-(q)$ is an odd function. We assume that both of these functions are smooth around $q = 0$, and we expand them in powers of q:

$$\begin{aligned}
\psi^+(q) &= \psi_2\,q^2 + \psi_4\,q^4 + \dots, \\
\psi^-(q) &= \psi_1\,q + \psi_3\,q^3 + \dots,
\end{aligned} \tag{6.21}$$

where the ψ_ns are given coefficients. We now express the rate function of j by the Gärtner-Ellis theorem:

$$I(j) = \sup_q \left[qj - \psi^{(j)}(q) \right] = q^* f - \psi^{(j)}(q^*), \tag{6.22}$$

where q^* is the solution of the equation $\psi'(q^*) = j$. Thus the average of j is of order δ. For fluctuating currents on the order of this average, one has

$$q^* = \frac{j - \psi_1\,\delta}{2\,\psi_2}, \tag{6.23}$$

which is also of order δ. Substituting this value into eq. (6.22) and keeping the lowest order in δ and j, we obtain

$$I(j) = \frac{(j - \psi_1\,\delta)^2}{4\,\psi_2}, \tag{6.24}$$

of order δ^2, neglecting terms of order δ^4 and higher. This argument can be generalized to cases where the system is brought out of equilibrium by several drivings, all assumed to be small.

Static and dynamic observables can be also introduced for continuous systems described by Langevin equations. In particular, the continuous version of empirical integrated currents are linear functionals of the increments dx of the process. From this perspective, the stochastic heat defined in eq. (3.88) is an empirical integrated current. The total entropy production defined in eq. (3.91) is also an empirical integrated current, provided that the system is in the steady state. The fact that these quantities are defined as Stratonovich products ensures that they are odd under time reversal. In any case, in this chapter we deal only with systems with discrete states for the sake of simplicity.

6.3 Large deviations and fluctuation relations

We now apply large deviation theory to the total entropy production s^{tot} produced during a time interval of duration $\mathcal{T}$ in a nonequilibrium steady state. We treat s^{tot} as an integrated empirical current of the form of eq. (6.14). We associate with s^{tot} an intensive current, called the **empirical entropy production rate**:

$$j_s^{\text{tot}} = \frac{s^{\text{tot}}}{\mathcal{T}}. \qquad (6.25)$$

Switching from the extensive variable s^{tot} to the intensive variable j_s^{tot}, the steady-state integral fluctuation relation (eq. (4.47)) becomes

$$\frac{p(j_s^{\text{tot}}; \mathcal{T})}{p(-j_s^{\text{tot}}; \mathcal{T})} = e^{\mathcal{T} j_s^{\text{tot}}/k_{\text{B}}}. \qquad (6.26)$$

The distribution $p(j_s^{\text{tot}}; \mathcal{T})$ satisfies a large deviation principle:

$$p(j_s^{\text{tot}}; \mathcal{T}) \asymp e^{-\mathcal{T} I(j_s^{\text{tot}})}, \qquad (6.27)$$

where the rate function $I(j_s^{\text{tot}})$ depends on the specific system. By expressing $p(j_s^{\text{tot}}; \mathcal{T})$ in terms of the rate function, eq. (6.26) entails the **asymptotic detailed fluctuation relation**

$$I(j_s^{\text{tot}}) - I(-j_s^{\text{tot}}) = -\frac{j_s^{\text{tot}}}{k_{\text{B}}}. \qquad (6.28)$$

If the distribution of j_s^{tot} is Gaussian, the rate function is a parabola, and eq. (6.28) yields

$$2k_{\text{B}} \langle j_s^{\text{tot}} \rangle = \tilde{\sigma}_{j_s^{\text{tot}}}^2, \qquad (6.29)$$

where $\tilde{\sigma}_{j_s^{\text{tot}}}^2 = \lim_{\mathcal{T} \to \infty} \mathcal{T} \sigma_{j_s^{\text{tot}}}^2$. Equation (6.29) is another example of a fluctuation-dissipation relation and can be seen as the intensive counterpart of eq. (3.66). If the rate function is not quadratic, i.e., if the distribution is not Gaussian, eq. (6.29) does not hold and eq. (6.28) does not impose any constraint between the mean and the scaled variance of the empirical entropy production rate.

The asymptotic detailed fluctuation relation, eq. (6.28), can be written in an even more compact form in terms of the scaled cumulant generating function. To do so, we apply the inverse form of the Gärtner-Ellis theorem, eq. (6.10):

$$\psi^{(j_s)}(q) = \sup_{j_s} \left[j_s q - I(j_s) \right] = \sup_{j_s} \left[j_s \left(q + \frac{1}{k_{\text{B}}} \right) - I(-j_s) \right]$$

$$= \psi^{(j_s)} \left(-q - \frac{1}{k_{\text{B}}} \right). \qquad (6.30)$$

The relation $\psi(q) = \psi(-q - 1/k_B)$, valid for intensive variables obeying an asymptotic detailed fluctuation relation, is often called the **Gallavotti-Cohen symmetry**.

6.4 Fluctuation theorem for currents (*)

In this section, we extend the network theory developed in section 3.9 to empirical currents. Our starting point is the the irreversibility relation (4.11), which we write in the form

$$\mathcal{P}_x \, e^{-s^{\mathrm{tot}}(x)/k_B} = \mathcal{P}_{\widehat{x}}, \tag{6.31}$$

where $\mathcal{P}_x$ is the probability of a trajectory x in the steady state, $\widehat{x}$ is the backward trajectory, and $s^{\mathrm{tot}}(x) = s^{\mathrm{res}}(x) + \Delta s^{\mathrm{sys}}(x)$ is the total entropy produced along the trajectory. We decompose a trajectory x of duration $\mathcal{T}$ into loops. A **loop** is an event in which the system returns to a previously visited state, having visited a sequence of two or more distinct states in between. Upon completion of a loop ℓ, the entropy of the reservoir changes by

$$s^{\mathrm{res}}(\ell) = \sum_\alpha c_\alpha(\ell) \frac{A_\alpha}{\mathcal{T}}, \tag{6.32}$$

where the sum runs over all the fundamental cycles and the A_αs are the affinities associated with these cycles; see section 3.9. The coefficient $c_\alpha(\ell)$ is equal to $+1$ if the loop goes through the chord of cycle α with the same orientation, is equal to -1 if it goes through the chord of cycle α with the opposite orientation, and vanishes if the loop does not go through the defining chord of cycle α. Therefore, the entropy change in the reservoir associated with a trajectory x can be expressed by

$$s^{\mathrm{res}}(x) = \sum_\ell s^{\mathrm{res}}(\ell) + s^{\mathrm{rem}}, \tag{6.33}$$

where the sum runs over all loops contained in x and the "remainder entropy" s^{rem} takes into account incomplete loops. Substituting eq. (6.32) into eq. (6.33), we obtain

$$s^{\mathrm{res}}(x) = \sum_\alpha \frac{A_\alpha}{\mathcal{T}} n_\alpha(x) + s^{\mathrm{rem}}(x), \tag{6.34}$$

where $n_\alpha(x)$ is the net number of loops in the trajectory x, including the chord α, counting $+1$ for the same orientation and -1 for the opposite orientation. Consequently, the total entropy produced is equal to

$$s^{\mathrm{tot}}(x) = \sum_\alpha \frac{A_\alpha}{\mathcal{T}} n_\alpha(x) + s^{\mathrm{rem}}(x) + \Delta s^{\mathrm{sys}}. \tag{6.35}$$

We now evaluate the scaled cumulant generating function of the cycle currents:

$$\psi^{(j)}(q) = \lim_{\mathcal{T} \to \infty} \frac{1}{\mathcal{T}} \ln \left\langle \exp\left(\sum_\alpha q_\alpha n_\alpha(x) \right) \right\rangle, \tag{6.36}$$

where $j = (j_\alpha)$ and $q = (q_\alpha)$. We thus have

$$
\psi^{(j)}\left(q - \frac{A}{k_{\mathrm{B}}T}\right) = \lim_{\mathcal{T}\to\infty} \frac{1}{\mathcal{T}} \ln \int \mathcal{D}x\, \mathcal{P}_x\, e^{\sum_\alpha q_\alpha n_\alpha(x) - s^{\mathrm{tot}}(x)/k_{\mathrm{B}}}
$$

$$
= \lim_{\mathcal{T}\to\infty} \frac{1}{\mathcal{T}} \ln \int \mathcal{D}\hat{x}\, \mathcal{P}_{\hat{x}}\, e^{\sum_\alpha q_\alpha n_\alpha(x)} \tag{6.37}
$$

$$
= \lim_{\mathcal{T}\to\infty} \frac{1}{\mathcal{T}} \ln \int \mathcal{D}\hat{x}\, \mathcal{P}_{\hat{x}}\, e^{-\sum_\alpha q_\alpha n_\alpha(\hat{x})} = \psi^{(j)}(-q),
$$

where $A = (A_\alpha)$. In the first line of eq. (6.37), we neglect contributions from $s^{\mathrm{rem}}(x)$ and Δs^{sys}, since they are both bounded functions of $\mathcal{T}$ and therefore do not contribute to the scaled cumulant generating function. We also use eq. (6.31) in going from the first to the second line. Thus $\psi^{(j)}(q)$ exhibits the Gallavotti-Cohen symmetry under the transformation $(q - A/k_{\mathrm{B}}T) \longleftrightarrow -q$.

We now show that the Onsager reciprocity relations (4.28) follow from this symmetry. The average cycle current J_α is expressed by

$$
J_\alpha(A) = \langle j_\alpha \rangle = \left. \frac{\partial \psi^{(j)}}{\partial q_\alpha} \right|_{q=0}, \tag{6.38}
$$

where $\psi^{(j)}(q)$ is evaluated with the given value of A. For small activities, we expand the current as in eq. (4.35), $J_\alpha(A) = \sum_\beta L_{\alpha\beta} A_\beta + \mathrm{o}(A)$, thus introducing the matrix $L = (L_{\alpha\beta})$ of kinetic coefficients. We have therefore

$$
L_{\alpha\beta} = \left. \frac{\partial^2 \psi^{(j)}}{\partial q_\alpha\, \partial A_\beta} \right|_{q=0, A=0}. \tag{6.39}
$$

Taking advantage of the Gallavotti-Cohen symmetry, we obtain

$$
\left. \frac{\partial^2 \psi^{(j)}}{\partial q_\alpha\, \partial A_\beta} \right|_{q,A} = \frac{1}{k_{\mathrm{B}}T} \left. \frac{\partial^2 \psi^{(j)}}{\partial q_\alpha\, \partial q_\beta} \right|_{-q-A/k_{\mathrm{B}}T,A} - \left. \frac{\partial^2 \psi^{(j)}}{\partial q_\alpha\, \partial A_\beta} \right|_{-q-A/k_{\mathrm{B}}T,A}. \tag{6.40}
$$

Setting $q = A = 0$, we obtain the Onsager reciprocity relations:

$$
L_{\alpha\beta} = \frac{1}{2k_{\mathrm{B}}T} \left. \frac{\partial^2 \psi^{(j)}}{\partial q_\alpha\, \partial q_\beta} \right|_{q=0, A=0} = L_{\beta\alpha}. \tag{6.41}
$$

Further relations can be obtained by considering higher-order terms in the expansion of J in powers of A. In this sense, one may interpret the Gallavotti-Cohen symmetry as the generalization of the Onsager reciprocity relations beyond linear response.

6.5 Tilting

We have seen that the scaled cumulant generating function of an observable such as an empirical current contains precious information about large deviations. Unfortunately, it is often difficult to compute the scaled cumulant generating function directly. A mathematical technique that substantially simplifies this task is **tilting**. To introduce tilting, we first define the generating function of a current j conditioned to a final state x:

$$\phi_x^{(j)}(q;t) = \left\langle e^{qTj} \middle| x(t) = x \right\rangle = \int \mathcal{D}\boldsymbol{x} \, \mathcal{P}_{\boldsymbol{x}} \, \delta_{x,x(t)}^{K} \, e^{qTj(\boldsymbol{x})}. \tag{6.42}$$

The standard generating function is retrieved by summing over the final state: $\phi^{(j)}(q;t) = \sum_x \phi_x^{(j)}(q;t)$. We explicitly write the trajectory probability $\mathcal{P}_{\boldsymbol{x}}$ in eq. (6.42):

$$\phi_x^{(j)}(q;t) = \int \mathcal{D}\boldsymbol{x} \, \delta_{x,x_n}^{K} \, e^{q\sum_{k=1}^{n} d_{x_k x_{k-1}}} \, k_{x x_{n-1}} e^{-k_{x_n}^{\text{out}}(t-t_{n-1})} \cdots$$

$$\cdots \times e^{-k_{x_1}^{\text{out}}(t_2-t_1)} k_{x_1 x_0} e^{-k_{x_0}^{\text{out}}(t_1-t_0)} \, p_{x_0}(t_0)$$

$$= \int \mathcal{D}\boldsymbol{x} \, e^{-k_x^{\text{out}}(t-t_{n-1})} \left(k_{x x_{n-1}} e^{q d_{x x_{n-1}}} \right) \cdots$$

$$\cdots \times \left(k_{x_1 x_0} e^{q d_{x_1 x_0}} \right) e^{-k_{x_0}^{\text{out}}(t_1-t_0)} \, p_{x_0}(t_0), \tag{6.43}$$

where we use the definition (6.14) of an integrated empirical current as a sum of distances associated with each observed jump. Equation (6.43) expresses the conditioned generating function of an empirical current in terms of a "modified" trajectory probability, where each jump rate $k_{xx'}$ is multiplied by a factor $e^{q d_{xx'}}$. As a consequence, the conditioned generating function evolves according to a **tilted master equation**:

$$\frac{d}{dt}\phi_x^{(j)}(q;t) = \sum_{x' (\neq x)} k_{xx'} \, e^{q \, d_{xx'}} \phi_{x'}^{(j)}(q;t) - k_x^{\text{out}} \phi_x^{(j)}(q;t). \tag{6.44}$$

The outflow terms in the master equation (6.44) are the same as in the original master equation, since in eq. (6.43) the factors corresponding to the dwell times are unaltered. Equations (6.43) and (6.44) embody the basic idea of tilting. The term *tilting* comes from the observation that increasing q creates an imbalance in the rates of eq. (6.44) similar to the effect of tilting an external potential. We write eq. (6.44) in a more compact form:

$$\frac{d}{dt}\phi_x^{(j)}(q;t) = \sum_{x'} L_{xx'}(q) \, \phi_{x'}^{(j)}(q;t), \tag{6.45}$$

where the **tilted generator** is given by

$$L_{xx'}(q) = \begin{cases} k_{xx'} \, e^{q d_{xx'}}, & \text{if } x \neq x'; \\ -k_x^{\text{out}}, & \text{if } x = x'. \end{cases} \tag{6.46}$$

In general, the tilted generator does not satisfy $\sum_x L_{xx'}(q) = 0$, and therefore the tilted master equation (6.44) does not preserve normalization. This reflects the fact that the conditioned generating function $\phi_x^{(j)}(q;t)$ does not need to be normalized. The formal solution of eq. (6.45) reads

$$\phi_x^{(j)}(q;t) = \sum_{x'} \left[\exp\left((t-t_0) L(q)\right) \right]_{xx'} P_{x'}(t_0). \tag{6.47}$$

For long times, the behavior of $\phi_x(q, t_f)$ is dominated by the largest eigenvalue $\lambda_{max}(q)$ of the tilted generator:

$$\phi_x^{(j)}(q;t) \asymp e^{(t-t_0)\,\lambda_{max}(q)}. \tag{6.48}$$

We now switch to the scaled cumulant generating function $\psi^{(j)}(q)$ introduced in eq. (6.8). Combining eqs. (6.48) and (6.8), we obtain

$$\psi^{(j)}(q) = \lim_{T \to \infty} \frac{1}{T} \ln \sum_{x} \phi_x^{(j)}(q;T) = \lambda_{max}(q). \tag{6.49}$$

Therefore, to find $\psi^{(j)}(q)$, we only need to compute the largest eigenvalue $\lambda_{max}(q)$ of the tilted generator. The Perron-Frobenius theorem ensures that this eigenvalue is always real and nondegenerate; see appendix A.5.

In practice, the largest eigenvalue $\lambda_{max}(q)$ can be obtained by diagonalizing the tilted generator as a function of q, at least when the number of states is small. However, this task can become computationally unfeasible for a large number of states. If we are only interested in the lowest-order scaled cumulants of the current, an explicit diagonalization of the tilted generator can be avoided in the following way. The characteristic equation associated with the tilted generator is

$$\det\left(L(q) - \lambda \mathcal{I}\right) = \sum_{r=0}^{N} a_r(q)\lambda^r = 0, \tag{6.50}$$

where $\mathcal{I}$ is the identity matrix. The maximum eigenvalue satisfies $\lambda_{max}(0) = 0$ since, for $q = 0$, the tilted generator reduces to the generator of the original master equation, for which the maximum eigenvalue vanishes due to the normalization condition. We therefore need to compute the derivatives with respect to q of the particular solution of eq. (6.50) satisfying this property, and evaluate them in $q = 0$. We can conveniently evaluate these derivatives by applying the implicit function theorem to the characteristic polynomial

$$g(\lambda, q) = \sum_{r=0}^{N} a_r(q)\lambda^r = 0. \tag{6.51}$$

For example, the first two scaled cumulants can be obtained from the expressions

$$\frac{d}{dq} g(\lambda(q), q) = \frac{\partial g}{\partial \lambda} \frac{\partial \lambda}{\partial q} + \frac{\partial g}{\partial q} = 0;$$

$$\frac{d^2}{dq^2} g(\lambda(q), q) = \frac{\partial^2 g}{\partial \lambda^2} \left(\frac{\partial \lambda}{\partial q}\right)^2 + \frac{\partial g}{\partial \lambda} \frac{\partial^2 \lambda}{\partial q^2} + 2\frac{\partial^2 g}{\partial \lambda \partial q} \frac{\partial \lambda}{\partial q} + \frac{\partial^2 g}{\partial q^2} = 0. \tag{6.52}$$

Both derivatives are equal to zero since the function $g(\lambda(q), q)$ vanishes for all values of q. Solving the first equation for $\partial\lambda/\partial q$ and the second for $\partial^2\lambda/\partial q^2$, and evaluating them for $\lambda = q = 0$, we obtain

$$
\begin{aligned}
\langle j \rangle &= \left.\frac{\partial\lambda}{\partial q}\right|_{\lambda=q=0} = -\frac{1}{a_1}\left.\frac{\partial a_0}{\partial q}\right|_{q=0}, \\
\tilde{\sigma}_j^2 &= \left.\frac{\partial^2\lambda}{\partial q^2}\right|_{\lambda=q=0} = -\frac{1}{a_1}\left(\left.\frac{\partial^2 a_0}{\partial q^2}\right|_{q=0} + 2\langle j\rangle\left.\frac{\partial a_1}{\partial q}\right|_{q=0} + 2\langle j\rangle^2 a_2\right),
\end{aligned}
\tag{6.53}
$$

where we evaluate derivatives using eq. (6.51). This trick allows us to evaluate cumulants without having to compute the maximum eigenvalue.

An alternative strategy is to evaluate $\lambda_{\max}(q)$ numerically by means of eq. (6.49). We define the matrix $\bar{L}_{xx'} = \delta_{xx'}^{\mathrm{K}} + \Delta t\, L_{xx'}$, with $\Delta t > 0$ and small enough so that the matrix $\bar{L}$ has nonnegative entries. Its largest eigenvalue $\bar{\lambda}_{\max}(q)$ is equal to $1 + \Delta t\,\lambda_{\max}(q)$. By repeatedly applying $\bar{L}$ to an arbitrary initial vector $\phi^{(0)} = (\phi_x^{(0)})$ with positive entries, we obtain, for large n,

$$
\bar{L}^n\phi^{(0)} \sim \left(\bar{\lambda}_{\max(q)}\right)^n \phi_{\max},
\tag{6.54}
$$

where $\phi_{\max}$ is the right eigenvector of $\bar{L}$ corresponding to $\bar{\lambda}_{\max}(q)$. Therefore,

$$
\bar{\lambda}_{\max}(q) = \lim_{n\to\infty}\left[\sum_x\left(\bar{L}^{n+1}\phi^{(0)}\right)_x \Big/ \sum_{x'}\left(\bar{L}^n\phi^{(0)}\right)_{x'}\right].
\tag{6.55}
$$

We can then evaluate the maximum eigenvalue of the tilted generator by the expression $\lambda_{\max}(q) = (\bar{\lambda}_{\max}(q) - 1)/\Delta t$. For large matrices, efficient numerical methods are available, like the Arnoldi-Lanczos method.

We apply the tilting method to specific problems in the next two sections.

6.6 Michaelis-Menten reaction scheme

In the stochastic Michaelis-Menten reaction scheme, a reactant A is transformed into a product B with the aid of a catalytic enzyme E. In chemical kinetics, complex formation is usually considered a reversible reaction, whereas product formation is often taken as irreversible. Here we consider a more general case in which both processes are reversible, i.e., the enzyme can reversibly form a complex E* with either A or B. This choice makes the model thermodynamically consistent. The full reaction scheme is

$$
A + E \underset{k_1^-}{\overset{k_1^+}{\rightleftharpoons}} E^* \underset{k_2^-}{\overset{k_2^+}{\rightleftharpoons}} B + E,
$$

where E* is the bound enzyme. We consider a scenario in which there is only one enzyme molecule and the average concentrations of A and B molecules are maintained constant by chemostats. The balance of the reaction is determined by the ratio

$$\frac{k_1^+ k_2^+}{k_1^- k_2^-} = e^{(-\epsilon + \mu_A - \mu_B)/k_B T} = e^{-\Delta G/k_B T}, \tag{6.56}$$

where ϵ is the energy difference between the reactant and the product, μ_A and μ_B are the chemical potentials of A and B, respectively, and $\Delta G = -\epsilon + \mu_A - \mu_B$ is the difference in the Gibbs free energy between one molecule of B and one molecule of A. This free-energy difference effectively drives the reaction: equilibrium is attained when $\Delta G = 0$. We parameterize the reaction rates by the intrinsic rates ω_1, ω_2 and the activation barrier ϵ^* for the reaction $A + E \rightarrow E^*$. Generalized detailed balance dictates

$$\frac{k_1^+}{k_1^-} = e^{(-\epsilon^* + \mu_A)/k_B T}; \qquad \frac{k_2^+}{k_2^-} = e^{(-\epsilon + \epsilon^* - \mu_B)/k_B T}. \tag{6.57}$$

We choose rates satisfying these conditions:

$$\begin{aligned}
k_1^+ &= \omega_1 \, e^{(2\mu_A - \epsilon^*)/2k_B T}; & k_1^- &= \omega_1 \, e^{\epsilon^*/2k_B T}; \\
k_2^+ &= \omega_2 \, e^{-(\epsilon - \epsilon^*)/2k_B T}; & k_2^- &= \omega_2 \, e^{(2\mu_B + \epsilon - \epsilon^*)/2k_B T}.
\end{aligned} \tag{6.58}$$

We also assume that the second reaction is much slower than the first, $\omega_2 \ll \omega_1$. We describe the evolution of the catalysis by the number n of B molecules synthesized since an arbitrary initial time. Since the reaction is reversible, n can become negative, i.e., at a given time there can be fewer B molecules than at the initial time. We denote the enzyme state by 0 if it is free and by 1 if it is bound. The master equation reads

$$\begin{aligned}
\frac{dp_{0n}}{dt} &= k_1^- p_{1n} + k_2^+ p_{1,n-1} - \left(k_1^+ + k_2^- \right) p_{0n}; \\
\frac{dp_{1n}}{dt} &= k_1^+ p_{0n} + k_2^- p_{0,n+1} - \left(k_1^- + k_2^+ \right) p_{1n}.
\end{aligned} \tag{6.59}$$

We call $p = \sum_n p_{1n}$ the probability that the enzyme is bound at a given time. From the master equation, this probability evolves according to

$$\frac{dp}{dt} = \left(k_1^+ + k_2^- \right) (1 - p) - \left(k_1^- + k_2^+ \right) p, \tag{6.60}$$

whose steady-state solution is

$$p^{st} = \left(1 + \frac{k_1^- + k_2^+}{k_1^+ + k_2^-} \right)^{-1}. \tag{6.61}$$

Using the generalized detailed balance conditions (6.56) and our assumption that $\omega_2 \ll \omega_1$, we approximate the steady-state probability by

$$p^{st} = \frac{e^{(\mu_A - \epsilon^*)/k_B T}}{1 + e^{(\mu_A - \epsilon^*)/k_B T}}. \tag{6.62}$$

Given the binding energy ϵ^* and the chemical potential μ_A of the ligand, the average production rate of B in the steady state is given by

$$\langle \dot{n}_B \rangle = k_2^+ p^{st} - k_2^-. \tag{6.63}$$

We now analyze the large deviations of the catalysis rate in the steady state by adapting the tilted dynamics introduced in section 6.5. We introduce a two-component generating function

$$\phi_\alpha(q,t) = \sum_{n=-\infty}^{+\infty} e^{q(n+\alpha/2)} p_{\alpha,n}(t), \tag{6.64}$$

where $\alpha \in \{0,1\}$, and we conventionally assign a half-integer value of n to the states where the enzyme is bound. From eq. (6.59), we obtain

$$\frac{d}{dt} \begin{pmatrix} \phi_0 \\ \phi_1 \end{pmatrix} = L(z) \begin{pmatrix} \phi_0 \\ \phi_1 \end{pmatrix}, \tag{6.65}$$

where $z = e^{q/2}$ and the tilted generator is given by

$$L(z) = \begin{pmatrix} -\left(k_1^+ + k_2^-\right), & z^{-1}k_1^- + zk_2^+ \\ zk_1^+ + z^{-1}k_2^-, & -(k_1^- + k_2^+) \end{pmatrix}. \tag{6.66}$$

Importantly, z only appears in the off-diagonal elements, and determines the eigenvalues of $L(z)$ only via their product:

$$\Pi(z) = z^2 k_1^+ k_2^+ + k_1^- k_1^+ + k_2^- k_2^+ + z^{-2} k_1^- k_2^-. \tag{6.67}$$

The eigenvalues of $L(z')$ are the same as those of $L(z)$ if z' is such that it exchanges the values of the first and last term:

$$z'^2 k_1^+ k_2^+ = z^{-2} k_1^- k_2^-, \tag{6.68}$$

$$z^2 k_1^+ k_2^+ = z'^{-2} k_1^- k_2^-, \tag{6.69}$$

from which we obtain

$$z'^2 = \frac{k_1^- k_2^-}{k_1^+ k_2^+ z^2}. \tag{6.70}$$

Taking into account eq. (6.56) and substituting $z = e^{q/2}$, we find that the eigenvalues are invariant under the transformation

$$q \longrightarrow q' = \frac{\Delta G}{k_B T} - q. \tag{6.71}$$

At equilibrium, where $\Delta G = 0$, this transformation reduces to $q \longrightarrow -q$.

We denote by $\lambda_{\max}(q)$ the leading eigenvalue of L. The generating function of the number of processed molecules n is a linear combination of ϕ_0 and ϕ_1 and therefore scales exponentially with $\mathcal{T}\lambda_{\max}(q)$:

$$\phi^{(n)}(q;t) = \sum_n e^{qn}[p_{0,n}(t) + p_{1,n}(t)] = \phi_0(q) + e^{-q/2}\phi_1(q) \asymp e^{\mathcal{T}\lambda_{\max}(q)}. \quad (6.72)$$

The number of products n grows linearly with $\mathcal{T}$ on average. We associate with it an intensive observable $\dot{n} = n/\mathcal{T}$, which we interpret as the empirical production rate. On the one hand, the scaled cumulant generating function of $\dot{n}$ is equal to

$$\psi^{(\dot{n})}(q) = \lim_{\mathcal{T}\to\infty} \frac{1}{\mathcal{T}} \ln \phi^{(\dot{n})}(q;t) = \lambda_{\max}(q). \quad (6.73)$$

On the other hand, we have seen that

$$\lambda_{\max}(q) = \lambda_{\max}\left(\frac{\Delta G}{k_B T} - q\right). \quad (6.74)$$

Equation (6.74) embodies the Gallavotti-Cohen symmetry for the intensive variable $\dot{n}$; see also eq. (6.30).

The explicit expression of the maximum eigenvalue is

$$\lambda_{\max}(q) = -\frac{1}{2}\Big[k_1^+ + k_1^- + k_2^+ + k_2^- \quad (6.75)$$
$$- \sqrt{(k_1^+ + k_1^- + k_2^+ + k_2^-)^2 + 4(1 - e^{-q})(k_1^+ k_2^+ e^q - k_1^- k_2^-)}\Big].$$

This expression behaves asymptotically as

$$\lambda_{\max}(q) \approx \begin{cases} \sqrt{k_1^+ k_2^+}\, e^{q/2}, & \text{for } q \to +\infty; \\ \sqrt{k_1^- k_2^-}\, e^{-q/2}, & \text{for } q \to -\infty. \end{cases} \quad (6.76)$$

We use this result to evaluate the average production rate $\langle \dot{n}\rangle$ via the relation

$$\langle \dot{n}\rangle = \left.\frac{d\lambda_{\max}}{dq}\right|_{q=0}. \quad (6.77)$$

An explicit evaluation of the derivative leads to the expression

$$\langle \dot{n}\rangle = \frac{k_1^+ k_2^+ - k_1^- k_2^-}{k_1^+ + k_1^- + k_2^+ + k_2^-}. \quad (6.78)$$

The average production rate $\langle \dot{n}\rangle$ vanishes at equilibrium, when eq. (6.56) is satisfied. In a similar way, one can obtain the scaled variance of $\dot{n}$ and scaled cumulants of high order. Since the derivative $d\lambda_{\max}/dq$ can take any real value for large enough $|q|$, the empirical production rate $\dot{n}$ is also unbounded. The Legendre-Fenchel transform of $\lambda_{\max}(q)$ is the rate function of $\dot{n}$:

$$p(\dot{n};\mathcal{T}) \asymp e^{-\mathcal{T} I(\dot{n})}, \quad (6.79)$$

where

$$I(\dot{n}) = \sup_q \left[\dot{n}\, q - \lambda_{\max}(q)\right]. \tag{6.80}$$

We obtain, for large values of $|\dot{n}|$,

$$I(\pm\dot{n}) \approx 2\,|\dot{n}| \left(\ln \frac{2\,|\dot{n}|}{\sqrt{k_1^{\pm}k_2^{\pm}}} - 1 \right). \tag{6.81}$$

By this technique, one can also evaluate the rate function of the entropy production rate j_s^{tot} and verify the asymptotic detailed fluctuation relation (6.28).

6.7   Fluctuation relations in a model of kinesin (*)

Molecular motors are important proteins that perform work inside cells by consuming chemical energy; see section 3.4. They are a paradigmatic example of biomolecules that can be studied with the tools of stochastic thermodynamics.

In this section, we study a model of kinesin, which is a molecular motor that walks on a microtubule—a hollow tubular structure that forms the cytoskeleton. The role of kinesin is to transport cargo along microtubules from the center of the cell to its periphery. The motion of kinesin proceeds in discrete, reversible steps. In the model, we subdivide each step along the microtubule into two states: a low-energy state $x = 0$, to which we conventionally assign an energy $\epsilon_0 = 0$, and a high-energy state $x = 1$, with energy $\epsilon_1 = \epsilon > 0$. We jointly represent the state of the motor and its position on the microtubule by an integer variable r. An increment of r by one corresponds to a half step of the motor, so that the motor is in state 0 at even-numbered coordinate r and in state 1 at odd-numbered coordinate r. We call d the physical distance on the microtubule corresponding to a half step.

There are four possible jumps connecting states 0 to states 1, each with its corresponding reverse jump, as shown in fig. 6.3a:

A: From a low-energy state, the motor makes a half step to the left, with rate $k_0^{\leftarrow}$, without consuming ATP. In the reverse jump, it moves a half step to the right from a high-energy state with rate $k_1^{\rightarrow}$.

B: From a low-energy state, the motor consumes one molecule of ATP and moves a half step to the left, with rate $k_0^{\nwarrow}$. In the reverse jump, it produces one molecule of ATP from a high-energy state and moves a half step to the right, with rate $k_1^{\searrow}$.

C: From a low-energy state, the motor consumes one molecule of ATP and moves a half step to the right, with rate $k_0^{\nearrow}$. In the reverse jump, the motor produces one molecule of ATP from a high-energy state and moves a half step to the left, with rate $k_1^{\swarrow}$.

D: From a low-energy state, the motor moves a half step to the right, with rate $k_0^{\rightarrow}$, without consuming ATP. In the reverse jump, it moves a half step to the left from a high-energy state, with rate $k_1^{\leftarrow}$.

Summarizing, jumps A and D represent spontaneous diffusion of the motor along the microtubule without the aid of ATP, whereas B and C are the corresponding jump rates with the use of ATP. In particular, jumps B and C are assisted by the same ATP

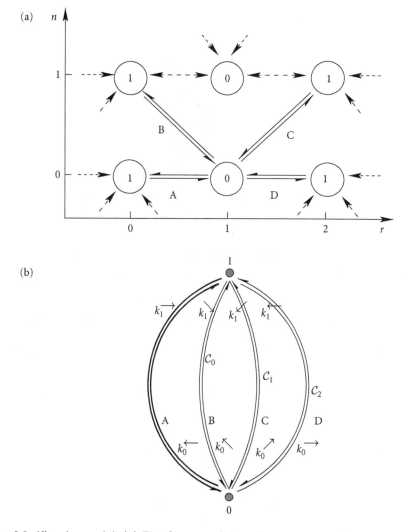

Figure 6.3. Kinesin model. (a) The four possible pathways connecting a state 0 to a state 1, represented in the (r, n) plane, where r denotes the position of the motor on the microtubule and n is the net number of ATP molecules consumed since the initial time. Equivalent jumps in different locations of the (r, n) plane are represented by dashed arrows. See Lacoste et al. [94]. (b) The jump network of the model. The spanning tree, which coincides with the A edge, is marked in bold. The cycle labels are placed close to the corresponding chords.

reservoir. We call $\Delta\mu$ the ATP-to-ADP chemical potential imbalance; see section 3.5. The motor is subject to a force f, considered positive in the sense of positive rs. The jump rates k_x^α ($x \in \{0, 1\}$, $\alpha \in \{\leftarrow, \nwarrow, \nearrow, \rightarrow, \searrow, \swarrow\}$) satisfy the generalized detailed balance conditions

$$\frac{k_0^\leftarrow}{k_1^\rightarrow} = e^{(-\epsilon - fd)/k_B T}; \qquad\qquad \frac{k_0^\nwarrow}{k_1^\searrow} = e^{(-\epsilon - fd + \Delta\mu)/k_B T}; \qquad (6.82a)$$

$$\frac{k_0^{\nearrow}}{k_1^{\swarrow}} = e^{(-\epsilon+fd+\Delta\mu)/k_{\mathrm{B}}T}; \qquad \frac{k_0^{\rightarrow}}{k_1^{\leftarrow}} = e^{(-\epsilon+fd)/k_{\mathrm{B}}T}. \qquad (6.82b)$$

In terms of these rates, the steady-state probability distribution is

$$p_0^{st} = \frac{\sum_\alpha k_1^\alpha}{\sum_{\alpha,x} k_x^\alpha};$$

$$p_1^{st} = \frac{\sum_\alpha k_0^\alpha}{\sum_{\alpha,x} k_x^\alpha}. \qquad (6.83)$$

We identify the fundamental cycles in the jump network following the method of section 3.9; see fig. 6.3b. We choose the edge A as the spanning tree. The corresponding fundamental cycles are outlined below.

$\mathcal{C}_0$: Formed by A and B. If run counterclockwise, it corresponds to moving a half step to the left with the consumption of one ATP molecule, and then a half step to the right without consuming ATP.

$\mathcal{C}_1$: Formed by A and C. Run counterclockwise, it corresponds to moving a half step to the right with the consumption of one ATP molecule, and then a half step again to the right without use of ATP.

$\mathcal{C}_2$: Formed by A and D. Run counterclockwise, it corresponds to a full step to the right without use of ATP.

We associate the affinities $A_{0,1,2}$ with these three cycles. Taking into account (6.82), we obtain

$$A_0 = k_{\mathrm{B}}T \ln \frac{k_0^{\nwarrow} k_1^{\rightarrow}}{k_0^{\leftarrow} k_1^{\searrow}} = \Delta\mu;$$

$$A_1 = k_{\mathrm{B}}T \ln \frac{k_0^{\nearrow} k_1^{\rightarrow}}{k_0^{\leftarrow} k_1^{\swarrow}} = -2fd + \Delta\mu; \qquad (6.84)$$

$$A_2 = k_{\mathrm{B}}T \ln \frac{k_0^{\rightarrow} k_1^{\rightarrow}}{k_0^{\leftarrow} k_1^{\leftarrow}} = -2fd.$$

Although there are three fundamental cycles, two currents are particularly interesting: the displacement velocity and the degradation rate of ATP. We evaluate the steady-state currents $J^{(r)} = \mathrm{d}\langle r \rangle /\mathrm{d}t$ and $J^{(n)} = \mathrm{d}\langle n \rangle /\mathrm{d}t$, where n is the number of ATP molecules consumed since the initial time. We express the steady-state currents in terms of the cycle currents J_α, $\alpha \in \{0, 1, 2\}$:

$$J^{(r)} = J_1 + J_2;$$

$$J^{(n)} = J_0 + J_1. \qquad (6.85)$$

We are now in a position to draw a phase diagram by identifying the lines in which the steady-state currents change sign (fig. 6.4). The most relevant working regimes are the following:

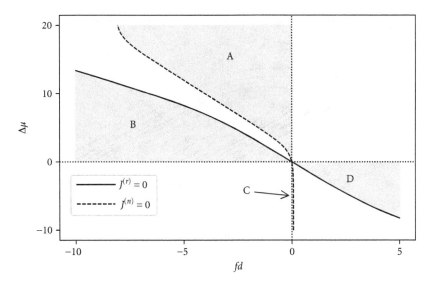

Figure 6.4. Phase diagram of the molecular motor in the (fd, $\Delta\mu$) plane, obtained by analytically evaluating the currents $J^{(r,n)}$ in the steady state. The axes (dotted) and the lines $J^{(r)} = 0$ (solid) and $J^{(n)} = 0$ (dashed) partition the plane into eight regions. The most relevant ones are A, where $J^{(r)} > 0$ and $J^{(n)} < 0$, with $f < 0$ and $\Delta\mu > 0$, and the system works as a motor, producing work via ATP hydrolysis; B, where it synthesizes ATP exploiting mechanical work; the narrow C region, where it exploits ADP in excess to perform mechanical work; and D, where the system produces ADP from molecular work. In the unshaded areas, the system acts as a dud. See Lacoste et al. [94].

A: In this regime, the system consumes ATP to produce work, therefore operating as a proper molecular motor. This regime is characterized by $f < 0$, $\Delta\mu > 0$, $J^{(n)} > 0$, and $J^{(r)} > 0$, i.e., the motor proceeds against the applied force.

B: In this regime, the motor exploits the applied force and uses up the mechanical work to produce ATP molecules, $J^{(r)} < 0$ and $J^{(n)} > 0$.

C: Here the motor exploits ADP in excess to perform mechanical work against a positive force. This regime occupies a narrow region of the phase diagram, bounded by the $f = 0$ and $J^{(n)} = 0$ lines.

D: In this regime, the motor produces ADP in excess from molecular work.

In the remaining four areas, unshaded in the figure, one has both $f J^{(r)} < 0$ and $\Delta\mu J^{(n)} < 0$, and therefore the motor does not produce anything useful and can be considered a dud.

In the steady state, the average entropy production rate is given by

$$T\dot{S} = A_0 J_0 + A_1 J_1 + A_2 J_2 = -2fd J^{(r)} + \Delta\mu J^{(n)}. \tag{6.86}$$

We characterize the fluctuations of the currents $j^{(r)}$, $j^{(n)}$ via the cumulant generating function

$$\psi^{(r,n)}(q_r, q_n) = \lim_{t_f \to \infty} \frac{1}{\mathcal{T}} \ln \left\langle \exp \left[\int_{t_0}^{t_f} dt \left(q_r j^{(r)}(t) + q_n j^{(n)}_t \right) \right] \right\rangle. \tag{6.87}$$

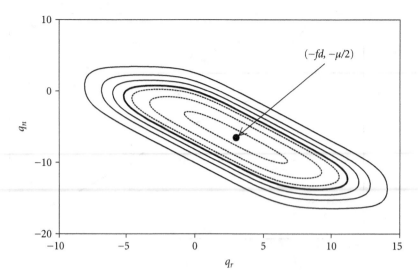

Figure 6.5. Contour plot of the cumulant generating function $\psi^{(j_r, j_n)}(q_r, q_n)$, exhibiting the Gallavotti-Cohen symmetry under the transformation (6.90). The invariant point of the transformation is shown. The zero-level contour is highlighted. We have set $fd = -3$, $\Delta\mu = 13$, and $k_B T = 1$.

We have seen in section 6.5 that $\psi^{(r,n)}(q_r, q_n)$ is equal to the maximum eigenvalue of the generator of the tilted dynamics. In our case, the generator is the matrix

$$L(q_r, q_n) = \begin{pmatrix} L_{00}, & L_{01} \\ L_{10}, & L_{11} \end{pmatrix}, \tag{6.88}$$

where

$$
\begin{aligned}
L_{00} &= -(k_0^{\leftarrow} + k_0^{\nwarrow} + k_0^{\nearrow} + k_0^{\rightarrow}); \\
L_{01} &= k_1^{\rightarrow} e^{q_x} + k_1^{\searrow} e^{-q_n + q_x} + k_1^{\swarrow} e^{-q_n - q_x} + k_1^{\leftarrow} e^{-q_x}; \\
L_{10} &= k_0^{\leftarrow} e^{-q_x} + k_0^{\nwarrow} e^{q_n - q_x} + k_0^{\nearrow} e^{q_n + q_x} + k_0^{\rightarrow} e^{q_x}; \\
L_{11} &= -(k_1^{\rightarrow} + k_1^{\searrow} + k_1^{\swarrow} + k_1^{\leftarrow}).
\end{aligned}
\tag{6.89}
$$

One can verify that, due to the generalized detailed balance conditions (6.82), the determinant of L, and therefore its leading eigenvalue, remains unchanged under the transformation $(q_r, q_n) \longrightarrow (q_r', q_n')$, where

$$q_r' = -\frac{2fd}{k_B T} - q_r; \qquad q_n' = -\frac{\Delta\mu}{k_B T} - q_n. \tag{6.90}$$

This transformation represents the Gallavotti-Cohen symmetry for this system, as shown in fig. 6.5.

6.8 Cloning (*)

In principle, observables of a master equation can be studied by stochastic simulations, for example, using the Gillespie method reviewed in appendix A.6 and sampling observables along the simulation. However, this approach is not practical if we are interested in large deviations: probabilities of large fluctuations become exponentially small over time, therefore simulations of any reasonable length are unlikely to sample them.

The tilting method introduced in section 6.5 provides an alternative to computing scaled cumulant generating functions and hence rate functions. However, extracting leading eigenvalues of very large tilted matrices can also be computationally prohibitive. An ideal way out would be to perform some kind of stochastic simulation of the tilted process. Unfortunately, the tilted dynamics introduced in eq. (6.44) does not conserve the normalization of $\phi_x(q;t)$. Therefore, it cannot be associated with a master equation nor with any other stochastic process that conserves probability.

A solution to this problem is **cloning**. Cloning algorithms simulate in parallel a population of N systems, which we call **clones**. Each clone evolves according to stochastic dynamics. During the dynamics, clones can be stochastically "pruned" or "copied," in a way to account for changes in statistical weights. To understand how cloning works, we provide the following definitions:

- The **tilted jump rate** $k_{xx'}(q) = k_{xx'}\, e^{qd_{xx'}}$;
- The **tilted escape rate** $k_x^{\text{out}}(q) = \sum_{x'\,(\neq x)} k_{x'x}(q)$;
- The **tilted dwell-time distribution** $\rho_x(t;q) = k_x^{\text{out}}(q)\, e^{-k_x^{\text{out}}(q)\,t}$;
- The **weights** $Y_x(t;q) = e^{(k_x^{\text{out}}(q) - k_x^{\text{out}})t}$.

Substituting these definitions into eq. (6.43), we rewrite the generating function of a current conditioned to a final state x as

$$
\phi_x^{(j)}(q,t) = \int \mathcal{D}x\, \delta_{x,x_n}^{K}\, \rho_{x_n}(t-t_n;q)\, Y_x(t-t_n;q)\, \frac{k_{x_n x_{n-1}}(q)}{k_{x_{n-1}}^{\text{out}}(q)}
$$

$$
\times\, \rho_{x_{n-1}}(t_n - t_{n-1};q)\, Y_{x_{n-1}}(t_n - t_{n-1};q)\, \cdots\, \frac{k_{x_2 x_1}(q)}{k_{x_1}^{\text{out}}(q)} \tag{6.91}
$$

$$
\times\, \rho_{x_1}(t_2 - t_1;q)\, Y_{x_1}(t_2 - t_1;q)\, \frac{k_{x_1 x_0}(q)}{k_{x_0}^{\text{out}}(q)}\, \rho_{x_0}(t_1;q)\, Y_{x_0}(t_1)\, p_{x_0}(t_0).
$$

The initial distribution $p_{x_0}(t_0)$ is usually chosen as the steady-state distribution, although this choice becomes irrelevant in the long time limit in which we are interested.

We numerically sample eq. (6.91) by introducing and evolving a population of N clones, where N is sufficiently large. At any given time t, each clone α is characterized by its state x^α and the time τ^α of its next jump. At the initial time t_0, the state x^α of each clone α is independently drawn from the initial probability distribution $p_x(t_0)$. We then draw, for each clone, the waiting time Δt_α to its next jump from the probability distribution $\rho_{x^\alpha}(\Delta t;q)$, and we set the time τ^α of the next jump to $t_0 + \Delta t_\alpha$. We then iterate the following procedure:

1. Identify the clone α^* with the smallest value of τ^α.
2. Advance the time t to τ^{α^*}.

3. Update the state x^{α^*} of the clone α^* to x, where x is drawn from the probability distribution $p_{x|x^{\alpha^*}} = k_{xx^{\alpha^*}}(q)/k_{x^{\alpha^*}}^{\text{out}}(q)$.

4. Set the time τ^{α^*} of the next jump of clone α^* to $t + \Delta t$, where Δt is drawn from the probability distribution $\rho_x(\Delta t; q)$.

5. Evaluate the weight $Y_x(\Delta t) = \exp\left[\left(k_x^{\text{out}}(q) - k_x^{\text{out}}\right)\Delta t\right]$.

6. Copy the clone into approximately $Y_x(\Delta t)$ other clones. More precisely:
   i. Evaluate $y = \lfloor Y_x(\Delta t) + \varepsilon \rfloor$, where ε is a uniformly distributed random number between 0 and 1 and $\lfloor z \rfloor$ is the largest integer not exceeding z.
   ii. If $y = 0$, remove clone α^* from the population. Otherwise, if $y > 1$, add $y - 1$ copies of clone α^* to the population. If $y = 1$, go to step 8.

7. Resize the population to N. More precisely: If $y = 0$, pick up a random clone and add a copy of it to the population. Otherwise, if $y > 1$, pick up at random (without replacement) N clones among the $N + y - 1$ available to form the new population.

8. Store the rescaling factor $X = N/(N + y - 1)$.

As the name suggests, the rescaling factor X is related to the fraction of clones other than α^* that are modified at each step in order to maintain the clone population size equal to N. In other words, if we were simply eliminating clones (when $y < 1$) and adding new clones (when $y \geq 1$) without replacing random clones, the population size would grow by a factor of X^{-1} at each jump. Therefore, to the leading order, the generating function grows with the inverse product of the rescaling factors

$$\phi_x^{(j)}(q, t) \asymp (X_1 X_2 \cdots X_n)^{-1}, \tag{6.92}$$

where the X_k are the factors X associated with each jump. It follows that the scaled cumulant generating function can be expressed as

$$\psi(q) = -\lim_{t \to \infty} \frac{1}{\mathcal{T}} \ln\left(X_1 X_2 \cdots X_n\right). \tag{6.93}$$

Scaled cumulant generating functions of static observables can also be estimated using cloning. We consider the generating function for an observable $\mathcal{A}(x) = \int_{t_0}^t dt'\, a_{x(t')}$, whose intensive version is $a = \mathcal{A}/\mathcal{T}$. As for dynamic observables, the strategy is to express the scaled cumulant generating function as an average over trajectory probabilities:

$$\phi_x^{(a)}(q; t) = \int \mathcal{D}x \, \mathcal{P}_x \, \delta_{x,x_n}^{\text{K}} \, e^{q\mathcal{A}(x)}$$

$$= \int \mathcal{D}x \, e^{-(k_x^{\text{out}} - a_x)(t - t_{n-1})} k_{xx_{n-1}} \cdots \tag{6.94}$$

$$\cdots \times k_{x_1 x_0} e^{-(k_{x_0}^{\text{out}} - a_{x_0})(t_1 - t_0)} p_{x_0}(t_0).$$

In this case, the tilted and the original jump rates are equal, whereas dwells of duration Δt in state x yield contributions $e^{q a_x \Delta t}$ to the generating function. Accordingly, we define the weights by

$$Y_x(\Delta t) = e^{q a_x \, \Delta t}. \tag{6.95}$$

By fixing q in the cloning procedure, we condition the system to exhibit exceptional values of a given observable, e.g., a current. In some cases, we are interested in averages of other observables in this exceptional ensemble. We might be tempted to evaluate them by averaging over the clone population that we obtain as time goes on. However, since the dynamics does not preserve normalization, the distribution of clones at a given time cannot be interpreted as the desired conditioned stationary distribution.

To obtain averages of observables at intermediate times t, we must therefore use a different method. We describe it for the example of static observables. The cloning algorithm is implemented as previously explained, but when the first jump time t_k after t is reached, the value a_{x^α} of the observable is attached to each clone α. This value is copied whenever the clone or its descendant is copied. Then we evaluate averages of a_{x^α} over the surviving clones at the end of the process. This may yield a rather noisy result, since the surviving clones may have only a few ancestors at time t. It is therefore preferable to evaluate integrated quantities, like $\left\langle \int_{t_0}^{t_f} dt \, a_{x(t)} \right\rangle / \mathcal{T}$, which are less noisy. These observables can be similarly obtained by attaching the accumulated value of the integral to each copy of a clone.

The scaled cumulant generating function $\psi^{(a)}(q)$ of a given observable a can be evaluated with the basic cloning algorithm. There is however a computationally more efficient procedure, called **thermodynamic integration**, based on the algorithm we just described. To introduce thermodynamic integration, we first define the **tilted average**

$$\langle a \rangle_q = \frac{d\psi^{(a)}(q)}{dq} \asymp \frac{\left\langle a \, e^{-q\mathcal{T} a} \right\rangle}{\left\langle e^{-q\mathcal{T} a} \right\rangle}. \tag{6.96}$$

Numerically, the tilted average is estimated by the mean of $a = \mathcal{A}/\mathcal{T}$ over the population of clones with a suitably large value of $\mathcal{T}$. From the tilted average, we can estimate $\psi(q)$ by

$$\psi(q) = \int_0^q dq' \, \langle a \rangle_{q'}. \tag{6.97}$$

The advantage of this procedure is that the integration smooths out the noise.

This method has been applied to study the large deviations in the asymmetric simple exclusion process with periodic boundary conditions. In this process, N particles are placed on a one-dimensional lattice of L sites, with $N < L$ and no more than one particle per site. Particles can jump to the next site to the right with rate k_R and to the left with rate k_L. Jumps can only take place if the target site is empty. Each jump contributes $+1$ to the current if it is to the right and -1 if it is to the left. The evaluation of the large deviation function of the current by the tilting method rapidly becomes unfeasible as L grows. In fact, tilting requires the evaluation of the leading eigenvalue of a matrix of size $\mathcal{N} \times \mathcal{N}$, where $\mathcal{N} = L!/((L-N)!N!)$ is the number of states. This number grows exponentially with L at constant N/L. Therefore, the cloning method is more advantageous for large values of L. In practice, by running calculations limited to a few hours on a desktop computer, one can reach $L \approx 20$ using tilting and $L \approx 400$ using cloning (fig. 6.6). Comparing the results for different values of L, we find small but significant differences due to the size dependence of the rate function.

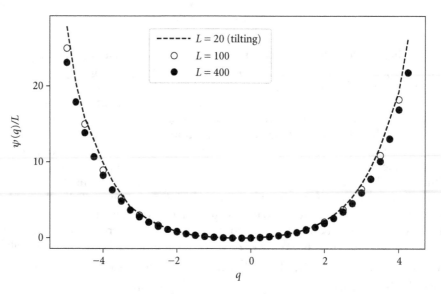

Figure 6.6. Scaled cumulant generating function $\psi(q)/L$ for the current in an asymmetric simple exclusion process with periodic boundary conditions. The jump rates are $k_R = 1.2$ and $k_L = 0.6$. Dashed line: Result of an evaluation by the tilting method for a system with $L = 20$ cases and $N = 10$ particles. Open circles: Result of a cloning simulation for $L = 100$ and $N = 50$. Filled circles: Result of a cloning simulation for $L = 400$ and $N = 200$. The function is symmetric around $q^* = \ln(k_L = k_R)/2 \approx -0.35$.

6.9   Levels of large deviations (*)

Stochastic processes can be characterized by quantities other than the static and dynamic observables introduced in section 6.2. A function of the instantaneous state of the system, such as the system energy $\epsilon_{x(t)}$ in stochastic thermodynamics, is one such example. More generally, one may consider observables that are functionals of long chunks of trajectories. To treat this broader set of observables, it is useful to introduce a hierarchy of **levels** of large deviations. Higher levels in this hierarchy correspond to a more detailed statistical description of trajectories:

- Level 1 includes the large deviation principles for functions of the instantaneous state of the system; for example, the energy ϵ_x.
- Level 2 describes the static observables introduced in section 6.2. The central observable of level 2 large deviations is the **empirical vector** $f = (f_x)$, where f_x is the empirical frequency defined by eq. (6.16).
- Level 3 encompasses the large deviation principles associated with observables defined on sequences of arbitrary length of the form $x = (x_t, x_{t+1}, \ldots, x_{t+k})$, made of successive states occupied by the system.

The reason for introducing a hierarchy of finer and finer large deviations lies in a fundamental result of large deviation theory, called the **contraction principle**. The contraction principle allows us to express the rate function $I(y)$ of a variable y, which is a continuous function of another variable, $y = f(x)$. Since this principle involves changing the independent variable of the rate function, we denote in this section the rate function, e.g., of the variable x by $I^{(x)}(x)$ to avoid misunderstandings. Now, if x satisfies a large deviation principle and the rate function $I^{(x)}(x)$ is known, the contraction principle states that the rate function for y is given by

$$I^{(y)}(y) = \inf_{x:f(x)=y} I^{(x)}(x). \tag{6.98}$$

The gist of this relation is that the asymptotic probability of a given value of y is dominated by the least improbable event x such that $f(x) = y$. Importantly, the contraction principle permits us to express lower-level rate functions in terms of higher-level ones.

For Markov processes, it is convenient to define an intermediate level between levels 2 and 3—level 2.5. A level 2.5 large deviation principle describes the joint asymptotic distribution of the empirical vector $f = (f_x)$ and the **empirical jump frequencies** $v = (v_{xx'})$. The empirical jump frequencies are the intensive counterparts of the empirical number of jumps introduced in eq. (6.15):

$$v_{xx'}(\boldsymbol{x}) = \frac{n_{xx'}(\boldsymbol{x})}{\mathcal{T}}. \tag{6.99}$$

Level 2.5 large deviations are of paramount relevance for Markov processes. The main reason is that, for master equations, this is the only level for which a relatively simple analytical expression for the rate functions is known. Moreover, rate functions for empirical currents can be obtained from the level 2.5 large deviation by means of the contraction principle. In particular, this procedure leads to expression for the rate function of important observables, such as the empirical entropy production rate.

We now derive the expression for the level 2.5 rate function. We consider a system described by a master equation with time-independent rates evolving at a steady state in a time interval $[t_0, t_f]$ of duration $\mathcal{T} = t_f - t_0$. Given a trajectory $\boldsymbol{x}$, we evaluate the empirical vector f_x and the empirical jump frequencies $v_{xx'}$. We make the additional assumption of **empirical stationarity**:

$$\sum_x v_{xx'} = \sum_x v_{x'x}. \tag{6.100}$$

Equation (6.100) states that the number of times that each state is reached is equal to the number of times that it is left. Any long trajectory of a master equation satisfies eq. (6.100) apart from a boundary term, stemming from the fact that the initial and final states can be different. This means that eq. (6.100) is a rather mild constraint for large $\mathcal{T}$. As $\mathcal{T}$ grows, f_x and $v_{xx'}$ converge to p_x^{st} and to $k_{xx'}p_{x'}^{st}$, respectively, for all x, x'. To obtain their probability densities, we introduce an auxiliary, fictitious process for which typical values of the empirical vector and number of jumps are equal to the observed ones. The jump rates of the auxiliary process are the **empirical jump rates** $\kappa_{xx'}$, defined by the relation

$$v_{xx'} = \kappa_{xx'}f_{x'}. \tag{6.101}$$

The escape rates in the auxiliary process are $\kappa_x^{\text{out}} = \sum_{x'} \kappa_{x'x}$. The probability $\mathcal{P}_x$ of a trajectory x in the original process is given by eq. (2.91). In the auxiliary process, the probability of the same trajectory is

$$
\mathcal{P}_x^* \, dt_n \cdots dt_1 = e^{-\kappa_{x_n}^{\text{out}}(t-t_n)} \kappa_{x_n x_{n-1}} \, dt_n \, e^{-\kappa_{x_{n-1}}^{\text{out}}(t_n - t_{n-1})} \cdots
$$
$$
\cdots \times \kappa_{x_1 x_0} \, dt_1 \, e^{-\kappa_{x_0}^{\text{out}}(t_1 - t_0)} p_{x_0}(t_0).
$$
(6.102)

Our goal is to compute the joint probability distribution $p(f, \nu; t)$. It is easier to first evaluate the probability densities $p(f, \kappa; t)$ and $p^*(f, \kappa; t)$ in the original and auxiliary processes, respectively, obtained by changing variables using eq. (6.99) and by summing $\mathcal{P}_x$ and $\mathcal{P}_x^*$ over all trajectories satisfying the condition (6.101). Since the auxiliary process is constructed to have f and ν as typical values for the corresponding observables, $p^*(f, \kappa; t)$ does not decrease exponentially with t for large t. Moreover, for all trajectories satisfying the condition (6.101), we can express the ratio of the two probability densities $\mathcal{P}_x$ and $\mathcal{P}_x^*$ in terms of f and κ only:

$$
\frac{\mathcal{P}_x}{\mathcal{P}_x^*} = \left(\prod_{i=1}^n \frac{k_{x_i x_{i-1}}}{\kappa_{x_i x_{i-1}}} \right) \left(\prod_{i=0}^n e^{-(k_{x_i}^{\text{out}} - \kappa_{x_i}^{\text{out}})(t_{i+1} - t_i)} \right)
$$
$$
= \exp\left[-\mathcal{T} \sum_{xx'} \left(\kappa_{xx'} \ln \frac{\kappa_{xx'}}{k_{xx'}} - \kappa_{xx'} + k_{xx'} \right) f_{x'} \right],
$$
(6.103)

where in the first line $t_{n+1} = t_f$. As a consequence, we obtain the large deviation principle

$$
p(f, \kappa; t) = p^*(f, \kappa; t) \left\langle \frac{\mathcal{P}_x}{\mathcal{P}_x^*} \middle| \text{conditions (6.100), (6.101)} \right\rangle \asymp \frac{\mathcal{P}_x}{\mathcal{P}_x^*}
$$
$$
\asymp e^{-\mathcal{T} I^{(f,\kappa)}(f,\kappa)},
$$
(6.104)

where the rate function is given by

$$
I^{(f,\kappa)}(f, \kappa) = \sum_{xx'} \left[\kappa_{xx'} \ln \frac{\kappa_{xx'}}{k_{xx'}} - \kappa_{xx'} + k_{xx'} \right] f_{x'}.
$$
(6.105)

This expression is the central result of this section. It can be used, for example, to formally express the rate function for the empirical vector f. Since the variables f_x and $\kappa_{xx'}$ satisfy a large deviation principle, f satisfies a large deviation principle by itself:

$$
P(f; \mathcal{T}) \asymp e^{-\mathcal{T} I^{(f)}(f)}.
$$
(6.106)

To obtain it, we apply the contraction principle:

$$
I^{(f)}(f) = \inf_\kappa I^{(f,\kappa)}(f, \kappa),
$$
(6.107)

where the infimum is taken among all $\kappa_{xx'}$s satisfying $\kappa_{xx'} \geq 0$ and $\sum_{x'} \kappa_{xx'} f_{x'} = \sum_{x'} \kappa_{x'x} f_x$ for all x, x'. Thus the rate function for the empirical vector f is the solution of a minimization problem, although its explicit expression can be hard to obtain.

We similarly express the rate function $I^{(f,j)}(f,j)$ by minimizing $I^{(f,\kappa)}(f,\kappa)$ with respect to κ, with the constraints $\kappa_{xx'} f_{x'} - \kappa_{x'x} f_x = j_{xx'}$. We impose the constraints via the Lagrange multipliers $\lambda_{xx'} = -\lambda_{x'x}$, and we minimize with respect to κ the quantity

$$\Phi(\kappa, k, \lambda) = \sum_{xx'} \left[I(f, \kappa) + \frac{1}{4} \lambda_{xx'} \left(\kappa_{xx'} f_{x'} - \kappa_{x'x} f_x \right) \right], \qquad (6.108)$$

where the factor $\frac{1}{4}$ is introduced for convenience. We obtain, with some algebra,

$$I^{(f,j)}(f,j) = I^{(f,\kappa)}(f,\kappa^*), \qquad (6.109)$$

with $\kappa_{xx'}^* = k_{xx'} e^{\lambda_{xx'}/2}$. The Lagrange multipliers are equal to

$$\lambda_{xx'} = 2 \ln \left[\frac{1}{2 k_{xx'} f_{x'}} \left(j_{xx'} + \sqrt{j_{xx'}^2 + \alpha_{xx'}^2} \right) \right], \qquad (6.110)$$

in which we define

$$\alpha_{xx'} = \sqrt{k_{xx'} k_{x'x} f_{x'} f_x} = \sqrt{\kappa_{xx'} \kappa_{x'x} f_{x'} f_x} = \alpha_{x'x}. \qquad (6.111)$$

We then obtain

$$I^{(f,j)}(f,j) = \frac{1}{4} \sum_{xx'} \lambda_{xx'} j_{xx'} - \frac{1}{2} \sum_{xx'} [t_{xx'}^*(f) - t_{xx'}(f)], \qquad (6.112)$$

where the ts are traffic observables given by

$$\begin{aligned} t_{xx'}(f) &= k_{xx'} f_{x'} + k_{x'x} f_x; \\ t_{xx'}^*(f) &= \kappa_{xx'}^* f_{x'} + \kappa_{x'x}^* f_x; \end{aligned} \qquad (6.113)$$

see also eq. (6.19). Also this result requires a stationarity condition, in this case for the currents:

$$\sum_{x'} j_{xx'} = 0, \quad \forall x. \qquad (6.114)$$

If the stationarity condition does not hold, one has $I^{(f,j)}(f,j) = +\infty$.

We finally obtain the rate function $I(j)$ of a current j by applying the contraction principle:

$$I^{(j)}(j) = \inf_f I^{(f,j)}(f,j). \qquad (6.115)$$

By choosing the steady-state distribution p^{st} for f, we obtain the bound

$$I^{(j)}(j) \leq I^{(f,j)}(p^{st}, j). \qquad (6.116)$$

This approach has been applied to derive general inequalities connecting the average value and the variance of steady-state currents with the entropy production rate, discussed in section 8.2.

6.10 Further reading

Theoretical physicists informally developed large deviation theory long before mathematicians established a unified theory, in particular following the work of Donsker and Varadhan. Indeed, some fundamental results by Boltzmann [24], and the theory of fluctuations established by Einstein [45], can be considered as results in large deviation theory. Touchette [168] reviews the large-deviation approach to statistical mechanics. Ellis [47], Den Hollander [43], and Dembo and Zeitouni [42] are useful textbooks.

Touchette [168, app. C-2] and Chetrite and Touchette [30] discuss the tilting method. The Arnoldi-Lanczos method is described, e.g., by Meyer [112, p. 651]. Koza [91] originally proposed the idea of using the implicit function theorem to compute scaled cumulants. Giardinà et al. [62] introduced cloning. The method is thoroughly discussed by Lecomte and Tailleur [99], including thermodynamic integration. Tizón-Escamilla et al. [167] discuss extension of the tilting method to Langevin processes.

The discussion in section 6.4 follows Andrieux and Gaspard [2, 3], and section 6.7 is inspired by Lacoste et al. [94]. For the definition of level 3 large deviation, see Ellis [47, §I.6]. Section 6.9 follows the derivation by Maes and Netočný [107]. Bertini et al. [17] show how one can apply level 2.5 large deviation theory to obtain the Gallavotti-Cohen symmetries for the rate functions of the empirical currents.

6.11 Exercises

6.1   Use the expression of the rate function for the binomial distribution (6.4) and the inverse Gärtner-Ellis theorem to compute the scaled cumulant generating function of the binomial distribution.

6.2   Consider a system for which the total entropy production is distributed according to a Gaussian and satisfies the integral fluctuation theorem. Compute explicitly the scaled cumulant generating function and verify that it satisfies the Gallavotti-Cohen symmetry.

6.3   Show that in a system described by a master equation, jumps between states that do not belong to any cycle do not contribute to the entropy production in the steady state.

6.4   A particle is dragged on a discrete ring with $N = 3$ states; see section 4.3. Consider a current equal to the number of jumps in the direction of the driving minus the number of jumps in the opposite direction. Compute the first two scaled cumulants of such a current as a function of the driving f using tilting. Check the form of the Gallavotti-Cohen symmetry for the cumulant generating function.

6.5   Consider the Mandal-Jarzynski machine described in section 5.5, but assume that the tape advances by one case at random time points, as in exercise 5.3.

Thus the system is described by a master equation with suitable rates, as discussed in section 5.8. Evaluate numerically the scaled cumulant generating function of the current between state 2 and state 1 by a cloning algorithm.

6.6   Compute the large deviation behavior of the total entropy production in the Michaelis-Menten reaction scheme introduced in section 6.6. Verify that the total entropy production satisfies the Gallavotti-Cohen symmetry.

CHAPTER 7

Experimental Applications

In this chapter, we discuss experiments that have tested predictions of stochastic thermodynamics or exploited them to measure quantities of physical and biological interests in mesoscopic systems. Perhaps the most celebrated experimental outcome of stochastic thermodynamics is the possibility of estimating equilibrium free-energy differences of biomolecules by means of nonequilibrium measurements and fluctuation theorems. Given the importance of this achievement, we explain in detail the idea underlying these experiments and different practical ways of estimating the free energy. A comprehensive review of the experimental aspects of stochastic thermodynamics falls beyond the scope of this book (and of our expertise). For this reason, this chapter is relatively synthetic and focuses on a limited set of experiments that we consider to be good representatives of the state of the art in stochastic thermodynamics.

7.1 The hairpin as a paradigm

A far-reaching consequence of stochastic thermodynamics is that equilibrium properties of mesoscopic systems can be measured by means of nonequilibrium experiments. This idea paved the way for novel methods to evaluate, for example, free-energy differences between different conformations of biological macromolecules.

A basic scheme of these experiments is sketched in fig. 7.1. A macromolecule, such as a stretch of DNA, RNA, or a protein, is anchored at one tip to a rigid, unmovable surface. A microscopic bead, made up of dielectric material, is connected to the other tip. The position of this bead, or the force applied to it, is controlled by means of an optical laser trap, also called **optical tweezers**.

In a classic "pulling" experiment, the initial position of the bead is relatively close to the surface. Under these conditions, polymers such as RNA tend to adopt a conformation called a **hairpin**, as sketched in fig. 7.1. This conformation is stabilized by hydrogen bonds between different nucleotides in the RNA chain. The bead is then pulled farther from the fixed surface, so that the hairpin is untied and the macromolecule conformation becomes elongated.

Ideally, in the limit of an infinitely slow pulling experiment, the thermodynamic transformation becomes reversible, and thus the work W made by the optical tweezers approaches the free-energy difference ΔF between the final and the initial configurations. This free-energy difference should be split into the free-energy change of the device and that of the system, i.e., the folding energy of the hairpin. Usually, it is

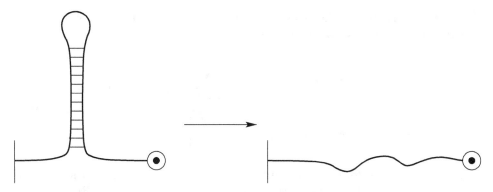

Figure 7.1. Sketch of a macromolecule pulling experiment. Narrow lines denote mono-mers connected by hydrogen bonds.

relatively easy to estimate the former contribution by independent means. This implies that experimental measurements of W can be used to infer the folding free energy of the hairpin.

The problem with this program is that adiabatic measurements on such mesoscopic systems are often not feasible. Free-energy differences of conformational changes of macromolecules are typically in the range of a few to a few hundred k_BT at physio-logical temperature, making their dynamics highly stochastic. The conclusion is that the folding/unfolding process is often not easy to manipulate in a quasi-static way. The counterintuitive fact that the Jarzynski equality (4.52) and the Crooks relation (4.53) relate equilibrium free energies with work distribution far from equilibrium have proved to be crucial to overcome these limitations of quasi-static measurement protocols.

7.2 A simpler model

Formulating a reasonably realistic model of the dynamics of a hairpin pulled by optical tweezers requires a number of concepts from polymer physics. Modeling the manipulation device is also not straightforward, as devices controlling the applied force or the position of the bead require slightly different descriptions. To avoid these digres-sions, we base our discussion on an idealized model. The model has a discrete number of states $x = 1, \ldots, n$, that represent the number of bonds formed by a hairpin. For $x = n$ the hairpin is completely formed, whereas for $x = 0$ it is unfolded. In the absence of a manipulation, the energy of the hairpin is simply proportional to the number of bonds,

$$\epsilon_x = -\epsilon x, \tag{7.1}$$

where the parameter ϵ represents the absolute value of the energy of each bond. We assume that the manipulation protocol λ "tilts" the energy landscape, so that the energy in the presence of manipulation reads

$$\epsilon_x = (\lambda - \epsilon)x. \tag{7.2}$$

A positive λ represents a state in which the manipulation is pulling the bead, therefore making the unfolded state $x = 0$ more favored. We consider a manipulation between an initial value of the manipulation parameter $\lambda = \lambda_0$ and a final value $\lambda = \lambda_f$. The free energy as a function of λ reads

$$
\begin{aligned}
F(\lambda) = -k_B T \ln Z(\lambda) &= -k_B T \ln \left[\sum_{x=0}^{n} \exp \left[\frac{(\epsilon - \lambda)x}{k_B T} \right] \right] \\
&= -k_B T \ln \left[\frac{1 - e^{(\epsilon - \lambda)(n+1)/k_B T}}{1 - e^{(\epsilon - \lambda)/k_B T}} \right].
\end{aligned}
\tag{7.3}
$$

Examples of trajectories of the hairpin model are shown in fig. 7.2a for a linear manipulation protocol. Parameter values are comparable with those of RNA hairpins commonly used in experiments. It is useful to compare "forward" trajectories with "backward" trajectories, i.e., with trajectories obtained by the backward manipulation protocol. This comparison shows that the folding/unfolding of the hairpin is characterized by **hysteresis**. In this context, *hysteresis* means that typical forward trajectories are markedly different than typical backward trajectories. Hysteresis is commonly observed in nonequilibrium experiments, as shown in fig. 7.2b.

7.3 Equilibrium free energies from nonequilibrium manipulations

In practice, the Jarzynski equality (4.52) is difficult to apply to systems very far from equilibrium due to the difficulty in sampling large deviations; see section 4.8. On the other hand, detailed fluctuation relations, such as the Crooks relation (4.53), provide more stringent predictions on the distribution of work in nonequilibrium protocols. It is important to establish the best way of exploiting these predictions and, consequently, how far we can push a system out of equilibrium if we want to reliably measure free-energy differences.

We consider an experiment in which a hairpin is brought from an initial folded state to a final elongated state by a protocol λ. The free-energy difference between the final and the initial equilibrium states is ΔF. In the experiment, we estimate the probability distribution of work $p_F(w) = p(w; \lambda)$, for example, by repeating the experiment many times and evaluating the work performed by the manipulation device in each realization. This may be achieved, e.g., by monitoring the applied force and the displacement of the tweezers as a function of time. We also assume to be able to perform the experiment backward, i.e., from the final state to the initial state via the backward protocol $\widehat{\lambda}$, and to estimate the distribution $p_B(w) = p(w; \widehat{\lambda})$ of the associated work.

We consider five methods to estimate the free-energy difference.

Average work. The simplest way to estimate the free energy is to assume that the transformation occurs close to equilibrium and therefore

$$
\Delta F \approx W = \langle w \rangle_{F,emp},
\tag{7.4}
$$

where we define the empirical average over $\mathcal{N}$ independent realizations of the experiment with the same forward protocol λ:

$$
\langle w \rangle_{F,emp} = \frac{1}{\mathcal{N}} \sum_{k=1}^{\mathcal{N}} w(\boldsymbol{x}_k).
\tag{7.5}
$$

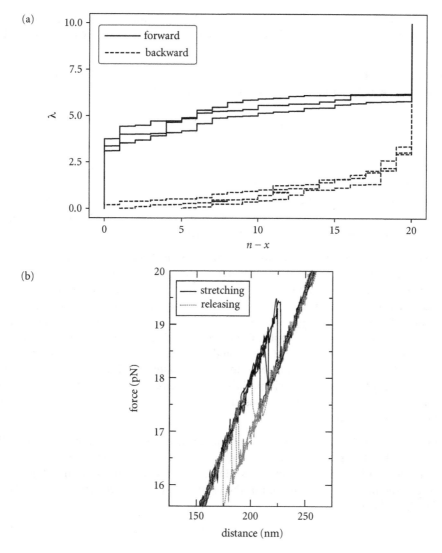

Figure 7.2. (a) Trajectories of the hairpin model. The force λ (in units of k_BT) is plotted against the number of open residues $n - x$. Parameters: $n = 20$, $\epsilon = 3\,k_BT$. The manipulation protocol is linear, $\lambda(t) = \alpha t\,k_BT$, bringing the system from the state $\lambda_0 = 0$ at initial time $t_0 = 0$ to $\lambda_f = 20\,k_BT$ at final time $t_f = 20/\alpha$. In this case, manipulation speed is $\alpha = 1$. Continuous lines are forward trajectories, dashed lines are backward trajectories. Forward trajectories are initialized at $x(t = 0) = n = 20$, whereas backward trajectories are initialized at $x(t = t_f) = 0$. (b) Experimental force-extension curves of an RNA hairpin experiment. Darker trajectories correspond to the forward (extension) protocol, while lighter ones correspond to the backward (refolding) one. (Adapted from Mossa et al. [113]. © SISSA Medialab Srl and IOP Publishing. Reproduced by permission of IOP Publishing. All rights reserved.)

Here x_k is the trajectory in the kth experimental realization. We expect this method to provide good results for slow transformations but poor approximations farther from equilibrium.

Jarzynski equality (JE). The free-energy difference can be directly estimated from the Jarzynski equality. In particular, the Jarzynski equality implies

$$\Delta F \approx -k_B T \ln \left\langle e^{-w/k_B T} \right\rangle_{F,emp}. \tag{7.6}$$

An advantage of this method is that it does not require performing experiments with the backward protocol, but only many replicates of the forward protocol. However, as we discussed in section 4.8, when the average dissipated work is large, i.e., when the protocol drives the system too far from equilibrium, the empirical average of $e^{-w/k_B T}$ can be very different from the theoretical one.

Crooks (crossing of forward/backward work distributions). This method requires us to evaluate both the work distribution $p_F(w)$ with the forward protocol and the corresponding $p_B(w)$ with the backward protocol, and therefore to perform both forward and backward experiments. A direct consequence of the Crooks relation (4.53) is that the distributions $p_F(w)$ and $p_B(-w)$ cross at the value $w = \Delta F$. This observation provides a simple graphical method to estimate the free energy (fig. 7.3b). However, this method necessitates enough realizations to resolve both distributions in the region where they intersect. Because of hysteresis, if the experiment is performed very far from equilibrium, the overlap between the two distributions occurs far on the tails, so that also in this case the required number of realizations may become prohibitively large.

Crooks (linear fit). A drawback of the estimate based on the crossing of the work distributions is that it exploits only the realizations of the experiments for which $w \approx \Delta F$ in the forward protocol and $w \approx -\Delta F$ in the backward protocol. To make use of the rest of the information gathered in the experiment, we rewrite the Crooks relation (4.53) in the form

$$\ln p_F(w) - \ln p_B(-w) = \frac{w - \Delta F}{k_B T}. \tag{7.7}$$

This means that the function $f(w) = \ln p_F(w) - \ln p_B(-w)$, as estimated from experiments, should look like a straight line with slope $(k_B T)^{-1}$, which crosses the line $y = 0$ for $w = \Delta F$. In practice, this relation is satisfied only in a limited range of values of w due to statistical limitations. The robustness of the result depends therefore on the quality of the fit. In any case, this method extends the previous one and, since it uses more of the data, is expected to be more robust.

Bennett-Crooks acceptance ratio (BC). The logic of the linear fit method can be generalized by introducing an arbitrary function of the work $f(w, z)$ that depends on an additional parameter z. We now consider a function

$$g(z) = \ln \left\langle f(-w, z) \right\rangle_B - \ln \left\langle f(w, z) \, e^{-w/k_B T} \right\rangle_F. \tag{7.8}$$

Using the Crooks relation to compute the averages in eq. (7.8) leads to the result $g(z) = \Delta F/k_B T$, independent of the choice of the function $f(w, z)$. This freedom

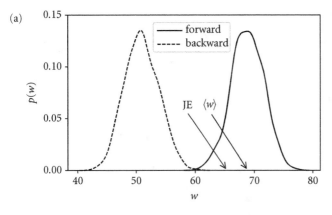

(a)

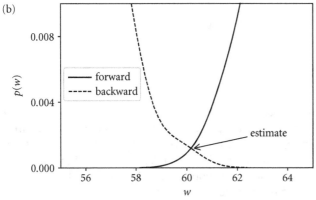

(b)

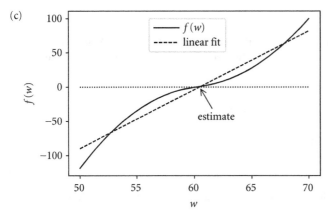

(c)

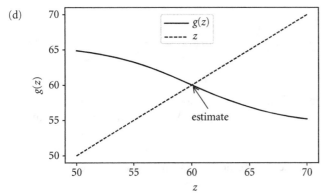

(d)

Figure 7.3. Methods for evaluating a free-energy difference ΔF. (a) The estimated probability distribution of the work w for the forward protocol and the corresponding distribution of $-w$ for the backward protocol. Work is plotted in units of $k_B T$. The mean $\langle w \rangle$ and the Jarzynski estimator JE are shown. (b) Close-up of the distributions $p_F(w)$ and $p_B(-w)$ near their crossing. (c) Linear fit to the function $f(w) = \ln p_F(w) - \ln p_B(-w) - w/k_B T$. The estimate corresponds to the intercept of this line with the horizontal line at 0. (d) The Bennett-Crooks estimator $g(z)$, defined in eq. (7.8), with $f(x, z)$ given in eq. (7.9), is plotted as a function of the parameter z. The best estimate is obtained at the crossing of $g(z)$ with z. The plots are based on 1000 replicates each of the forward and backward protocols on the model introduced in section 7.2, with $\alpha = 0.04$. The actual free-energy difference is $\Delta F = 60\,k_B T$. A comparison of results obtained by the different methods is shown in fig. 7.4.

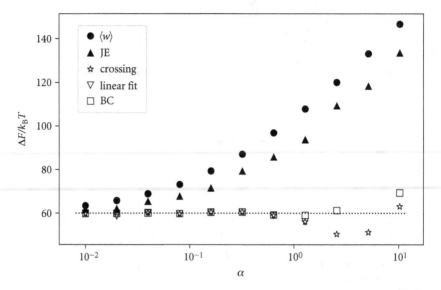

Figure 7.4. Estimates of the free-energy difference ΔF with different methods for the hairpin model as a function of the protocol speed α, which controls the distance from equilibrium. In all cases, the estimate is based on 1000 replicates of the forward protocol. In addition, methods based on the Crooks relation use 1000 replicates of the backward protocol. For large α, the linear fit method cannot be applied because the empirical distributions $p_F(w)$ and $p_B(-w)$ do not sufficiently overlap.

permits us to pick a function $f(w, z)$ and a range of z that minimizes the expected error on the estimate of ΔF. Bennett showed that the best estimate is obtained by choosing

$$f(w) = \frac{1}{e^{-w/k_B T} + e^{-z/k_B T}}, \tag{7.9}$$

where z is chosen such that

$$g(z) = z. \tag{7.10}$$

A derivation of this result is sketched in appendix A.11. We therefore obtain

$$\Delta F \approx k_B T \left[\ln \left\langle \frac{e^{z/k_B T}}{1 + e^{(w+z)/k_B T}} \right\rangle_{\text{B,emp}} - \ln \left\langle \frac{1}{1 + e^{(w-z)/k_B T}} \right\rangle_{\text{F,emp}} \right], \tag{7.11}$$

with $z \approx \Delta F$. If the numbers of forward ($\mathcal{N}_F$) and backward ($\mathcal{N}_B$) samples are not equal, the best choice of z becomes $\Delta F - k_B T \ln(\mathcal{N}_F/\mathcal{N}_B)$, and eq. (7.10) must be changed accordingly.

A numerical comparison of these five methods of estimating a free-energy difference is shown in fig. 7.4.

7.4 Maxwell demons

In chapter 5, we discussed apparent violations of the second law of thermodynamics that occur when manipulating a mesoscopic system using information about its state. In stochastic thermodynamics, information-based manipulation can be studied within concrete models, making the correspondence between free energy and information more explicit and resolving apparently paradoxical scenarios. Placing information manipulation in the context of concrete models has the additional merit of laying the groundwork for experimental tests. In particular, a crucial test of thermodynamics of information is to extract work from a thermal reservoir via feedback control.

We discuss an experiment in which a mesoscopic dimeric particle is pinned on a surface but is allowed to rotate freely on the surface plane. The particle is subject to an electric field, which applies a constant torque f in the negative angle direction, in addition to a potential $\epsilon(\phi, \lambda)$, where λ is the manipulation parameter. Moreover, the particle undergoes rotational diffusion due to the interactions with a heat reservoir. The potential is periodic, $\epsilon(\phi + \pi) = \epsilon(\phi)$. The manipulation parameter can assume two discrete values, $\lambda \in \{0, 1\}$, which change the phase of the periodic potential, $\epsilon(\phi, \lambda) = \epsilon(\phi + \lambda\pi/2)$.

The periodic potential is manipulated according to the result of measurements. Initially, the control parameter is set to λ. At a given time, the position of the particle is measured. If it is found in the shaded region shown in fig. 7.5a, after a delay τ the parameter λ is switched to $1 - \lambda$, otherwise it is left unaltered. The shaded region is chosen so that the particle, after switching, is likely to be found in a well at a higher energy than that at the beginning of the interval. The whole procedure is periodically repeated at intervals of time equal to 44 ms.

For small values of τ, the device operates as a Maxwell demon, i.e., it uses the information provided by the measurement to extract heat from the heat reservoir and perform work, on average, against the external torque f; see fig. 7.5b. The apparent violation of the second law of thermodynamics agrees with the Sagawa-Ueda theory presented in section 5.4, even though in this case the system is periodically manipulated, whereas in section 5.4 we considered a manipulation in a finite time interval where the system starts and ends at thermodynamic equilibrium. In fact, in the present case, the relaxation time of the system is small enough so that the system can relax to thermodynamic equilibrium in each manipulation, at least for small values of τ.

7.5 Landauer principle

According to the Landauer principle, erasing information in a microscopic or mesoscopic device requires an amount of work of at least $k_B T \ln 2$ per bit; see sections 5.1 and 5.5. The Landauer principle is crucial to reconcile the workings of the Maxwell demon and the Szilard engine with the second law of thermodynamics, a point made by Penrose and, independently, by Bennett. In recent years, it has been possible to test this principle, thanks to novel methods to closely monitor the position of Brownian particles in precisely controllable potentials.

To test the Landauer principle, one needs a Brownian particle in a time-dependent bistable potential. Experimentally, the potential can be generated by a highly focused laser beam that is rapidly moved between two locations. A different approach uses feedback loops to apply a force that depends on the particle position. In the latter approach, the spatial dependence of the force mimics a virtual potential $U(x, t)$ chosen by the

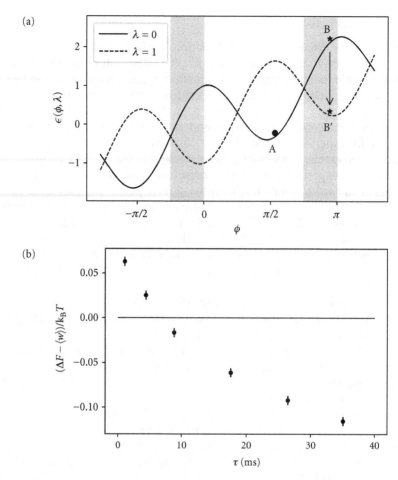

Figure 7.5. Maxwell demon experiment. (a) Scheme of the feedback manipulation protocol. The parameter λ is set to 0. If, upon measurement, the particle is found in the unshaded areas, as in A, the parameter remains unchanged. If it is found in the shaded areas, as near point B, the parameter switches to 1 and the particle is located near the local minimum in B′, which is at a higher level than A. In this way, the manipulation allows the particle to "climb the staircase." (b) Experimental results. The net work per measurement cycle $(\Delta F - \langle w \rangle)/k_B T$ is plotted against the time interval τ from the measurement to the control. The first two points correspond to the "Maxwell demon" regime. (Data from Toyabe et al. [169].)

experimenter. The two wells of the potential are thought of as the two states of 1 bit of memory. In order to act as a memory for a certain time, the potential barrier between the two minima should be so large that the probability that the particle switches location during this time, due to thermal fluctuations, is negligible. The particle is initially in one minimum (say, -1) or the other (say, $+1$), with a certain probability p_x, $x \in \{-1, 1\}$. The maximum Shannon entropy of $\ln 2$ (1 bit) is reached when $p_x = 1/2$.

To reset the memory, one manipulates the system so that, at the end of the manipulation, the particle is found in a reference minimum, say, $x = +1$, with probability close to 1, regardless of the initial state. A possible manipulation scheme is presented in fig. 7.6.

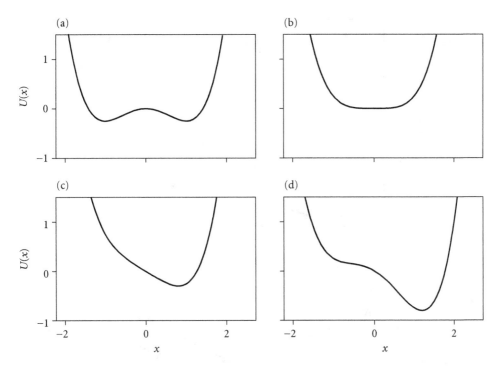

Figure 7.6. Scheme of the potential manipulation in a full erasure ($p = 1$) protocol. (a) Initial potential $U(x)$ (arbitrary units). (b) The central barrier is lowered. (c) A symmetry-breaking force is applied. (d) The central barrier is raised. At the final step, symmetry is restored and the initial form of the potential is recovered.

Starting from the initial bistable potential shown in (a), the central barrier is first lowered (b), leading to a potential with a single minimum. Then a force directed toward positive x is applied (c), displacing the minimum to the right. At this point the central barrier is raised again (d), and finally the symmetry-breaking force is removed (not shown) to recover the initial bistable potential. With this protocol, the probability p that the particle is found in the minimum at 1 at the end of the manipulation is very close to 1. We compare this case with a no-erasure protocol in which the symmetry-breaking force is not applied. This protocol randomly reshuffles the memory, so that the probability p of finding the particle at the end of the manipulation in the minimum at 1 is close to $1/2$. In this situation, the change in the system entropy vanishes and so does the minimum required dissipated work.

We now discuss more closely a specific experimental setup, where the manipulation is carried out via a feedback trap. The trap acquires an image of the particle diffusing in an aqueous solution and evaluates its position. Then voltages are applied across two sets of electrodes, creating electrical forces that move the particle. The forces mimic those that would originate by the desired potential $U(x, t)$ at the given time and at the measured position of the particle. The feedback loop is updated on a timescale that is much faster than the timescale of the potential dynamics, set by the local relaxation within the wells.

With a manipulation potential such as the one shown in fig. 7.6, we expect the Landauer bound $k_B T \ln 2 \approx 0.69 \, k_B T$ to be reached as the manipulation time τ goes to

(a)

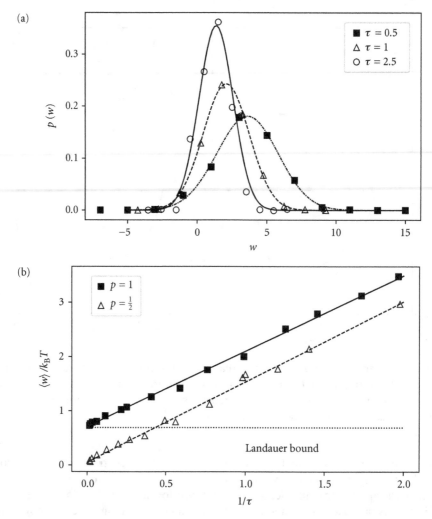

(b)

Figure 7.7. (a) Histogram of the work w measured in a full erasure protocol, for different values of the protocol duration τ (measured in units of τ_0). The curves are Gaussian fits. (b) Average work $\langle w \rangle / k_B T$ plotted against the inverse manipulation time $1/\tau$ (measured in units of τ_0). The Landauer bound $k_B T \ln 2$ is also shown. (Adapted from Jun et al. [84], with permission.)

infinity. Therefore, the average work $\langle w \rangle_\tau$ obtained in a manipulation of duration τ is fit to a law of the form

$$\langle w \rangle_\tau \approx \langle w \rangle_\infty + \frac{K}{\tau}. \tag{7.12}$$

The results of the experiment are shown in fig. 7.7. The fit (lines) is compatible with the Landauer bound for $p = 1$ (erasure) and with 0 for $p = 1/2$ (no erasure). One obtains $\langle w \rangle_\infty = 0.71\, k_B T$ and $K = 1.39\, \tau_0$ for $p = 1$, and $\langle w \rangle_\infty = 0.05\, k_B T$ and $K = 1.48\, \tau_0$ for $p = 1/2$, where $\tau_0 \approx 15\,\text{s}$ is a timescale corresponding to the time needed by the Brownian particle to freely diffuse a distance equal to the distance between the two minima and thus forget the initial state of the memory.

7.6 Further reading

Early experiments measuring equilibrium free energies via nonequilibrium protocols used the Jarzynski equality and were therefore necessarily performed close to equilibrium. In particular, Liphardt et al. [104, 103] measured in this way the folding free energy of RNA molecules that can be folded/unfolded in a quasi-static way. Collin et al. [31] showed that using the Crooks relations and the variants described in this chapter allows one to obtain reliable measurements in experiments much farther away from equilibrium. This observation greatly extended the range of macromolecules that can be studied using these techniques. Crooks [35] proposed to use in this context the Bennett-Crooks estimator [11].

Toyabe et al. [169] performed the Maxwell demon experiment. The Landauer principle was discussed by Penrose [126, ch. VI, §3] and Bennett [10]. Jun et al. [84] performed the experiment reported in section 7.5. Other similar experiments have been performed on colloidal particles (see, e.g., Bérut et al. [27]) and on technologically relevant systems, such as nanomagnetic memory bits (Hong et al. [76]). Koski et al. [89, 90] have carried out closely related experiments in single-electron boxes.

CHAPTER 8

Developments

The core theory of stochastic thermodynamics is nowadays rather well established. In recent years, there have been rapid developments in several directions, with a growing body of surprisingly general results. In this chapter, we discuss some of these developments. Progress in this field is growing by the day, and therefore we make no attempt at being exhaustive.

8.1 Stochastic efficiency

An original motivation for the development of thermodynamics has been to understand the efficiency of heat engines; see section 2.2. The concept of efficiency also plays an important role for the mesoscopic systems that are the subject of stochastic thermodynamics. In particular, the thermal Carnot bound also limits the maximum average work that can be extracted from a mesoscopic heat engine. The reason is that this bound is a direct consequence of the fundamental laws of thermodynamics, and we have seen that these laws hold for average observables in stochastic thermodynamics.

Throughout this book, we have analyzed mesoscopic engines that perform different kinds of energy conversion. Prominent examples are molecular motors that convert chemical energy into work. Information can also fuel a mesoscopic engine, as discussed in section 5.5. To include all these cases in a comprehensive framework, *efficiency* is defined as the ratio between input and output entropy change $\eta_S = -S^{\text{in}}/S^{\text{out}}$, as discussed in section 2.2. The efficiency is limited by the Carnot bound $\eta_S \leq \eta_C = 1$.

Since entropy is a fluctuating quantity in stochastic thermodynamics, it is natural to introduce the **stochastic efficiency**

$$\eta = -\frac{s^{\text{in}}}{s^{\text{out}}}. \tag{8.1}$$

The stochastic efficiency exhibits surprising universal properties. To study them, we consider the joint probability distribution $p(s^{\text{in}}, s^{\text{out}})$ of the input and output entropy production, with $s^{\text{tot}} = s^{\text{in}} + s^{\text{out}}$. The two entropy productions s^{in} and s^{out} satisfy a detailed fluctuation relation

$$\frac{p(s^{\text{in}}, s^{\text{out}})}{p(-s^{\text{in}}, -s^{\text{out}})} = e^{(s^{\text{in}} + s^{\text{out}})/k_B}. \tag{8.2}$$

Equation (8.2) derives from the fact that both s^{in} and s^{out} are odd under time reversal:

$$
\begin{aligned}
p(s^{\text{in}}, s^{\text{out}}) &= \int \mathcal{D}x\, \delta(s^{\text{in}} - s^{\text{in}}(x))\, \delta(s^{\text{out}} - s^{\text{out}}(x))\, \mathcal{P}(x) \\
&= \int \mathcal{D}\widehat{x}\, \delta(s^{\text{in}} - s^{\text{in}}(x))\, \delta(s^{\text{out}} - s^{\text{out}}(x))\, e^{s^{\text{tot}}(x)}\, \mathcal{P}(\widehat{x}) \\
&= e^{s^{\text{tot}}/k_{\text{B}}} \int \mathcal{D}\widehat{x}\, \delta(s^{\text{in}} + s^{\text{in}}(\widehat{x}))\, \delta(s^{\text{out}} + s^{\text{out}}(\widehat{x}))\, \mathcal{P}(\widehat{x}) \\
&= e^{s^{\text{tot}}/k_{\text{B}}}\, p(-s^{\text{in}}, -s^{\text{out}}).
\end{aligned}
\tag{8.3}
$$

This implies that, in the long time limit, the joint rate function of the empirical entropy production rates $j_s^{\text{in}} = s^{\text{in}}/\mathcal{T}$ and $j_s^{\text{out}} = s^{\text{out}}/\mathcal{T}$ satisfies

$$
I(j_s^{\text{in}}, j_s^{\text{out}}) - I(-j_s^{\text{in}}, -j_s^{\text{out}}) = \frac{j_s^{\text{in}} + j_s^{\text{out}}}{k_{\text{B}}}.
\tag{8.4}
$$

By the contraction principle, eq. (6.98), the rate function of the efficiency can be expressed as the minimum of $I(j_s^{\text{in}}, j_s^{\text{out}})$ compatible with a given value of η:

$$
I(\eta) = \min_{j_s^{\text{in}}, j_s^{\text{out}} \mid -j_s^{\text{in}}/j_s^{\text{out}} = \eta} I(j_s^{\text{in}}, j_s^{\text{out}}) = \min_{j_s^{\text{in}}} I(j_s^{\text{in}}, -j_s^{\text{in}}/\eta).
\tag{8.5}
$$

We analyze the qualitative behavior of the rate function $I(\eta)$ using a contour plot of the rate function $I(j_s^{\text{in}}, j_s^{\text{out}})$ (fig. 8.1). For a given value of η, $I(\eta)$ can be graphically computed by looking for the minimum in the contour plot along the straight line $j_s^{\text{out}} = -j_s^{\text{in}}/\eta$ (see dotted and dashed lines in the figure). The rate function of the entropy production rates vanishes at the average values, $I(J_s^{\text{in}}, J_s^{\text{out}}) = 0$. It follows that the rate function of the stochastic efficiency also vanishes at the macroscopic efficiency $\eta_S = -J_s^{\text{in}}/J_s^{\text{out}}$.

$$
I(\eta_S) = \min_{j_s^{\text{in}}} I\left(j_s^{\text{in}}, j_s^{\text{out}} = -j_s^{\text{in}}/\eta_S\right) = 0,
\tag{8.6}
$$

since the corresponding line intercepts the minimum (dotted line and black dot in fig. 8.1a).

The function $I(\eta)$ attains its maximum at the Carnot efficiency $\eta_C = 1$; see fig. 8.1b. This means that the Carnot efficiency is, in general, the *least likely efficiency*, at least in the infinite time limit governed by large deviation theory. This fact is not a peculiarity of the particular rate function $I(j_s^{\text{in}}, j_s^{\text{out}})$ shown in the figure but holds in general. Indeed, eq. (8.4) for $j_s^{\text{in}} + j_s^{\text{out}} = 0$ implies $I(j_s^{\text{in}}, j_s^{\text{out}}) = I(-j_s^{\text{in}}, -j_s^{\text{out}})$. It follows that the function $I(j_s^{\text{in}}, -j_s^{\text{in}}/\eta_C)$ is a convex even function of j_s^{in}. Its minimum is therefore achieved for $j_s^{\text{in}} = 0$:

$$
I(\eta_C) = I(j_s^{\text{in}}, j_s^{\text{out}})\big|_{j_s^{\text{in}} = j_s^{\text{out}} = 0}.
\tag{8.7}
$$

On the other hand, from eq. (8.5) we have

$$
I(\eta) = \min_{j_s^{\text{in}}} I(j_s^{\text{in}}, -j_s^{\text{in}}/\eta) \leq I(j_s^{\text{in}}, -j_s^{\text{in}}/\eta_C)\big|_{j_s^{\text{in}} = 0} = I(\eta_C),
\tag{8.8}
$$

(a)

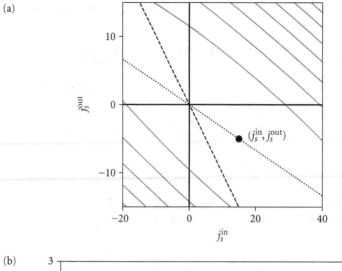

(b)

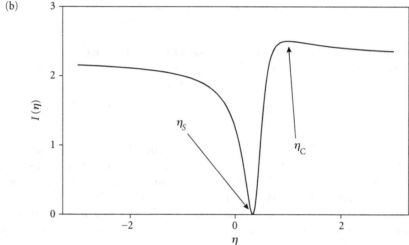

Figure 8.1. (a) Contour plot of the rate function $I(j_s^{in}, j_s^{out})$. In this example, the rate function is a quadratic form, corresponding to a case where j_s^{in} and j_s^{out} are two correlated Gaussian random variables with scaled variances $\tilde{\sigma}_{in}^2 = 50$, $\tilde{\sigma}_{out}^2 = 10$, and correlation $\tilde{c}_{in,out} = -20$. The same qualitative picture holds for a general convex rate function. The dotted line marks the macroscopic efficiency $-j_s^{in}/j_s^{out} = -J_s^{in}/J_s^{out} = \eta_S$. The dashed line marks the maximal efficiency $-j_s^{in}/j_s^{out} = \eta_C = 1$. (b) The corresponding rate function for the efficiency, obtained from eqs. (8.6) and (8.10). See, e.g., Verley et al. [172].

i.e., $I(\eta)$ reaches its maximum for $\eta = \eta_C$. For $|\eta| \gg 1$, the rate function $I(\eta)$ tends to a finite limit, corresponding to the minimum of $I(j_s^{in}, j_s^{out})$ for $j_s^{in} = 0$. It is important to stress in this context that large deviation theory only describes the leading behavior for large t. The distribution of η for finite time might significantly differ due to the role of subleading terms.

To explicitly compute the rate function of the stochastic efficiency in concrete examples, we exploit the inverse Gärtner-Ellis theorem to work with the scaled cumulant generating function rather than with the rate functions:

$$\psi^{(j_s^{in}, j_s^{out})}(q_{in}, q_{out}) = \lim_{\mathcal{T} \to \infty} \frac{1}{\mathcal{T}} \ln \left\langle e^{q_{in}s^{in} + q_{out}s^{out}} \right\rangle. \tag{8.9}$$

Combining eqs. (6.9) and (8.6), we express the rate function as

$$
\begin{aligned}
I(\eta) &= \inf_{j_s^{in}} \sup_{q_{in}, q_{out}} \left[q_{in}j_s^{in} - q_{out}(\eta j_s^{in}) - \psi^{(j_s^{in}, j_s^{out})}(q_{in}, q_{out}) \right] \\
&= \inf_{j_s^{in}} \sup_q \left[q j_s^{in} - f(q) \right] \\
&= -\sup_{j_s^{in}} [q' j_s^{in} - \sup_q [q j_s^{in} - f(q)]] \qquad \text{for} \quad q' = 0 \\
&= -f(0) = -\inf_{q_{out}} \psi^{(j_s^{in}, j_s^{out})}(q_{out}\eta, q_{out}),
\end{aligned}
\tag{8.10}
$$

where in the second equality we substitute $q = q - q_{out}\eta$ and define the convex function $f(q) = \inf_{q_{out}} \psi^{(j_s^{in}, j_s^{out})}(q + q_{out}\eta, q_{out})$. In the third equality, we introduce a fictitious variable q' (to be then set to 0) and change sign to cast the expression into the Legendre transform of $f(q)$. The last equality is a consequence of the involution property of the Legendre transformation.

We apply eq. (8.10) to an example in which the empirical entropy production rates j_s^{in} and j_s^{out} are correlated Gaussian random variables. In this case, the scaled cumulant generating function is a quadratic form,

$$
\begin{aligned}
\psi^{(j_s^{in}, j_s^{out})}(q_{in}, q_{out}) &= \frac{1}{2} \left(\tilde{\sigma}_{in}^2 q_{in}^2 + \tilde{\sigma}_{out}^2 q_{out}^2 + 2\tilde{c}_{in,out} \, q_{in}q_{out} \right) \\
&\quad + j_s^{in} q_{in} + j_s^{out} q_{out},
\end{aligned}
\tag{8.11}
$$

where $\tilde{\sigma}_{in}^2$, $\tilde{\sigma}_{out}^2$ are the scaled variances of s^{in} and s^{out}, respectively, and we define the scaled covariance $\tilde{c}_{in,out} = \lim_{\mathcal{T} \to \infty} (\langle s^{in}(\mathcal{T})s^{out}(\mathcal{T}) \rangle - \langle s^{in}(\mathcal{T}) \rangle \langle s^{out}(\mathcal{T}) \rangle)/\mathcal{T}$. Evaluating eq. (8.11) for $q_{in} = \eta \, q_{out}$ and minimizing with respect to q_{out} yields $q_{out} = (j_s^{in}\eta + j_s^{out})/(\tilde{\sigma}_{in}\eta^2 + 2\tilde{c}_{in,out}\eta + \tilde{\sigma}_{out})$. Substituting again into eq. (8.10) yields

$$I(\eta) = \frac{1}{2} \frac{(j_s^{in}\eta + j_s^{out})^2}{\tilde{\sigma}_{in}^2 \eta^2 + 2\tilde{c}_{in,out}\eta + \tilde{\sigma}_{out}^2}. \tag{8.12}$$

We use eq. (8.4) to further simplify the expression. The Gallavotti-Cohen symmetry associated with this detailed fluctuation relation reads $\psi^{(j_s^{in}, j_s^{out})}(q_{in}, q_{out}) = \psi^{(j_s^{in}, j_s^{out})}(-1/k_B - q_{in}, -1/k_B - q_{out})$. By imposing this condition for the Gaussian case of eq. (8.11), we express the average empirical entropy production rates in terms of the scaled variances and covariances, $2j_s^{in} = \tilde{\sigma}_{in}^2 + \tilde{c}_{in,out}$ and $2j_s^{out} = \tilde{\sigma}_{out}^2 + \tilde{c}_{in,out}$. In this way, the rate function of the stochastic efficiency can be expressed in terms of the scaled variances and covariances alone:

$$I(\eta) = \frac{1}{8} \frac{\left[\eta\tilde{\sigma}_{in}^2 + (1+\eta) \, \tilde{c}_{in,out} + \tilde{\sigma}_{out}^2 \right]^2}{\tilde{\sigma}_{in}^2 \eta^2 + 2\tilde{c}_{in,out}\eta + \tilde{\sigma}_{out}^2}. \tag{8.13}$$

8.2   Uncertainty relations

Uncertainty relations are universal inequalities relating the average value of an arbitrary current, its variance, and the average entropy production rate. In particular, most uncertainty relations set lower bounds on the **coefficient of variation** $\tilde{\sigma}_j/\left|\langle j\rangle\right|$ of an empirical current j, bounds that are expressed in terms of the entropy production rate. The coefficient of variation measures the amplitude of fluctuations relative to the mean. In a nutshell, uncertainty relations state that achieving very accurate currents, with a very small coefficient of variation, requires in general a minimal cost in terms of entropy production. In fact, a number of several interconnected results have been obtained, all broadly termed *uncertainty relations*. They differ in their underlying hypotheses and provide tighter or looser bounds in different situations.

The original thermodynamic uncertainty relation represents a remarkable achievement in stochastic thermodynamics. Its nontrivial proof exploits the properties of the level 2.5 large deviations introduced in section 6.9. For pedagogical reasons, we start our discussion from a more recent thermodynamic uncertainty relation, which holds in a nonequilibrium steady state and has the advantage of being relatively easy to prove. Later in this section, we present a more general version of the original uncertainty relation involving several currents, and we sketch its proof in appendix A.13. We conclude with some weaker inequalities that are useful for experimental applications.

We consider a system obeying a master equation as in eq. (2.77) in the steady state p_x^{st}. We focus on the joint probability distribution $p(s^{\mathrm{tot}}, \mathcal{J})$ of the total entropy production s^{tot} and a given integrated empirical current $\mathcal{J}$ being odd under time reversal. Both these quantities are evaluated over a time interval of duration $\mathcal{T}$. The joint probability distribution satisfies a detailed fluctuation relation:

$$p(s^{\mathrm{tot}}, \mathcal{J}) = e^{s^{\mathrm{tot}}/k_{\mathrm{B}}}\, p(-s^{\mathrm{tot}}, -\mathcal{J}). \tag{8.14}$$

The proof of eq. (8.14) goes along the same lines of eq. (8.3):

$$
\begin{aligned}
p(s^{\mathrm{tot}}, \mathcal{J}) &= \int \mathcal{D}\boldsymbol{x}\, \delta(s^{\mathrm{tot}} - s^{\mathrm{tot}}(\boldsymbol{x}))\, \delta(\mathcal{J} - \mathcal{J}(\boldsymbol{x}))\, \mathcal{P}(\boldsymbol{x}) \\
&= \int \mathcal{D}\boldsymbol{x}\, \delta(s^{\mathrm{tot}} - s^{\mathrm{tot}}(\boldsymbol{x}))\, \delta(\mathcal{J} - \mathcal{J}(\boldsymbol{x}))\, e^{s^{\mathrm{tot}}(\boldsymbol{x})}\, \mathcal{P}(\widehat{\boldsymbol{x}}) \\
&= e^{s^{\mathrm{tot}}/k_{\mathrm{B}}} \int \mathcal{D}\widehat{\boldsymbol{x}}\, \delta(s^{\mathrm{tot}} + s^{\mathrm{tot}}(\widehat{\boldsymbol{x}}))\, \delta(\mathcal{J} + \mathcal{J}(\widehat{\boldsymbol{x}}))\, \mathcal{P}(\widehat{\boldsymbol{x}}) \\
&= e^{s^{\mathrm{tot}}/k_{\mathrm{B}}}\, p(-s^{\mathrm{tot}}, -\mathcal{J}).
\end{aligned}
\tag{8.15}
$$

We introduce a probability density normalized over the *positive* values of the total entropy production:

$$p^{+}(s^{\mathrm{tot}}, \mathcal{J}) = \left(1 + e^{-s^{\mathrm{tot}}/k_{\mathrm{B}}}\right) p(s^{\mathrm{tot}}, \mathcal{J}). \tag{8.16}$$

One has indeed, by eq. (8.14),

$$\int_{0}^{\infty} \mathrm{d}s^{\mathrm{tot}} \int_{-\infty}^{+\infty} \mathrm{d}\mathcal{J}\, p^{+}(s^{\mathrm{tot}}, \mathcal{J}) = \int_{-\infty}^{\infty} \mathrm{d}s^{\mathrm{tot}} \int_{-\infty}^{+\infty} \mathrm{d}\mathcal{J}\, p(s^{\mathrm{tot}}, \mathcal{J}) = 1. \tag{8.17}$$

We then express the average of our extensive current as

$$\langle \mathcal{J} \rangle = \int_{-\infty}^{+\infty} ds^{\mathrm{tot}} \int_{-\infty}^{+\infty} d\mathcal{J} \, \mathcal{J} \, p(s^{\mathrm{tot}}, \mathcal{J})$$

$$= \int_{0}^{\infty} ds^{\mathrm{tot}} \int_{-\infty}^{+\infty} d\mathcal{J} \, \mathcal{J} \, p(s^{\mathrm{tot}}, \mathcal{J}) \left(1 - e^{-s^{\mathrm{tot}}/k_{\mathrm{B}}} \right)$$

$$= \int_{0}^{\infty} ds^{\mathrm{tot}} \int_{-\infty}^{+\infty} d\mathcal{J} \, \mathcal{J} \, p^{+}(s^{\mathrm{tot}}, \mathcal{J}) \frac{1 - e^{-s^{\mathrm{tot}}/k_{\mathrm{B}}}}{1 + e^{-s^{\mathrm{tot}}/k_{\mathrm{B}}}} \qquad (8.18)$$

$$= \left\langle \mathcal{J} \tanh \left(\frac{s^{\mathrm{tot}}}{2k_{\mathrm{B}}} \right) \right\rangle^{+},$$

where we define

$$\langle \cdots \rangle^{+} = \int_{0}^{\infty} ds^{\mathrm{tot}} \int_{-\infty}^{+\infty} d\mathcal{J} \, \cdots \, p^{+}(s^{\mathrm{tot}}, \mathcal{J}). \qquad (8.19)$$

The average entropy production can be similarly related to the average over positive values only:

$$\langle s^{\mathrm{tot}} \rangle = \left\langle s^{\mathrm{tot}} \tanh \left(\frac{s^{\mathrm{tot}}}{2k_{\mathrm{B}}} \right) \right\rangle^{+}. \qquad (8.20)$$

By applying the Cauchy-Schwarz inequality (see appendix A.12) to eq. (8.18), we obtain

$$\langle \mathcal{J} \rangle^{2} = \left(\left\langle \mathcal{J} \tanh \left(\frac{s^{\mathrm{tot}}}{2k_{\mathrm{B}}} \right) \right\rangle^{+} \right)^{2} \leq \langle \mathcal{J}^{2} \rangle^{+} \left\langle \tanh^{2} \left(\frac{s^{\mathrm{tot}}}{2k_{\mathrm{B}}} \right) \right\rangle^{+}. \qquad (8.21)$$

For any nonnegative variable x, one has

$$\langle \tanh^{2} x \rangle \leq \tanh \langle x \rangle. \qquad (8.22)$$

Indeed, if $x \geq 0$, one has $\tanh^{2} x \leq \tanh x$, and moreover $\langle \tanh x \rangle \leq \tanh \langle x \rangle$ by the Jensen inequality, since $\tanh x$ is concave for nonnegative x. We obtain therefore the following **thermodynamic uncertainty relation**

$$\frac{\sigma_{\mathcal{J}}^{2}}{\langle \mathcal{J} \rangle^{2}} \geq \frac{2}{e^{S^{\mathrm{tot}}/k_{\mathrm{B}}} - 1}, \qquad (8.23)$$

where $\sigma_{\mathcal{J}}^{2} = \langle \mathcal{J}^{2} \rangle - \langle \mathcal{J} \rangle^{2}$ and we take into account that $\langle \mathcal{J}^{2} \rangle = \langle \mathcal{J}^{2} \rangle^{+}$.

The average of $\mathcal{J}$ is proportional to the duration $\mathcal{T}$ of the time interval: $\langle \mathcal{J} \rangle = \mathcal{T} J^{\mathrm{st}}$, where J^{st} is the steady-state average of the associated intensive current. The same holds for the total entropy production: $\langle s^{\mathrm{tot}} \rangle = \mathcal{T} \dot{S}^{\mathrm{tot}}$. Also the variance of $\mathcal{J}$ is extensive, $\sigma_{\mathcal{J}}^{2} = \tilde{\sigma}_{j}^{2} \mathcal{T}$, where $\tilde{\sigma}_{j}^{2}$ is the scaled variance of j. Substituting in (8.23) and taking the small $\mathcal{T}$ limit, we obtain

$$\frac{\tilde{\sigma}_j^2}{(J^{st})^2} \ge \frac{2\,k_B}{\dot{S}^{tot}}, \qquad \mathcal{T} \ll k_B/\dot{S}^{tot}, \tag{8.24}$$

where we expand $e^{\dot{S}^{tot}/k_B} \approx 1 + \mathcal{T}\dot{S}^{tot}/k_B$ for small $\mathcal{T}$. Equation (8.24) implies that the coefficient of variation of the current can be reduced only at the expense of an increased entropy production rate, as we had anticipated. This bound holds in fact for any value of $\mathcal{T}$, as we discuss later.

The thermodynamic uncertainty relation can be generalized to multiple currents $\mathcal{J}_\alpha(\boldsymbol{x})$. To this aim, we define $\mathcal{J}(\boldsymbol{x})$ as a linear combination of the $\mathcal{J}_\alpha$'s:

$$\mathcal{J}(\boldsymbol{x}) = \sum_\alpha \lambda_\alpha \mathcal{J}_\alpha(\boldsymbol{x}), \tag{8.25}$$

where the coefficients λ_α are arbitrary. We then write eq. (8.23) in the form

$$\sum_{\alpha\beta} \lambda_\alpha \lambda_\beta \left(C_{\alpha\beta} - \frac{1}{K} \langle \mathcal{J}_\alpha \rangle \langle \mathcal{J}_\beta \rangle \right) \ge 0, \tag{8.26}$$

where we introduce the correlation matrix

$$C_{\alpha\beta} = \langle \mathcal{J}_\alpha \mathcal{J}_\beta \rangle - \langle \mathcal{J}_\alpha \rangle \langle \mathcal{J}_\beta \rangle \tag{8.27}$$

and the shorthand notation

$$K = \frac{1}{2} \left(e^{\langle \dot{s}^{tot} \rangle / k_B} - 1 \right). \tag{8.28}$$

If we now choose the coefficients λ by

$$\lambda_\alpha = \sum_\gamma C_{\alpha\gamma}^{-1} \langle \mathcal{J}_\gamma \rangle, \tag{8.29}$$

where C^{-1} is the matrix inverse of C, we obtain

$$\sum_{\alpha\beta} \langle \mathcal{J}_\alpha \rangle C_{\alpha\beta}^{-1} \langle \mathcal{J}_\beta \rangle - \frac{1}{K} \left(\sum_{\alpha\beta} \langle \mathcal{J}_\alpha \rangle C_{\alpha\beta}^{-1} \langle \mathcal{J}_\beta \rangle \right)^2 \ge 0, \tag{8.30}$$

which implies the inequality

$$\sum_{\alpha\beta} \langle \mathcal{J}_\alpha \rangle C_{\alpha\beta}^{-1} \langle \mathcal{J}_\beta \rangle \le K. \tag{8.31}$$

Switching to the currents $J_\alpha^{st} = \lim_{\mathcal{T} \to 0^+} \langle \mathcal{J}_\alpha \rangle / \mathcal{T}$ and to the corresponding scaled variances and covariances $\tilde{\sigma}^2 = (\tilde{\sigma}_{\alpha\beta}^2) = \lim_{\mathcal{T} \to 0^+} C_{\alpha\beta} / \mathcal{T}$, we obtain the **multidimensional uncertainty relation**

$$\sum_{\alpha\beta} J_\alpha^{st} (\tilde{\sigma}^2)_{\alpha\beta}^{-1} J_\beta^{st} \le \frac{\dot{S}^{tot}}{2k_B}. \tag{8.32}$$

We now discuss an important extension of uncertainty relations beyond the second scaled cumulant. The rate function $I(j)$ for large deviations of any current j satisfies the bound

$$I(j) \le \sum_{x' < x} \frac{(j_{xx'} - J_{xx'}^{st})^2}{4k_B (J_{xx'}^{st})^2} \dot{S}_{xx'}^{tot}, \tag{8.33}$$

in which $J_{xx'}^{st} = k_{xx'} p_{x'}^{st} - k_{x'x} p_x^{st}$ is the average current along the xx' edge in the steady state and $\dot{S}_{xx'}^{tot}$ is the contribution of the xx' edge to the average entropy production rate, given by

$$\dot{S}_{xx'}^{tot} = 2k_B J_{xx'}^{st} \sinh^{-1} \frac{J_{xx'}^{st}}{\alpha_{xx'}} = J_{xx'}^{st} \frac{A_{xx'}}{T}, \tag{8.34}$$

where $\alpha_{xx'} = 2\sqrt{k_{xx'} k_{x'x} p_x^{st} p_{x'}^{st}}$. Equation (8.33) is proved in appendix A.13.

Near equilibrium, fluctuations of the empirical entropy production rate j_s^{tot} are Gaussian (see section 6.2) and satisfy eq. (6.29). In keeping with eq. (8.34), we define the empirical current $j_{s,xx'}^{tot} = j_{xx'} A_{xx'} / T$, representing the contribution of the xx' edge to the empirical entropy production rate. With this definition, we have $\sum_{x' < x} j_{s,xx'}^{tot} = j_s^{tot}$ and $\dot{S}_{xx'}^{tot} = \langle j_{s,xx'}^{tot} \rangle$. We also denote by $I_{LR}(j_s^{tot})$ the rate function of j_s^{tot} in the linear response regime. We obtain

$$\begin{aligned} I_{LR}(j_s^{tot}) &= \sum_{x' < x} \frac{(j_{s,xx'}^{tot} - \dot{S}_{xx'}^{tot})^2}{4k_B \dot{S}_{xx'}^{tot}} = \sum_{x' < x} \frac{(j_{xx'} - J_{xx'}^{st})^2}{4k_B J_{xx'}^{st}} \frac{A_{xx'}}{T} \\ &= \sum_{x' < x} \frac{(j_{xx'} - J_{xx'}^{st})^2}{4k_B (J_{xx'}^{st})^2} J_{xx'}^{st} \frac{A_{xx'}}{T}, \end{aligned} \tag{8.35}$$

which is equal to the right-hand side of eq. (8.33). This result provides additional physical insight into eq. (8.33). Since $I(j) \le I_{LR}(j)$, eq. (8.33) effectively means that large fluctuations of an arbitrary empirical current in the steady state are always more likely than predicted by linear response theory.

From eq. (8.33), we derive a weaker but handier inequality, which holds for any current j:

$$I(j) \le I_{WLR}(j) = \frac{(j - J^{st})^2}{4k_B (J^{st})^2} \dot{S}^{tot}. \tag{8.36}$$

This bound depends only on the average current J_d^{st} and the average entropy production rate $\dot{S}^{tot}$. As a corollary, the bound of eq. (8.24) holds for any value of $\mathcal{T}$, as anticipated.

Choosing in particular the coefficients $d_{xx'} = A_{xx'}/T$, we obtain a weak linear response bound for the entropy production rate j_s^{tot} itself:

$$I(j_s^{\text{tot}}) \leq I_{\text{WLR}}(j_s^{\text{tot}}) = \frac{(j_s^{\text{tot}} - \dot{S}^{\text{tot}})^2}{4k_B \dot{S}^{\text{tot}}}. \tag{8.37}$$

The right-hand side of eq. (8.37) is the parabolic rate function of the entropy production rate predicted by linear response theory; see section 4.2. The thermodynamic uncertainty relation implies that linear response theory provides a lower bound for fluctuations of the entropy production rate.

8.3   Applications of uncertainty relations

In this section, we apply thermodynamic uncertainty relations to two models of molecular motors.

We first use eq. (8.23) to bound the efficiency of a continuous molecular motor. The molecular motor moves with an average speed $v = \langle \dot{r} \rangle$, where r is its position along a microtubule. The motor carries a load that exerts on the motor a force f. The average entropy production rate is given by

$$T \dot{S}^{\text{tot}} = \dot{W}^{\text{chem}} - f v, \tag{8.38}$$

where $\dot{W}^{\text{chem}}$ is the average rate of consumption of the chemical free energy. We express $\dot{W}^{\text{chem}}$ in terms of the average ATP consumption rate $J_n = \langle \dot{n} \rangle$ and the ATP chemical potential imbalance $\Delta\mu$ by

$$\dot{W}^{\text{chem}} = \Delta\mu \, \langle \dot{n} \rangle . \tag{8.39}$$

The macroscopic efficiency of the motor is

$$\eta_S = \frac{fv}{\dot{W}^{\text{chem}}} = \frac{fv}{fv + T \dot{S}^{\text{tot}}}. \tag{8.40}$$

By substituting (8.24), we obtain

$$\eta_S \leq \frac{1}{1 + 2v \, k_B T / (\tilde{\sigma}_j^2 f)}. \tag{8.41}$$

The bound provided by eq. (8.41) is particularly useful since all quantities appearing in the right-hand side are in principle accessible to experiments.

Our second example is in the context of the kinesin model introduced in section 6.7. This model is characterized by three independent currents. We focus on two of them: the displacement $\langle \dot{r} \rangle$ and the rate of ATP consumption $\langle \dot{n} \rangle$. Applying the relation (8.32), we obtain the bounds

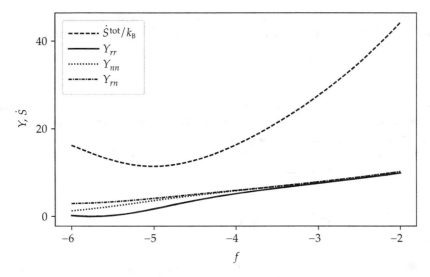

Figure 8.2. Multidimensional uncertainty relation for a model of a molecular motor. The model is defined in section 6.7. The entropy production rate $\dot{S}^{\text{tot}}$, in units of k_B, is plotted against the applied load $f = Fd/k_BT$ for a chemical potential imbalance $\Delta\mu = 13\,k_BT$. The time unit is arbitrary. The quantities Y_{rr}, Y_{nn}, and Y_{rn} defined in eq. (8.42) are all bounded by $\dot{S}^{\text{tot}}/k_B$ by the multidimensional uncertainty relation (8.32). The bound is tightest for Y_{rn}.

$$Y_{rr} = \frac{2\,\langle \dot{r} \rangle^2}{\tilde{\sigma}_{rr}^2} \leq \frac{\dot{S}}{k_B};$$

$$Y_{nn} = \frac{2\,\langle \dot{n} \rangle^2}{\tilde{\sigma}_{nn}^2} \leq \frac{\dot{S}}{k_B}, \qquad\qquad (8.42)$$

$$Y_{rn} = 2(\langle \dot{r} \rangle, \langle \dot{n} \rangle) \cdot \left(\tilde{\sigma}^2\right)^{-1} \cdot \begin{pmatrix} \langle \dot{r} \rangle \\ \langle \dot{n} \rangle \end{pmatrix} \leq \frac{\dot{S}}{k_B},$$

where $\dot{S}$ is the average entropy production rate. The bound on Y_{rn} is the most significant, since this quantity is the largest in our example, as shown in fig. 8.2.

8.4 First-passage times

Large deviation theory describes the steady-state distribution of an integrated current $\mathcal{J}$ for long time intervals $\mathcal{T}$; see chapter 6. Here we address a complementary problem of characterizing the distribution of the **first-passage time** $\mathcal{T}_{\text{fp}}(J_{\text{thr}})$ of a current. As the name says, the first-passage time is the time at which the integrated current first crosses a given positive or negative threshold J_{thr}. In the steady state and for large $|J_{\text{thr}}|$, an uncertainty relation also holds for the first passage time:

$$\frac{\sigma_{t_{\text{fp}}}^2}{\langle t_{\text{fp}} \rangle} \geq \frac{2k_B}{\dot{S}^{\text{tot}}}, \qquad\qquad (8.43)$$

where we define the intensive counterpart of the first-passage time, $t_{fp} = \mathcal{T}_{fp}/|J_{thr}|$. To prove eq. (8.43), we exploit the large deviation principle obeyed by currents:

$$p(\mathcal{J}; \mathcal{T}) \asymp e^{-\mathcal{T} I(\mathcal{J}/\mathcal{T})}, \tag{8.44}$$

where $I(x)$ is the rate function. The average of the first-passage time $\mathcal{T}_{fp}$ is approximately proportional to $|J_{thr}|$ for large $|J_{thr}|$. Therefore, $\mathcal{T}_{fp}$ also satisfies a large deviation principle. The behavior of the first-passage time for a positive (+) or negative (−) threshold J_{thr} can in principle be different. We denote with $I_+(t_{fp})$ and $I_-(t_{fp})$ the rate functions corresponding to these two cases:

$$p(\mathcal{T}_{fp}; J_{thr}) \asymp \begin{cases} e^{-J_{thr} I_+(\mathcal{T}_{fp}/J_{thr})}, & \text{if } J_{thr} > 0; \\ e^{J_{thr} I_-(-\mathcal{T}_{fp}/J_{thr})}, & \text{if } J_{thr} < 0. \end{cases} \tag{8.45}$$

We use the short notation $I_\pm(t_{fp})$ for either $I_+(t_{fp})$ or $I_-(t_{fp})$, depending on the sign of J_{thr}. The rate functions for the first-passage times are related to the rate function for the current:

$$I_\pm^{(t)}(t_{fp}) = t_{fp}\, I(\pm 1/t_{fp}). \tag{8.46}$$

We first prove eq. (8.46) in the case $J_{thr} > 0$. We express the probability density of $\mathcal{T}_{fp}$ in terms of the probability density of the trajectories:

$$p(\mathcal{T}_{fp} = t_{fp} J_{thr}) \approx \int \mathcal{D}x\, \delta(\mathcal{T}_{fp} - t_{fp} J_{thr})\, \mathcal{P}(x). \tag{8.47}$$

Equation (8.47) is approximate since we did not impose that the threshold is passed for the first time. However, for large J_{thr}, the times of all crossings are quite likely to be relatively close to each other, so that eq. (8.47) provides the correct leading order in J_{thr}. The trajectories that contribute to the integral in eq. (8.47) are those for which the current is equal to J_{thr}. Thus the integral boils down to the probability $p(J_{thr}; \mathcal{T}_{fp})$ of having $\mathcal{J} = J_{thr}$ at time $\mathcal{T}_{fp}$. Using the large deviation form of this probability, we obtain

$$p(\mathcal{T}_{fp} = t_{fp} J_{thr}) \asymp e^{-\mathcal{T}_{fp} I(J_{thr}/\mathcal{T}_{fp})} = e^{-J_{thr} t_{fp} I(1/t_{fp})}, \tag{8.48}$$

from which (8.46) follows. The case $J_{thr} < 0$ can be treated in the same way.

We now extend the relation (8.46) between the rate functions of the current and the first-passage time to the scaled cumulant generating functions, defined by

$$\psi^{(j)}(q) = \lim_{\mathcal{T} \to \infty} \frac{1}{\mathcal{T}} \ln \langle e^{qJ} \rangle; \qquad \psi_\pm^{(t_{fp})}(q) = \lim_{|J_{thr}| \to \infty} \frac{1}{|J_{thr}|} \ln \langle e^{q\, t_{fp}} \rangle, \tag{8.49}$$

where $\pm$ corresponds to the sign of J_{thr}. We now split $\psi(q)$ in one branch (+) with positive slope and one (−) with negative slope. Remarkably, the scaled cumulant generating functions are related by

$$\psi_\pm^{(t_{fp})}(q) = -\frac{1}{\psi_\pm^{(j)}(-q)}. \tag{8.50}$$

To prove this relation, we express the generating functions from the corresponding rate function by means of the inverse Gärtner-Ellis theorem. Setting $J_{thr} > 0$ for definiteness, we obtain

$$\psi^{(j)}(q) = q j^* - I(j^*), \qquad\qquad j^*: \quad I'(j^*) = q; \qquad\qquad (8.51a)$$

$$\psi_+^{(t_{fp})}(q) = q\, t_{fp}^* - I_+(t_{fp}^*), \qquad\qquad t^*: \quad I'_+(t_{fp}^*) = q. \qquad\qquad (8.51b)$$

We evaluate $\psi^{(j)}(-\psi_+^{(t_{fp})}(q))$, taking into account that t_{fp}^* corresponds to $1/j^*$, obtaining

$$\psi\left(-\psi_+^{(t_{fp})}(q)\right) = -\left(q\, t_{fp}^* - I_+(t_{fp}^*)\right) j^* - I(j^*)$$

$$= -q + \frac{1}{t_{fp}^*} I_+(t_{fp}^*) - I\left(\frac{1}{t_{fp}^*}\right) = -q, \qquad\qquad (8.52)$$

where we use eq. (8.46). This proves (8.50). The relations between these functions are shown in fig. 8.3. Evaluating $\langle j \rangle, \langle t_{fp} \rangle$ and the respective scaled variances by taking the derivatives of the scaled cumulant generating functions around $q = 0$ results in

$$|\langle j \rangle| = \frac{1}{\langle t_{fp} \rangle}; \qquad \tilde{\sigma}_j^2 = |\langle j \rangle| \,\tilde{\sigma}_{t_{fp}}^2 = \frac{\tilde{\sigma}_{t_{fp}}^2}{\langle t_{fp} \rangle}. \qquad\qquad (8.53)$$

Substituting these expressions in the uncertainty relation (8.24), we obtain the uncertainty relation (8.43) for the first-passage time.

Some currents, such as the entropy production rate j_s^{tot}, exhibit the Gallavotti-Cohen symmetry; see section 6.4. In particular, the Gallavotti-Cohen symmetry for j_s^{tot} is expressed by eq. (6.30) for the scaled cumulant generating function. The corresponding symmetry for the scaled cumulant generating functions of the first-passage time is expressed by

$$\psi_+^{(t_{fp})}(q) = \psi_-^{(t_{fp})}(q) - \frac{1}{k_B}. \qquad\qquad (8.54)$$

Applying once more the Gärtner-Ellis theorem, we find that the rate functions for positive and negative thresholds differ by a constant:

$$I_+(t_{fp}) = I_-(t_{fp}) - \frac{1}{k_B}. \qquad\qquad (8.55)$$

This relation implies that, up to the leading order, the probability of reaching a given positive value s^{tot} of the total entropy produced in a given time $\mathcal{T}$ exceeds the probability of reaching the opposite value $-s^{tot}$ by a factor $\exp(s^{tot}/k_B)$.

Taking into account the weak linear response bound for the rate function $I(j)$ expressed by eq. (8.36), we obtain the inequality

$$I_+(t_{fp}) \leq I_{WLR}(t_{fp}) = \frac{(t - \langle t \rangle)^2}{4k_B t}\, \dot{S}^{tot}, \qquad\qquad (8.56)$$

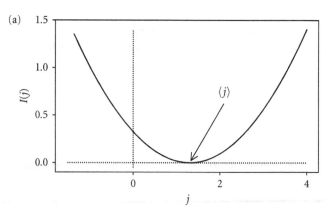

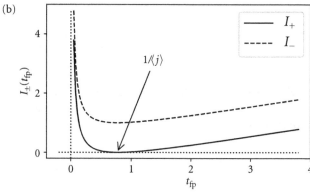

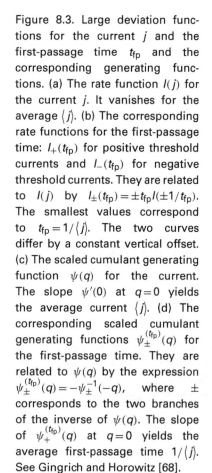

Figure 8.3. Large deviation functions for the current j and the first-passage time t_{fp} and the corresponding generating functions. (a) The rate function $I(j)$ for the current j. It vanishes for the average $\langle j \rangle$. (b) The corresponding rate functions for the first-passage time: $I_+(t_{fp})$ for positive threshold currents and $I_-(t_{fp})$ for negative threshold currents. They are related to $I(j)$ by $I_{\pm}(t_{fp}) = \pm t_{fp} I(\pm 1/t_{fp})$. The smallest values correspond to $t_{fp} = 1/\langle j \rangle$. The two curves differ by a constant vertical offset. (c) The scaled cumulant generating function $\psi(q)$ for the current. The slope $\psi'(0)$ at $q=0$ yields the average current $\langle j \rangle$. (d) The corresponding scaled cumulant generating functions $\psi_{\pm}^{(t_{fp})}(q)$ for the first-passage time. They are related to $\psi(q)$ by the expression $\psi_{\pm}^{(t_{fp})}(q) = -\psi_{\pm}^{-1}(-q)$, where $\pm$ corresponds to the two branches of the inverse of $\psi(q)$. The slope of $\psi_+^{(t_{fp})}(q)$ at $q=0$ yields the average first-passage time $1/\langle j \rangle$. See Gingrich and Horowitz [68].

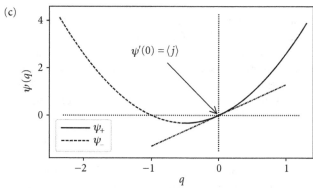

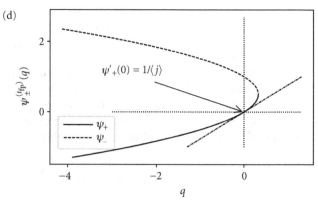

which shows how the first-passage time fluctuations are bounded by a function of the entropy production rate, i.e., of the dissipation. The probability distribution corresponding to the right-hand side is an inverse Gaussian:

$$p_{\text{WLR}}(t_{\text{fp}}) = \sqrt{\frac{J_{\text{thr}}\,\dot{S}^{\text{tot}}}{4k_{\text{B}}\pi t^3}}\,\exp\left[-\frac{J_{\text{thr}}(t_{\text{fp}} - \langle t_{\text{fp}}\rangle)^2\,\dot{S}^{\text{tot}}}{4k_{\text{B}}t_{\text{fp}}}\right]. \tag{8.57}$$

Interestingly, this is the same distribution that we would obtain by simply treating the process as a one-dimensional diffusion process with constant drift $\langle j\rangle = 1/\langle t_{\text{fp}}\rangle$ and diffusion coefficient $D_j = k_{\text{B}}\langle j\rangle^2/\dot{S}^{\text{tot}}$.

8.5 Fully irreversible processes

The irreversibility relation (4.12) connects the total entropy $s^{\text{tot}}(x)$ produced in a trajectory x to the ratio of the probability density of the trajectory x in the forward protocol λ to that of the backward trajectory $\widehat{x}$ with the backward protocol $\widehat{\lambda}$. A crucial underlying hypothesis is that the manipulation is slow enough that the reservoir remains always at equilibrium. Microscopic reversibility is a consequence of this hypothesis: whenever a jump $x \longrightarrow x'$ is allowed (its rate is nonvanishing), the reverse transition $x' \longrightarrow x$ is also allowed.

In some situations, this hypothesis is not satisfied because the manipulation protocol is too fast or the work involved too large. As an example, we consider a particle that diffuses in a finite one-dimensional system. Its discrete coordinate x is in the range $1 \le x \le L_0$, where $L_0 > 1$ is the coordinate of a removable wall. At equilibrium, the probability distribution is uniform on the allowed states, and the system entropy is given by $S_0 = k_{\text{B}}\ln L_0$. At a time t_c between t_0 and t_f, the wall at L_0 is suddenly removed and the particle is allowed to wander over a range $1 \le x \le L_f$, with $L_f > L_0$. Since no work is performed by removing the wall, we have

$$\left\langle e^{-w/k_{\text{B}}T}\right\rangle = 1. \tag{8.58}$$

The equilibrium entropy of the final state is given by $S_f = k_{\text{B}}\ln L_f > S_0$, while the internal energy has remained constant, and therefore the free energy changes by

$$\Delta F = -T\,\Delta S = -k_{\text{B}}T\ln\frac{L_f}{L_0} < 0. \tag{8.59}$$

The Jarzynski equality $\left\langle e^{(\Delta F - w)/k_{\text{B}}T}\right\rangle = 1$ is therefore violated. To interpret this result, we must have a closer look at the consequences of breaking microscopic reversibility.

We say that a trajectory x is **allowed** if its probability density $\mathcal{P}_x$ does not vanish. When microscopic reversibility (cf. section 2.6) does not hold, there exist allowed backward trajectories $\widehat{x}$ under the backward protocol $\widehat{\lambda}$ whose reverse x are not allowed, i.e., $\mathcal{P}_x(\lambda) = 0$. In our example, this is the case for all the trajectories $\widehat{x}$ in which the particle is found at time $\widehat{t_c}$ at a site x in the interval $L_0 < x \le L_f$. We call **reversible** those trajectories $\widehat{x}$ that are allowed under the reverse protocol $\widehat{\lambda}$ and whose reverse x is allowed under the forward protocol λ. With these definitions, we generalize the irreversibility relation (4.12) to

$$\mathcal{P}_x(\lambda)\, e^{-s^{\text{tot}}(x)/k_B} = \begin{cases} \mathcal{P}_{\widehat{x}}(\widehat{\lambda}), & \text{if } \widehat{x} \text{ is reversible;} \\ 0, & \text{otherwise.} \end{cases} \tag{8.60}$$

We define p^{rev} as the total probability of reversible trajectories with the backward protocol. Integrating eq. (8.60), we obtain the **modified Jarzynski equality**

$$\left\langle e^{-s^{\text{tot}}(x)} \right\rangle_F = \int \mathcal{D}x\, \mathcal{P}_x(\lambda)\, e^{-s^{\text{tot}}(x)/k_B} = p^{\text{rev}}. \tag{8.61}$$

In our specific example, the modified Jarzynski equality yields the correct result,

$$\int \mathcal{D}x\, \mathcal{P}_x(\lambda)\, e^{-s^{\text{tot}}(x)/k_B} = e^{-\Delta S/k_B} = \frac{L_0}{L_f}, \tag{8.62}$$

since in this case $p^{\text{rev}} = L_0/L_f$.

A similar reasoning applies to steady states. For example, we consider a system with L states arranged on a ring. Jumps $x \longrightarrow (x+1)$ and $(x+1) \longrightarrow x$ are all allowed for $1 \leq x < L$. The jump $L \to 1$ is also allowed, but the jump rate $1 \longrightarrow L$ vanishes. In this case, we say that the jump $1 \longrightarrow L$ is **fully** or **absolutely** irreversible. We assume that all the nonvanishing jump rates are equal. Then all the jumps $x \longleftrightarrow (x+1)$ with $1 \leq x < L$ do not produce entropy. Entropy is produced only in the $L \longrightarrow 1$ transition, which has no reverse. Therefore, if the trajectory x is reversible, i.e., if its reverse $\widehat{x}$ is allowed, the relation

$$\frac{\mathcal{P}_x}{\mathcal{P}_{\widehat{x}}} = e^{s^{\text{tot}}(x)/k_B} \tag{8.63}$$

is satisfied. Integrating over all reversible trajectories taking place from $t = t_0$ to $t = t_f$, we obtain

$$\int_{t=t_0}^{t=t_f} \mathcal{D}x\, \mathcal{P}_x\, e^{-s^{\text{tot}}(x)/k_B} = p^{\text{rev}}(t_f), \tag{8.64}$$

where also in this case $p^{\text{rev}}(t_f)$ is defined as the probability that an allowed trajectory x is reversible. The rate of decrease of p^{rev} is given by the total rate at which fully irreversible jumps take place in the steady state. In our model, this rate is equal to

$$\frac{1}{p^{\text{rev}}(t)}\frac{dp^{\text{rev}}}{dt} = -k_{1L}\, p_L^{\text{st}}. \tag{8.65}$$

Systems that possess **absorbing states** provide a common example of fully irreversible processes. In these cases, we characterize the dynamics by the **survival probability** $p^{\text{surv}}(t)$, i.e., the probability of *not* having been absorbed at time t. We consider a system characterized by absorbing states during a time interval from t_0 to t_f. If a trajectory x ends up in an absorbing state before t_f, its reverse trajectory has a vanishing probability. This also holds in reverse. As a consequence, we have

$$\int \mathcal{D}x\, \mathcal{P}_x\, e^{-s^{\text{tot}}(x)/k_B} = p^{\text{surv}}(t_f), \tag{8.66}$$

where the integral extends only over the reversible trajectories that run from $t = t_0$ to $t = t_f$.

8.6 Optimal protocols

Many experiments in stochastic thermodynamics involve manipulation of a meso-scopic system. We consider a manipulation that is specified by several control param-eters $\lambda(t) = (\lambda_\alpha(t))$, with $\alpha \in \{1, \ldots, r\}$. The experiment is carried out from an initial equilibrium state, characterized by values $\lambda_\alpha(t_0)$ of the manipulation parameters to a final state characterized by other values $\lambda_\alpha(t_f)$. In principle, there are infinite ways of performing this manipulation, specified by the functions $\lambda_\alpha(t)$ at fixed initial and final values. Among these infinite possibilities, it is often desirable to choose the **optimal manipulation protocol**, defined as the protocol that minimizes the average dissipated work

$$W^{\mathrm{diss}} = W - F(\lambda_f) + F(\lambda_0) = W - \Delta F. \tag{8.67}$$

If $\mathcal{T}$ is very large and the manipulation is performed very slowly, the amount of work can be made equal to the free-energy difference ΔF and therefore $W^{\mathrm{diss}} = 0$. For finite $\mathcal{T}$, the average dissipated work is positive, and determining the optimal protocol is not always easy. This problem allows for an elegant solution if the manipulation is still rela-tively slow and smooth, so that the system is never too far from equilibrium. The average instantaneous power spent on the system at time t is

$$\dot{W} = \frac{dF}{dt} + \dot{W}^{\mathrm{diss}}(t). \tag{8.68}$$

The last term in eq. (8.68) represents the average dissipation rate. It turns out that it can be expressed in the form

$$\dot{W}^{\mathrm{diss}}(t) = \sum_{\alpha\beta} g_{\alpha\beta}(\lambda(t)) \frac{d\lambda_\alpha}{dt} \frac{d\lambda_\beta}{dt}, \tag{8.69}$$

in which $g = (g_{\alpha\beta}(\lambda))$ is a **generalized friction coefficient**. The generalized friction coefficient is a symmetric and positive semidefinite matrix, which depends smoothly on λ, except possibly at macroscopic phase transitions. Its specific form depends on the system at hand. The matrix g can be expressed in terms of observables $X_{\alpha,x}$, defined by

$$X_{\alpha,x} = \frac{\partial \epsilon_x(\lambda)}{\partial \lambda_\alpha}, \tag{8.70}$$

where $\epsilon_x(\lambda)$ is the energy of state x. We have in fact

$$g_{\alpha\beta}(\lambda) = \frac{1}{k_{\mathrm{B}}T} \int_0^\infty dt\, C_{\alpha\beta}(t; \lambda), \tag{8.71}$$

in which we introduce the equilibrium correlation function

$$C_{\alpha\beta}(t; \lambda) = \left\langle \left(X_{\alpha,x}(t) - \langle X_{\alpha,x} \rangle^{\mathrm{eq}} \right) \left(X_{\beta,x}(0) - \langle X_{\beta,x} \rangle^{\mathrm{eq}} \right) \right\rangle^{\mathrm{eq}}, \tag{8.72}$$

where the averages are evaluated at fixed λ.

Before deriving eq. (8.69), we interpret it in the language of differential geometry with the aim of finding the optimal protocol. The symmetric and positive semidefinite matrix $g_{\alpha\beta}(\lambda)$ defines a metric in the space spanned by the manipulation variables λ. If the λ_αs change by $d\lambda_\alpha$, the corresponding traveled distance in this space is

$$ds = \sqrt{\sum_{\alpha\beta} g_{\alpha\beta}\, d\lambda_\alpha\, d\lambda_\beta}. \tag{8.73}$$

We express the average dissipated work $\dot{W}^{\mathrm{diss}}$ in terms of the speed ds/dt in the manipulation space:

$$\dot{W}^{\mathrm{diss}} = \sum_{\alpha\beta} g_{\alpha\beta} \frac{d\lambda_\alpha}{dt} \frac{d\lambda_\beta}{dt} = \left(\frac{ds}{dt}\right)^2. \tag{8.74}$$

This is the same expression of the kinetic energy of a free particle of mass equal to 2 moving on a space with metric $g_{\alpha\beta}$. We define the total **dissipation length** of a trajectory λ:

$$\mathcal{L}(\lambda) = \int_{t_0}^{t_f} dt \sqrt{\sum_{\alpha\beta} g_{\alpha\beta}(\lambda(t)) \frac{d\lambda_\alpha}{dt} \frac{d\lambda_\beta}{dt}}. \tag{8.75}$$

One can verify that this expression is invariant under reparameterizations $\lambda(t) \longrightarrow \lambda'(\tau)$, where $\lambda'(\tau(t)) = \lambda(t)$, with $\tau(t)$ a monotonically increasing function for $t \in [t_0, t_f]$. We assign to the same **path** $[\lambda]$ all trajectories λ that can be mapped one onto the other by reparameterization. The dissipation length is the same for all these trajectories, hence it can be considered as a function of the path $\mathcal{L} = \mathcal{L}([\lambda])$. Given a path $[\lambda]$, the average dissipated work W^{diss} on any trajectory of duration $\mathcal{T} = t_f - t_0$ associated with it satisfies the bound

$$W^{\mathrm{diss}}\, \mathcal{T} \geq \mathcal{L}^2. \tag{8.76}$$

This bound follows from the Cauchy-Schwarz inequality: we write $\mathcal{L}$ in the form

$$\mathcal{L} = \int_{t_0}^{t_f} dt\, 1 \cdot \left(\sum_{\alpha\beta} g_{\alpha\beta}(\lambda(t)) \frac{d\lambda_\alpha}{dt} \frac{d\lambda_\beta}{dt}\right)^{1/2}. \tag{8.77}$$

We then have

$$\mathcal{L}^2 \leq \int_{t_0}^{t_f} dt\, 1^2 \times \int_{t_0}^{t_f} dt \sum_{\alpha\beta} g_{\alpha\beta}(\lambda(t)) \frac{d\lambda_\alpha}{dt} \frac{d\lambda_\beta}{dt} = \mathcal{T} \cdot W^{\mathrm{diss}}. \tag{8.78}$$

One can verify that the bound is saturated if the protocol satisfies the condition

$$\sum_{\alpha\beta} g_{\alpha\beta}(\lambda(t)) \frac{d\lambda_\alpha}{dt} \frac{d\lambda_\beta}{dt} = \mathrm{const.} \tag{8.79}$$

This corresponds to imposing that the average dissipated power is constant during the manipulation.

We summarize the above results by a prescription for the optimal manipulation protocol λ^*, given the generalized friction coefficient $g_{\alpha\beta}(\lambda)$:

1. Identify the **geodesic path** $[\lambda^*]$, i.e., the path of minimal dissipation length $\mathcal{L}$ connecting the two endpoints λ_0 and λ_f.
2. Find the trajectory λ that runs along the path $[\lambda^*]$ and satisfies the condition (8.79).

We can associate with a trajectory $\lambda = (\lambda(t))$ other trajectories $\lambda_{\mathcal{T}}$ of arbitrary duration $\mathcal{T}$ that follow the same path $[\lambda]$, by linearly reparameterizing the time t. The average dissipated work W^{diss} in these trajectories decreases like $\mathcal{T}^{-1}$ as $\mathcal{T}$ becomes large.

We now sketch the derivation of (8.69). From (8.68), we obtain

$$\dot{W}^{\text{diss}} = \sum_\alpha \frac{d\lambda_\alpha}{dt} \left(\langle X_\alpha(t; \lambda(t)) \rangle_\lambda - \langle X_\alpha(\lambda(t)) \rangle_{\lambda(t)}^{\text{eq}} \right), \tag{8.80}$$

where $\langle \cdots \rangle_\lambda^{\text{eq}}$ is the equilibrium average with the given value of λ and we make explicit the dependence of X_α on λ. We now exploit the linear response theory developed in section 3.11. We obtain, to first order in the perturbation and for a given protocol λ such that $\lim_{t \to -\infty} \lambda(t) = \lambda^{(0)}$,

$$\langle X_\alpha(t; \lambda(t)) \rangle_\lambda - \langle X_\alpha(\lambda(t)) \rangle_{\lambda(t)}^{\text{eq}} = \sum_\beta \int_{-\infty}^t dt' \, K_{\alpha\beta}(t - t'; \lambda^{(0)}) \left(\lambda_\beta(t') - \lambda^{(0)} \right), \tag{8.81}$$

where

$$K_{\alpha\beta}(t; \lambda) = -\frac{\theta(t)}{k_B T} \frac{d}{dt} C_{\alpha\beta}(t; \lambda), \tag{8.82}$$

and $C_{\alpha\beta}(t; \lambda)$ is the correlation function defined in (8.72). We substitute this expression into eq. (8.81) and integrate by parts. The boundary terms vanish, one trivially and the other assuming that the manipulation starts from equilibrium. We thus obtain

$$\dot{W}^{\text{diss}}(t) = \frac{1}{k_B T} \sum_{\alpha\beta} \frac{d\lambda_\alpha}{dt} \int_{-\infty}^t dt' \, C_{\alpha\beta}(t - t'; \lambda(t)) \frac{d\lambda_\beta(t')}{dt'}. \tag{8.83}$$

Changing the variable to $\tau = t - t'$, eq. (8.83) takes the form

$$\dot{W}^{\text{diss}}(t) = \frac{1}{k_B T} \sum_{\alpha\beta} \frac{d\lambda_\alpha}{dt} \int_0^\infty d\tau \, C_{\alpha\beta}(\tau; \lambda(t)) \left. \frac{d\lambda_\beta(t')}{dt'} \right|_{t'=t-\tau}. \tag{8.84}$$

We now expand in τ, obtaining

$$\left. \frac{d\lambda_\beta(t')}{dt'} \right|_{t'=t-\tau} = \frac{d\lambda_\beta}{dt} - \tau \frac{d^2\lambda_\beta}{dt^2} + o(\tau). \tag{8.85}$$

Assuming that the manipulation is slow enough so that the change in $d\lambda/dt$ in a time on the order of the correlation time can be neglected, we keep only the first term and obtain

$$\dot{W}^{\mathrm{diss}}(t) = \sum_{\alpha\beta} \frac{d\lambda_\alpha}{dt} g_{\alpha\beta}(\lambda(t)) \frac{d\lambda_\beta}{dt}, \tag{8.86}$$

where $g_{\alpha\beta}(\lambda)$ is given by (8.71), as anticipated.

An interesting, exactly solvable example is the case of a Brownian particle in a one-dimensional harmonic potential of strength κ whose minimum is placed at ℓ. We describe the dynamics by a Langevin equation with mobility μ_P:

$$\frac{dx}{dt} = -\mu_P \kappa \, (x - \ell) + \xi(t), \tag{8.87}$$

where $\langle \xi(t) \rangle = 0, \langle \xi(t) \xi(t') \rangle = 2\mu_P \, k_B T \, \delta(t - t')$ by the Einstein relation. The manipulation parameters are ℓ and κ, and the associated observables are

$$\frac{\partial \epsilon(x)}{\partial \ell} = -\kappa \, (x - \ell); \qquad \frac{\partial \epsilon(x)}{\partial \kappa} = \frac{1}{2} \, (x - \ell)^2. \tag{8.88}$$

Since the equation of motion is linear, the correlation function $C_{\alpha\beta}(t; \lambda)$, with $\lambda = (\ell, \kappa)$, can be analytically obtained:

$$C(t; \ell, \kappa) = \begin{pmatrix} \kappa \, k_B T \, e^{-\mu_P \kappa |t|}, & 0 \\ 0, & (k_B T/\kappa)^2 e^{-2\mu_P \kappa |t|}/2 \end{pmatrix}, \tag{8.89}$$

and therefore the generalized friction coefficient is given by

$$g(\ell, \kappa) = \begin{pmatrix} 1/\mu_P, & 0 \\ 0, & k_B T/4\mu_P \kappa^3 \end{pmatrix}. \tag{8.90}$$

The matrix $g_{\alpha\beta}$ is diagonal, and its entries are independent of each other. Thus the optimization problem boils down to two separate ones, for ℓ and κ, respectively. The solution for ℓ is $d\ell/dt = \mathrm{const}$. For κ, we obtain

$$\frac{\dot{\kappa}^2}{\kappa^3} = \mathrm{const.} \tag{8.91}$$

or, equivalently,

$$\frac{d\kappa^{-1/2}}{dt} = \mathrm{const.}, \tag{8.92}$$

which can be used to impose the boundary conditions.

The theory developed in this section was exploited to experimentally design energetically efficient nonequilibrium processes in the folding and unfolding of a DNA hairpin. The generalized friction coefficient $g(\lambda)$ was estimated by measuring the equilibrium autocorrelation function of the applied force at different values of the control parameter λ, which, in the case of the DNA hairpin, corresponds to a fixed value of the separation of the optical traps to which the DNA molecule is tethered. The distribution of the force

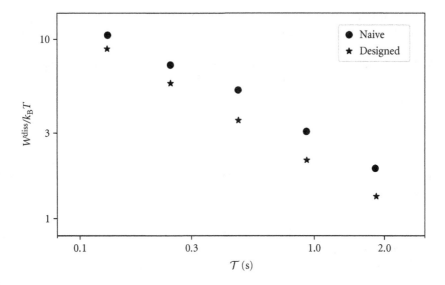

Figure 8.4. Average dissipated work with a naive and a designed protocol. The average dissipated work W^{diss} (in units of $k_B T$) is plotted against the duration of the protocol $\mathcal{T}$. In the naive protocol, the separation of the optical traps is increased linearly. In the designed protocol, the speed is inversely proportional to the square root of the generalized friction coefficient, eq. (8.79). Both axes are logarithmic. (Data from Tafoya et al. [164].)

is unimodal at small and large separations but is bimodal at intermediate values, showing how the molecule vacillates from a partially folded to a partially unfolded state. From knowledge of $g(\lambda)$, the optimal protocol is evaluated by imposing the condition (8.79). In this way, one obtains a substantially smaller average dissipation, as shown in fig. 8.4.

8.7 Martingales

Martingales are a powerful concept in the theory of stochastic processes. We consider a stochastic process $x(t)$, where the variable x can be either discrete or continuous. We estimate $\langle x(t)\rangle$ based on past observations $x(t_1), x(t_2), x(t_3), \ldots$, with $t > t_1 > t_2 > t_3 \cdots$. A **martingale** is a process in which this conditioned average is equal to the last observed value,

$$\langle x(t)|x(t_1), x(t_2), x(t_3), \ldots\rangle = x(t_1), \tag{8.93}$$

regardless of the number of past observations, their times, or their outcomes. In a nutshell, a martingale is a stochastic process with no average drift. As is the case in much of probability theory, the concept of a martingale finds its origin in gambling: if $x(t)$ represents a gambler's fortune at a certain time t, $x(t)$ is a martingale if the game is fair, i.e., if the gambler should expect neither to lose nor to win money on average.

To understand the usefulness of martingales, we introduce the concept of a **stopping time** t_{stop}. A stopping time is a random time determined by some conditions on the trajectory. First-passage times are an example of stopping times, determined by the condition of reaching a given state. Stopping times are however more general: a stopping time could be, for example, the Nth time at which a given state has been visited, or the first time that a state has been visited after having visited another state. The only restriction is that a stopping time cannot depend on the future, i.e., on events or conditions determined after t_{stop}.

A major result of the theory of martingales is Doob's **optional stopping theorem**, which states that, under certain assumptions, the average of a martingale at a stopping time is equal to its initial value:

$$\langle x \rangle_{t_{\text{stop}}} = x(t_0). \tag{8.94}$$

There are several versions of Doob's theorem in the literature, based on slightly different assumptions. A common version requires the stopping time to satisfy *at least one* of the following three properties:

1. The stopping time is bounded, $t_{\text{stop}} \leq c$, where $c < \infty$ is a constant.
2. The average stopping time is finite, $\langle t_{\text{stop}} \rangle < \infty$, and the increments of the process are bounded, $|x(t + \Delta t) - x(t)| \leq c$ with $c < \infty$.
3. The process itself is bounded, $|x(t)| \leq c$ for $t \leq t_{\text{stop}}$ and $c < \infty$.

An example of a stopping time that does not satisfy any of these conditions is the first-passage time for a one-dimensional unbiased random walk. Indeed, an unbiased one-dimensional random walk is a martingale, but its first-passage time does not satisfy eq. (8.94): the average of x at the first-passage time is equal to position x of the absorbing state, which is in general different from the initial condition $x(t_0)$.

The meaning (and importance) of Doob's theorem is best understood by returning to our example of gambling. Strategic players have long tried to profit from fair games by devising strategies to leave the gambling table when certain conditions are met, for example, when their fortune surpasses a given value. Doob's theorem confirms the intuition that these strategies never lead to an average profit. However, a gambler willing to play for an infinite time and having access to infinite economic resources can break all three conditions of Doob's theorem and therefore profit, on average, from a fair game.

The link between martingale theory and stochastic thermodynamics comes from the fact that, in any steady-state mesoscopic system, the quantity e^{-s^{tot}/k_B} is a martingale. We now verify that this is the case. Since we consider Markov processes, conditioning to a succession of past events is equivalent to conditioning to the last event only:

$$\left\langle e^{-s^{\text{tot}}(t)/k_B} \middle| e^{-s^{\text{tot}}(t_1)/k_B}, e^{-s^{\text{tot}}(t_2)}, \cdots \right\rangle = \left\langle e^{-s^{\text{tot}}(t)/k_B} \middle| e^{-s^{\text{tot}}(t_1)/k_B} \right\rangle. \tag{8.95}$$

We now use the expression of entropy production in terms of trajectory probabilities, eq. (4.12). Given a trajectory x defined between t_0 and t, we have

$$e^{s^{\text{tot}}(x)/k_B} = \frac{\mathcal{P}_x}{\mathcal{P}_{\widehat{x}}} = \frac{p_{x_0} \mathcal{P}_{x|x_0}}{p_{x_t} \mathcal{P}_{\widehat{x}|x_t}}. \tag{8.96}$$

We call x_1 the part of the trajectory in the time interval $[t_0, t_1)$ and x_2 the part of the trajectory in the time interval $[t_1, t]$, including t_1. This means that the sum over the state of the system at time t_1 is included in $\mathcal{D}x_2$ but not in $\mathcal{D}x_1$. We also decompose the entropy production $s^{\text{tot}}(t)$ into two contributions, $s^{\text{tot}}(t) = s_1 + s_2(x_2)$, where s_1 and s_2 are the entropies produced in the time intervals $[t_0, t_1)$ and $[t_1, t]$, respectively. Since we condition on the value of s_1, we do not explicitly write its dependence on x_1. These two contributions have the same form of eq. (8.96) thanks to the fact that the system is in a steady state. We now write the conditional average in eq.(8.95) explicitly:

$$\left\langle e^{-s^{\text{tot}}(t)/k_B} \middle| e^{-s^{\text{tot}}(t_1)/k_B} \right\rangle = \frac{\int \mathcal{D}x_1 \int \mathcal{D}x_2 \, \mathcal{P}_x \, e^{-s^{\text{tot}}(t)/k_B} \, \delta(s^{\text{tot}}(t_1) - s_1)}{\int \mathcal{D}x_1 \int \mathcal{D}x_2 \, \mathcal{P}_x \, \delta(s^{\text{tot}}(t_1) - s_1)}. \tag{8.97}$$

The delta function selects the trajectories characterized by the given entropy production s_1 in the first time interval. Since the function to be averaged is constant, summing over x_1 leaves us with the probability of a state x_{t_1} conditioned to having produced an entropy s_1 in the first interval:

$$\left\langle e^{-s^{\text{tot}}(t)/k_B} \middle| e^{-s^{\text{tot}}(t_1)/k_B} \right\rangle = e^{-s^{\text{tot}}(t_1)/k_B} \frac{\int \mathcal{D}x_2 \, \mathcal{P}_{x_2|x_{t_1}} \, e^{-s_2(x_2)/k_B} p_{x_{t_1}|s_1}}{\int \mathcal{D}x_2 \, \mathcal{P}_{x_2}}, \tag{8.98}$$

where

$$p_{x_{t_1}|s_1} = \frac{\int \mathcal{D}x_1 \, \mathcal{P}_{x_1} \, \delta(s^{\text{tot}}(t_1) - s_1) \, \delta^K_{x(t_1),x_{t_1}}}{\int \mathcal{D}x_1 \, \mathcal{P}_{x_1} \delta(s^{\text{tot}}(t_1) - s_1)}. \tag{8.99}$$

The denominator on the right-hand side of eq. (8.98) is equal to 1 because of the normalization. Using the definition of entropy production, eq. (8.96), we obtain

$$\left\langle e^{-s^{\text{tot}}(t)/k_B} \middle| e^{-s^{\text{tot}}(t_1)/k_B} \right\rangle = e^{-s^{\text{tot}}(t_1)/k_B} \int \mathcal{D}x_2 \, p_{x_t} \mathcal{P}_{\hat{x}_2|x_t} \frac{p_{x_{t_1}|s_1}}{p_{x_{t_1}}}$$

$$= e^{-s^{\text{tot}}(t_1)/k_B} \sum_{x_{t_1}} p_{x_{t_1}|s_1} = e^{-s^{\text{tot}}(t_1)/k_B}, \tag{8.100}$$

as anticipated.

We now explore the physical consequences of the fact that, in the steady state, e^{-s^{tot}/k_B} is a martingale. First of all, at the initial time t_0, s^{tot} vanishes, and therefore $e^{-s^{\text{tot}}(t_0)/k_B} = 1$. This implies that, for any stopping time t_{stop} satisfying at least one of the conditions above, one has

$$\left\langle e^{-s^{\text{tot}}/k_B} \right\rangle_{t_{\text{stop}}} = 1. \tag{8.101}$$

Equation (8.101) is the **integral fluctuation relation at stopping times**. For a fixed t_{stop}, eq. (8.101) reduces to the steady-state integral fluctuation theorem, eq. (4.15).

An interesting nontrivial example is the stopping time t_{stop} defined as the first time for which $e^{-s^{\text{tot}}/k_B} \geq c$ with $c > 1$. If in a given trajectory e^{-s^{tot}/k_B} never reaches c, then $t_{\text{stop}} = \infty$. Before the stopping time, one has

$$0 < \left\langle e^{-s^{\text{tot}}/k_B} \right\rangle_{t_{\text{stop}}} \leq c, \tag{8.102}$$

and therefore the stopping time satisfies condition 3 of Doob's theorem. We call p_c the probability that $e^{-s^{\text{tot}}/k_B} \geq c$ at a finite time. Since for long times $e^{-s^{\text{tot}}/k_B} \to 0$, we obtain from eq. (8.101)

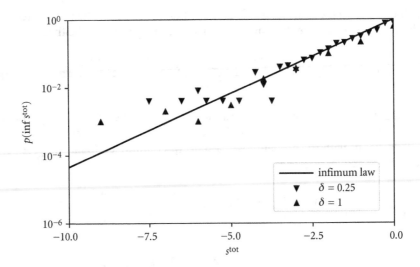

Figure 8.5. Infimum law. Points are obtained from 10^4 simulations of a dragged particle on a ring; see section 4.3. The parameter δ represents the amount of external work per step in the positive direction in units $k_B T$. The probability density $p(\inf s^{\text{tot}})$ is plotted against s^{tot}, measured in units of k_B. The line is the infimum law given by eq. (8.105). The vertical axis is logarithmic.

$$p_c \left\langle e^{-s^{\text{tot}}/k_B} \right\rangle_c = 1, \tag{8.103}$$

where $\left\langle e^{-s^{\text{tot}}/k_B} \right\rangle_c$ is the average value of e^{-s^{tot}/k_B} at the time of passing the threshold c, conditioned on passing it. If the trajectories of s^{tot} are continuous, then $\left\langle e^{-s^{\text{tot}}/k_B} \right\rangle_c = c$, and we obtain $p_c = 1/c$. Changing variables, p_c represents the probability that s^{tot} attains values smaller than or equal to $\bar{s} = -k_B \ln c$ at a finite time. This means that, for continuous processes,

$$p \left(s^{\text{tot}} \leq \bar{s} \text{ for some finite } t \right) = e^{\bar{s}/k_B}. \tag{8.104}$$

Equation (8.104) can be seen as the cumulative distribution of the **infimum** of s^{tot} along an infinite trajectory. We obtain the probability density of the infimum by taking the derivative with respect to $\bar{s}$:

$$p \left(\inf s^{\text{tot}} = \bar{s} \right) = \frac{1}{k_B} e^{\bar{s}/k_B}. \tag{8.105}$$

Equation (8.105) is the **infimum law** (fig. 8.5). It states that, for any steady-state continuous process, the probability of observing a "negative record" $\bar{s}$ of entropy production is exponentially distributed with average k_B. For discontinuous trajectories, such as in discrete processes, one has in general $\left\langle e^{-s^{\text{tot}}/k_B} \right\rangle_c \leq c$, since in principle discontinuous trajectories can jump below the threshold without hitting it exactly. Therefore, in the discontinuous case, the distribution (8.104) is an upper bound:

$$p\left(s^{\text{tot}} \leq \bar{s} \text{ for some finite } t\right) \leq e^{\bar{s}/k_B}. \tag{8.106}$$

The importance of the infimum law is that it provides a very simple bound on the probability of observing negative values of the entropy production. Mathematically, it is also a very good example of the usefulness of martingales and Doob's theorem: using them allows us to prove some simple and surprising general results in very few steps. Following a similar logic, we can prove that several statistical properties of s^{tot} related to more complex first-passage times are universal, and compute them explicitly.

8.8 Random time

Entropy production in continuous systems can be studied by a technique of stochastic processes called a **random time transformation**. To introduce this idea, we start with an instructive exercise in stochastic calculus. In section 3.13, we derived stochastic thermodynamics observables following the rules of the Stratonovich convention. We now derive the same observables in the Ito convention, which we use throughout this section. The energy change is expressed by

$$\frac{d}{dt}\epsilon(x(t), t) = \frac{\partial}{\partial t}\epsilon(x, t) + \frac{\partial}{\partial x}\epsilon(x, t) \cdot \frac{dx}{dt} + D\frac{\partial^2}{\partial x^2}\epsilon(x, t), \tag{8.107}$$

where the last term comes from applying the Ito formula (2.128). Similarly, the work change is expressed by

$$\frac{d}{dt}w(x(t), t) = \frac{\partial}{\partial t}\epsilon(x, t) + f(x, t) \cdot \frac{dx}{dt} + D\frac{\partial}{\partial x}f(x, t). \tag{8.108}$$

We obtain the heat change by applying the first law of stochastic thermodynamics:

$$\frac{d}{dt}q(x(t), t) = \frac{d}{dt}w(x, t) - \frac{d}{dt}\epsilon(x, t) = \mathcal{F}(x, t) \cdot \frac{dx}{dt} + D\frac{\partial}{\partial x}\mathcal{F}(x, t), \tag{8.109}$$

where $\mathcal{F}(x, t) = -(\partial_x \epsilon + f)$ is defined in eq. (3.82). Equations (8.107), (8.108), and (8.109) are equivalent to the expressions (3.86), (3.87), and (3.88), respectively, in the Stratonovich convention. We now proceed to compute the entropy production rate. The rate of system entropy change is given by

$$\frac{d}{dt}s^{\text{sys}} = -k_B \left\{ \frac{1}{p(x; t)}\frac{\partial}{\partial t}p(x; t) + \frac{1}{p(x; t)}\frac{\partial}{\partial x}p(x; t) \cdot \frac{dx}{dt} \right.$$
$$\left. - \frac{D}{p^2(x; t)}\left[\frac{\partial}{\partial x}p(x; t)\right]^2 + \frac{D}{p(x; t)}\frac{\partial^2}{\partial x^2}p(x; t) \right\}$$

$$= k_B \left\{ -\frac{2}{p(x;t)}\frac{\partial}{\partial t}p(x;t) - \frac{\mathcal{F}(x,t)}{k_B T}\cdot\frac{dx}{dt} + \frac{J(x,t)}{Dp(x;t)}\cdot\frac{dx}{dt} \right.$$

$$\left. + \frac{D}{p^2(x;t)}\left[\frac{\partial}{\partial x}p(x;t)\right]^2 - \frac{1}{p(x;t)}\frac{\partial}{\partial x}[\mu_P\mathcal{F}(x,t)p(x;t)] \right\}, \tag{8.110}$$

where in the last equality we use $J(x,t) = \mu_P\mathcal{F}(x,t)p(x;t) - D\,\partial_x p(x;t)$ and the Fokker-Planck equation. We obtain the total entropy production rate by applying the Ito formula once more:

$$\frac{d}{dt}s^{tot} = \frac{d}{dt}s^{env}(x(t),t) + \frac{d}{dt}s^{sys}(x(t),t)$$

$$= k_B \left\{ -\frac{2}{p(x;t)}\frac{\partial}{\partial t}p(x;t) + \mu_P\frac{\partial}{\partial x}\mathcal{F}(x,t) + \frac{J(x,t)}{Dp(x;t)}\cdot\frac{dx}{dt} \right.$$

$$\left. + \frac{D}{p^2(x;t)}\left[\frac{\partial}{\partial x}p(x;t)\right]^2 - \frac{1}{p(x;t)}\frac{\partial}{\partial x}[\mu_P\mathcal{F}(x,t)p(x;t)] \right\}$$

$$= k_B \left\{ -\frac{2}{p(x;t)}\frac{\partial}{\partial t}p(x;t) + \frac{J(x,t)}{Dp(x;t)}\cdot\frac{dx}{dt} \right.$$

$$\left. + \frac{D}{p^2(x;t)}\left[\frac{\partial}{\partial x}p(x;t)\right]^2 - \frac{\mu_P\mathcal{F}(x,t)}{p(x;t)}\frac{\partial}{\partial x}p(x;t) \right\} \tag{8.111}$$

$$= k_B \left[-\frac{2}{p(x;t)}\frac{\partial}{\partial t}p(x;t) + \frac{J(x,t)}{Dp(x;t)}\cdot\frac{dx}{dt} - \frac{J(x,t)}{p^2(x;t)}\frac{\partial}{\partial x}p(x;t) \right]$$

$$= k_B \left[-\frac{2}{p(x;t)}\frac{\partial}{\partial t}p(x;t) + \frac{J(x,t)^2}{Dp^2(x;t)} + \frac{\sqrt{2}J(x,t)}{\sqrt{D}p(x;t)}\xi(t) \right].$$

It is convenient to define

$$v_s(x,t) = \frac{J^2(x,t)}{Dp^2(x;t)}. \tag{8.112}$$

In terms of v_s, the entropy production rate is expressed by

$$\frac{ds^{tot}}{dt} = k_B \left[-2\frac{\partial}{\partial t}\ln p(x;t) + v_s(x,t) + \sqrt{2v_s(x,t)}\xi(t) \right]. \tag{8.113}$$

Equation (8.113) is most useful in the steady state, where $v_s(x)$ can be interpreted as the expected rate of entropy production in state x:

$$\frac{ds^{tot}}{dt} = k_B v_s(x) + k_B\sqrt{2v_s(x)}\,\xi(t). \tag{8.114}$$

The drift term of the Langevin equation (8.114) is equal to the diffusion term. This property is related to the fact that e^{-s^{tot}/k_B} is a martingale. In the steady state, we have in fact

$$\frac{\mathrm{d}}{\mathrm{d}t} e^{-s^{\mathrm{tot}}/k_B} = -e^{-s^{\mathrm{tot}}/k_B} \sqrt{2 v_s(x)}\, \xi(t). \tag{8.115}$$

Equation (8.115) is a Langevin equation (in the Ito interpretation) without drift: the drift term cancels out with the Ito term in the chain rule, precisely because the drift and diffusion terms in eq. (8.114) are equal. Since the average of the noise term in the Ito equation always vanishes, the future average of $e^{-s^{\mathrm{tot}}/k_B}$ is equal to the initial value, i.e., $e^{-s^{\mathrm{tot}}/k_B}$ is a martingale.

A convenient way to analyze eq. (8.114) is to introduce the **random time** t_{rnd}, such that

$$\mathrm{d}t_{\mathrm{rnd}} = v_s(x(t))\, \mathrm{d}t. \tag{8.116}$$

As the name suggests, the transformation between the "real" time t and the random time t_{rnd} is stochastic, i.e., it depends on the trajectory via the quantity $v_s(x(t))$. The random time transformation contracts or dilates time depending on whether the local entropy production rate is respectively large or small. In terms of the random time, eq. (8.114) becomes

$$\frac{\mathrm{d}}{\mathrm{d}t_{\mathrm{rnd}}} s^{\mathrm{tot}} = k_B + k_B \sqrt{2}\, \xi'(t_{\mathrm{rnd}}), \tag{8.117}$$

where $\xi'(t_{\mathrm{rnd}})$ is white noise. Equation (8.117) shows that the dynamics of entropy production is system-independent when measured in terms of the random time. This implies that first-passage probabilities of entropy production are universal, since these probabilities do not depend on the "speed" of the clock. For example, the infimum law introduced in eq. (8.105) can be derived by studying the probability of a trajectory of eq. (8.117) to cross a given negative threshold. Other universal properties of continuous systems can be similarly derived by considering different first-passage problems.

8.9 Population genetics

There exists a deep analogy between population genetics and thermodynamics. R. A. Fisher [55], one of the fathers of population genetics, already pointed out this analogy in the 1930s. In the context of the fundamental theorem of natural selection, according to which the rate of increase in reproductive value (fitness) of an organism is proportional to its variance due to genetic factors, he states:

> It will be noticed that the fundamental theorem proved above bears some remarkable resemblances to the second law of thermodynamics. Both are properties of population, or aggregates, true irrespective of the nature of the units which compose them; both are statistical laws; each requires the constant increase of a measurable quantity, in the one case the entropy of a physical system and in the other the fitness, measured by m, of a biological population. (p. 36)

This analogy extends to stochastic thermodynamics, and several works have exploited this idea to obtain new results in population genetics. We focus here on one approach, which allows us to obtain information on selection pressures in populations from genealogical data.

The reproduction rate of an individual in a population depends on many features that we call **traits**. Some of these traits are inheritable. In population genetics, it is important to identify traits and understand how they affect reproduction rates. This goal can be achieved by relating two distinct probability distributions of traits.

We consider a population initially made up of $N^{\text{tot}}(t_0) \gg 1$ asexually reproducing individuals. Each individual carries a trait specified by a variable x. Individuals reproduce at a constant rate r_x, depending on the trait and possibly on the environment. In a reproduction event, the "mother" cell divides into two "daughters." We momentarily consider the simplest scenario in which the two daughters carry the same value of the trait as their mother. At a given final time t_f, the population consists of $N^{\text{tot}}(t_f)$ individuals. We characterize each individual i by its trait x_i and by its **lineage**, i.e., by the collection of identities of its ancestors, from its mother up to its ancestor in the initial population at t_0. In particular, the lineage permits us to determine the number ρ_i of cell divisions between the initial ancestor and the individual at time t_f. If the model includes birth events only, the population grows exponentially with time. However, the size of most natural populations is limited by factors such as resource availability. A simple modeling choice in this case is to maintain the population size N constant: at every cell division, one other individual is removed. For example, in the **Moran model** of population genetics, the individual to be removed is chosen at random.

In these cases, the N individuals constituting the population at time t_f are the descendants of a smaller number $N(t_0)$ of individuals in the initial population. In the following, we focus on the individuals present at time t_f and all their ancestors as specified by their lineages. The number of individuals belonging to this set of lineages decreases monotonically from $N(t_f)$ to $N(t_0)$ as we go back in time. We assume that N is always large enough, so that we can neglect the effects of sampling fluctuations.

We momentarily focus on a single **clone**, i.e., on the family of descendants of a single individual in the initial population. Lineages within a clone can be sampled in two ways. One possibility is **retrospective sampling**. In retrospective sampling, we trace back the ancestors of the final population. Specifically, we randomly pick up one individual i in the final population with probability $p_i^{\text{ret}} = 1/N(t_f)$, where $N(t_f)$ is the size of the clone, and sample its lineage and its phenotypic trait x_i. An alternative is **chronological sampling**. In chronological sampling, we start from the ancestor and randomly pick up one of the two daughters at each division with equal probabilities. In this case, the lineage of an individual i that underwent ρ_i divisions since the initial time is sampled with probability $p_i^{\text{chr}} = 2^{-\rho_i}$. The probability p_i^{chr} depends only on the number ρ_i of divisions of the lineage of individual i. In contrast, the probability p_i^{ret} also depends on the reproductive performance of the other lineages, via the total number $N^{\text{tot}}(t_f)$ of individuals present at time t_f. The discrepancy between the two sampling probabilities reflects the variability in the number of divisions—i.e., in reproductive success—among the different lineages. Our aim is to exploit this discrepancy to evaluate the effect of the trait on the division rate.

We now generalize this idea to a heterogeneous population made up of individuals with diverse traits and possibly belonging to different clones. We assume to know the lineage, phenotypic trait x_i, and number of divisions ρ_i of each individual i in the final population. We then evaluate the joint retrospective distribution

$$p_{\rho,x}^{\text{ret}} = \frac{n_{\rho,x}}{N(t_f)}, \tag{8.118}$$

where $n_{\rho,x}$ is the number of individuals at time t_f with a lineage with ρ divisions and phenotypic trait x. We also define the joint chronological distribution

$$p_{\rho,x}^{\text{chr}} = \frac{2^{-\rho}}{N(t_0)} n_{\rho,x}. \tag{8.119}$$

The population growth rate is defined by

$$\Lambda = \frac{1}{\mathcal{T}} \ln \frac{N(t_{\text{f}})}{N(t_0)}, \tag{8.120}$$

where $\mathcal{T} = t_{\text{f}} - t_0$. We then have the relation

$$p_{\rho,x}^{\text{ret}} = e^{\mathcal{T} (\tilde{h}_\rho - \Lambda)} p_{\rho,x}^{\text{chr}}, \tag{8.121}$$

where

$$\tilde{h}_\rho = \frac{1}{\mathcal{T}} \rho \ln 2. \tag{8.122}$$

Equation (8.121) bears a formal similarity to fluctuation relations, with the two distributions $p_{\rho,x}^{\text{ret}}$ and $p_{\rho,x}^{\text{chr}}$ respectively playing the roles of the forward and backward trajectory probabilities. The quantity $\mathcal{T} (\tilde{h}_x - \Lambda)$ is therefore analogous to a total entropy production. As with fluctuation theorems, it is possible to derive variants of eq. (8.121) by choosing different forward and backward distributions.

We follow this approach to estimate **fitness**. The fitness of a trait x measures the reproductive success of the individuals carrying it. It is formally defined as the expected number of offspring of an individual with the given trait. In practice, fitness is rather difficult to evaluate. By comparing retrospective and chronological sampling, we can at least obtain a related quantity. We introduce the marginals

$$p_x^{\text{ret}} = \sum_\rho p_{\rho,x}^{\text{ret}}; \qquad p_x^{\text{chr}} = \sum_\rho p_{\rho,x}^{\text{chr}}. \tag{8.123}$$

We define the **fitness landscape** h_x as a measure of the dependence of the effective division rate on a trait x:

$$h_x = \frac{1}{\mathcal{T}} \ln \frac{N(t_{\text{f}}) p_x^{\text{ret}}}{N(t_0) p_x^{\text{chr}}} = \Lambda + \frac{1}{\mathcal{T}} \ln \frac{p_x^{\text{ret}}}{p_x^{\text{chr}}}. \tag{8.124}$$

Inverting this relation, we obtain

$$p_x^{\text{ret}} = e^{\mathcal{T} (h_x - \Lambda)} p_x^{\text{chr}}. \tag{8.125}$$

The fitness landscape specifies the dependence of the division rate on the trait x. If genealogical data on a population is available, the fitness landscape h_x can be obtained by sampling both p_x^{ret} and p_x^{chr}. In the Moran model, if some individuals in the initial population do not have descendants at t_{f}, the fitness landscape differs from the reproduction rate by an additive constant, dependent on the probability of such events. This phenomenon can be understood using the formalism of section 8.5: lineages having no descendant are analogous to irreversible trajectories in stochastic thermodynamics.

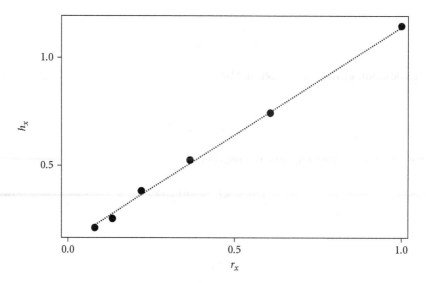

Figure 8.6. Estimated fitness landscape h_x plotted against division rate r_x for a population of $N = 10,000$ individuals evolving according to a Moran model for time $\mathcal{T} = 5$, where the unit of time is the doubling time of the fastest clones. The trait x can take values from 0 to 5, with division rate $r_x = e^{-x/2}$. The dotted line is a fit to $h = r + \text{const}$. The average number of divisions per lineage in this run is 5.64.

Substituting the relation

$$p_{\rho|x}^{\text{chr}} = \frac{p_{\rho,x}^{\text{chr}}}{p_x^{\text{chr}}} \tag{8.126}$$

in eq. (8.124) and expressing $p_{\rho,x}^{\text{ret}}$ by

$$p_{\rho,x}^{\text{ret}} = \frac{N(t_0)}{N(t_f)} 2^{\rho} p_{\rho,x}^{\text{chr}}, \tag{8.127}$$

we obtain the following alternative expression of the fitness landscape:

$$h_x = \frac{1}{\mathcal{T}} \ln \sum_{\rho} 2^{\rho} p_{\rho|x}^{\text{chr}}. \tag{8.128}$$

The expression (8.128) provides a more convenient way to estimate the fitness landscape from lineage distributions. Given lineage data from a large enough population, this procedure can be carried out by looking at a relatively small number of generations (fig. 8.6).

We now generalize our theory to population dynamics characterized by **phenotypic switching**. Phenotypic switching is a process that changes the trait x of an individual, and can originate from different biological mechanisms. For example, a strong evolutionary pressure can lead to rapid adaptation of a population, with a consequential shift of individual traits within a few generations. More recently, it has been observed that even genetically identical populations often diversify individual phenotypes. In particular, individuals belonging to these populations may stochastically switch their trait x in the course of their lifetimes.

We model the latter case assuming that individuals switch from phenotype x' to x with constant rate $k_{xx'}$. Phenotypic switching and duplication occur independently. Switching can be costly for individuals. We denote by $e^{d_{xx'}}$ the survival probability of an individual switching its phenotype, where $d_{xx'} \leq 0$. We define the **phenotypic trajectory** of a lineage as

$$x = ((x_0, t_0), (x_1, t_1), \dots, (x_n, t_n), t_f). \tag{8.129}$$

The expected number of descendants of an individual with phenotypic trajectory x is

$$\mathcal{N}_x = e^{r_{x_n}(t_f - t_n)} \prod_{\ell=0}^{n-1} \left(e^{r_{x_\ell}\,(t_{\ell+1} - t_\ell) + d_{x_{\ell+1} x_\ell}} \right)$$

$$= \exp\left[\sum_x r_x \tau_x(x) + \sum_{x \neq x'} d_{xx'} n_{xx'}(x) \right], \tag{8.130}$$

where $\tau_x(x)$ are the empirical dwell times and $n_{xx'}(x)$ the empirical jump numbers associated with the phenotypic trajectory x, respectively defined by eqs. (6.13) and (6.99). The probability of a phenotypic trajectory is also given by the expression (2.91) of the trajectory probability $\mathcal{P}_x$ of a master equation with jump rates $k_{xx'}$. We therefore identify $\mathcal{N}_x$ as the probability ratio of the phenotypic trajectory x in the retrospective sampling to that in the chronological sampling. The expected total population size at time t_f is thus given by

$$N^{\text{tot}}(t_f) = \int \mathcal{D}x \, \mathcal{N}_x \, \mathcal{P}_x \, N^{\text{tot}}(t_0). \tag{8.131}$$

The population growth rate (8.120) is therefore

$$\Lambda = \lim_{\mathcal{T} \to \infty} \frac{1}{\mathcal{T}} \ln \langle \mathcal{N}_x \rangle$$

$$= \lim_{\mathcal{T} \to \infty} \frac{1}{\mathcal{T}} \ln \left\langle \exp\left[\sum_x r_x \tau_x(x) + \sum_{x \neq x'} d_{xx'} n_{xx'}(x) \right] \right\rangle, \tag{8.132}$$

where the average is taken with respect to the probability density $\mathcal{P}_x$ of the phenotypic trajectories. Equation (8.132) shows that the population growth rate is the scaled cumulant generating function of the dwell-time and jump-number distribution. By the Gärtner-Ellis theorem, we express Λ as the Legendre-Fenchel transform of the rate function $I(f, \kappa)$ of the empirical jump rates κ and the empirical vector f:

$$\Lambda = \psi^{(f, \kappa)}(r, d) = \sup_{f, \kappa} \left[\sum_x r_x f_x + \sum_{x \neq x'} d_{xx'} \kappa_{xx'} f_{x'} - I(\kappa, f) \right]. \tag{8.133}$$

The rate function $I(f, \kappa)$ is given by eq. (6.105), and the supremum is taken over all f, κ satisfying the empirical stationarity condition; see section 6.9. We compute the supremum (f^*, κ^*) using tilting. The generator of the tilted dynamics reads

$$L_{xx'} = \left(r_x - k_x^{\mathrm{out}}\right) \delta_{xx'}^{\mathrm{K}} + e^{d_{xx'}} k_{xx'}, \tag{8.134}$$

where $k_x^{\mathrm{out}} = \sum_{x'} k_{x'x}$. We obtain

$$\kappa_{xx'}^* = \frac{u_x}{u_{x'}} \left(L_{xx'} - \Lambda \, \delta_{xx'}^{\mathrm{K}}\right), \qquad f_x^* = v_x, \tag{8.135}$$

where $u = (u_x)$ and $v = (v_x)$ are respectively the left and right eigenvectors of $L = (L_{xx'})$ belonging to its maximum eigenvalue Λ. This procedure bears an analogy with the cloning algorithm described in section 6.8 to evaluate the scaled cumulant generating function of a current.

The rates κ^* define the **retrospective process** associated with our population dynamics. The retrospective process is obtained by picking up a random individual in the present population and tracing its phenotypic trajectory backward to a very remote time when it becomes a stationary process. Its stationary distribution f^* is called the **ancestral distribution**. Its trajectories x are the phenotypic histories, with an associated trajectory probability density $\mathcal{P}_x^*$. The variational principle (8.133) permits us to obtain the change of the population growth rate Λ under a change of the division rate r, the switch cost d, or the switch rate k. Taking the derivative of the evolution operator L with respect to these quantities and taking into account the properties of κ^*, f^*, we obtain

$$\delta \Lambda = \sum_x f_x^* \, \delta r_x + \sum_{x \neq x'} \kappa_{xx'}^* f_x^* \, \delta d_{xx'} + \sum_{x \neq x'} \left(\frac{\kappa_{xx'}^* f_{x'}^*}{k_{xx'}} - f_{x'}^*\right) \delta k_{xx'}. \tag{8.136}$$

In this way, the analogy with stochastic thermodynamics leads to fluctuation relations valid for populations evolving in time-varying environments.

8.10 Further reading

Our discussion of stochastic efficiency is mainly based on Verley et al. [172, 175]. Discussions of the finite time behavior of stochastic efficiency can be found in Gingrich et al. [71] and Polettini et al. [131].

The thermodynamic uncertainty relations were first proposed by Barato and Seifert [9]. Gingrich et al. [69] rigorously proved the original thermodynamic uncertainty relation using large deviation theory. The first part of section 8.2 on uncertainty relations follows Hasegawa and Van Vu [74]. Pietzonka et al. [127] apply them to study the efficiency of molecular motors. Horowitz and Gingrich [78] point out that the finite time uncertainty relation follows from large deviation theory relations applied to large ensembles. Dechant and Sasa [39] (see also [40, 38]) generalize the uncertainty relations to multidimensional currents and establish their connection with the Cramér-Rao inequality. Seifert [150] and Horowitz and Gingrich [79] review general aspects of uncertainty relations.

The statistics of first-passage times was first considered by Saito and Dhar [141] and Garrahan [60]. There are several further theoretical studies, and experiments are reported by Singh et al. [156]. We follow in particular Gingrich and Horowitz [68]. Murashita et al. [114] discuss fully irreversible processes. This approach leads to an interesting way of discussing the Gibbs paradox [Murashita and Ueda, 115]. The discussion of the optimal protocol in section 8.6 follows Sivak and Crooks [157], and the experiment mentioned in the section is due to Tafoya et al. [164]. An alternative approach is due to Schmiedl and Seifert [145] and is developed by Aurell et al. [6, 5], among others. Our discussion on the application of martingales in stochastic thermodynamics is based on Neri et al. [117, 118]. The random time transformation for the entropy production is introduced in the work by Pigolotti et al. [128].

Leibler and Kussell [102] compare retrospective and chronological sampling in population dynamics. Mustonen and Lässig [116] propose a fluctuation relation involving fitness in evolving populations. Kobayashi and Sughiyama [88] and Sughiyama et al. [161] establish and develop the analogy with stochastic thermodynamics, obtaining several fluctuation relations. García-García et al. [58] further discuss these relations and their applications. Our exposition follows Nozoe et al. [119].

8.11 Exercises

8.1   A particle is dragged on a ring with N states as in section 4.3. Evaluate its entropy production rate in the steady state and verify that, for any value of the driving δ, the current satisfies the uncertainty bounds in section 8.2.

8.2   Evaluate the rate function $I(j)$ for the model of enzyme-mediated reaction introduced in section 4.4, by applying the tilted dynamics and the Gärtner-Ellis theorem. Compare with the weak linear response bound given in eq. (8.36).

8.3   A Brownian particle is subject to a constant force f in one dimension. Verify analytically and numerically that the empirical one-dimensional current saturates the uncertainty bound (8.24). Provide a heuristic explanation of why this is the case.

8.4   Following the logic of section 8.7, prove that e^{-s^a/k_B} is a martingale, where s^a is the adiabatic entropy production introduced in section 4.9.

8.5   Use the random time form (8.117) of the entropy production rate to prove the infimum law (8.105). (Hint: Use the backward Kolmogorov equation (2.106).)

CHAPTER 9

Perspectives

Stochastic thermodynamics is nowadays an established branch of nonequilibrium statistical physics, and there is a growing interest in extending its application to a broader range of systems. Prominent examples come from the field of active matter, which seeks to describe, for example, scenarios in which the chemical reservoir is itself a nonequilibrium system fueled by chemical energy. Other theoretical challenges arise when applying stochastic thermodynamics to the quantum world. Information-processing biological systems constitute another interesting venue, also thanks to the steady improvement of quantitative techniques in experimental biophysics. Before discussing these future directions, we touch upon more fundamental aspects that are related to how stochastic thermodynamics fits into the grand problem of understanding nonequilibrium physical systems.

In short, stochastic thermodynamics aims at relating a mesoscopic physical system, described in terms of a stochastic process, with the macroscopic laws of thermodynamics. This goal is sketched in fig. 1.1, at the very beginning of this book. We have seen throughout this book that this goal is achieved in a relatively simple and elegant way. Compare, for example, the compact expressions of fluctuation theorems with the intricacies of deriving irreversibility from kinetic theory. A main reason for this simplification is that the mesoscopic dynamics which constitutes the starting point of stochastic thermodynamics, is already irreversible, so that the fundamental problem of irreversibility is not solved but rather swept under the rug; see section 1.2. It is probably fair to conclude that stochastic thermodynamics aims at a simpler goal than kinetic theory and succeeds in providing simpler answers.

A more concerning and subtle problem is the following. Let us imagine being able to derive the macroscopic thermodynamic behavior of a nonequilibrium physical system from its microscopic dynamics. In parallel, we also describe the system via a coarse-grained stochastic dynamics, and therefore also obtain the macroscopic behavior using the tools of stochastic thermodynamics. Can we expect the two macroscopic descriptions to be identical? In other words, are the two paths from the microscopic to the macroscopic world in fig. 1.1 in general equivalent?

Unfortunately, theoretical arguments suggest that this equivalence does not hold in general [72]. The discrepancy between the two descriptions originates in the definition of entropy in statistical mechanics. One definition is the so-called **Boltzmann entropy**. One associates with a microscopic state ξ of a large system the empirical distribution $p(\vec{r}, \vec{p}_r)$ of the positions and momenta of all particles. The set of microstates compatible with the same $p(\vec{r}, \vec{p}_r)$ identifies a **macrostate** $M(\xi)$. We denote by $\Gamma(M(\xi))$ the region of the phase space spanned by these microstates and by $|\Gamma(M(\xi))|$ its volume. The Boltzmann entropy is proportional to the logarithm of the phase-space volume of this region:

$$S^{(B)} = k_B \ln |\Gamma(M(\xi))| . \tag{9.1}$$

The Boltzmann entropy is a fluctuating quantity, since it depends on the microstate ξ. Alternatively, we can describe a macrostate by an *ensemble* $p(\xi)$, as is customarily done in equilibrium statistical mechanics. The entropy associated with this ensemble is called the **Gibbs entropy**:

$$S^{(G)} = -k_B \int d\xi \, p(\xi) \ln p(\xi). \tag{9.2}$$

At equilibrium, these two definitions are equivalent, in the sense that the average of $S^{(B)}$ is equal to $S^{(G)}$ (at least for gases) and its fluctuations are small. Out of equilibrium, this is not necessarily the case. A classic example is a gas prepared in an atypical state, for example, compressed into a corner of a container having adiabatic walls, and then allowed to relax to equilibrium. For such a system, the Boltzmann entropy defined in eq. (9.1) increases as expected during relaxation. On the contrary, due to the Liouville theorem, the Gibbs entropy defined in eq. (9.2) always remains constant since the system is isolated. This observation, combined with the fact that eq. (9.1) is defined for a single macroscopic system whereas eq. (9.2) requires the definition of an ensemble, leads us to conclude that the Boltzmann entropy is the proper choice to characterize irreversibility in isolated thermodynamic systems.

We should remark that this conclusion is not universally accepted. Defenders of eq. (9.2) claim that both eqs. (9.1) and (9.2) require a coarse graining of phase space, since it is impossible to measure positions and momenta with infinite precision. In the presence of coarse graining, the Gibbs entropy becomes a random quantity that increases with time, on average. However, following this line of thought, one finds that the increase with time of the Gibbs entropy depends on the level of coarse graining, at variance with that of the Boltzmann entropy [52]. It is our contention that the relaxation to equilibrium of a large isolated system is objective, i.e., independent of the precision of any conceivable microscopic measurement.

On the other hand, stochastic thermodynamics is firmly based on ensembles and consequently on the ensemble definition of entropy; see section 3.6. Stochastic thermodynamics works because of the presence of a large heat reservoir always at equilibrium, which makes the system dynamics irreversible. In this case, the ensemble perspective is justified by the fact that the system is small and therefore its dynamics is not deterministic like the dynamics of macroscopic systems. In short, macroscopic thermodynamics and stochastic thermodynamics are consistent theories if taken individually. However, it is not obvious whether one can invoke a "correspondence principle" and recover macroscopic nonequilibrium thermodynamics by taking a limit in which a mesoscopic system becomes large.

Another subtle issue with the program of stochastic thermodynamics concerns the separation between the mesoscopic system and the heat reservoir surrounding it. Besides the separation of timescales, an underlying assumption is that the interaction energy between the system and the reservoir should be negligible compared with the system energy. If this assumption does not hold, the use of projection techniques, such as those mentioned in section 4.15, requires more care. Stochastic thermodynamics has been generalized to these situations [83], although the interpretation of thermodynamic quantities has generated some controversy [165, 160].

Besides these fundamental aspects, stochastic thermodynamics has the potential to be applied to a broader range of systems than those studied so far. For example, "traditional" stochastic thermodynamics assumes the presence of equilibrium reservoirs in contact with the nonequilibrium system under study. However, many interesting physical and biological systems are embedded in fluids that are steadily kept out of equilibrium by the consumption of energy. For example, the cytosol, i.e., the liquid found inside biological cells, contains several chemical species undergoing chemical reactions. These reactions are kept out of equilibrium by other chemical reservoirs. Fluid (or solid) nonequilibrium media collectively fall under the denomination of **active matter** [109]. The phenomenology of mesoscopic nonequilibrium systems in contact with active reservoirs is rather rich. For example, work can be extracted from a single active heat reservoir without necessarily violating the second law of thermodynamics. There have been interesting attempts to extend stochastic thermodynamics to include active reservoirs. Such attempts are still at relatively early stages at the time of writing this book.

Quantum systems represent another important field of application. The governing equations of quantum mechanical systems coupled to a heat reservoir, formulated either in terms of the Schrödinger equation or in terms of a density matrix, display symmetries similar to those of classical systems. In particular, the unitary nature of quantum dynamics, which preserves normalization of the wave function, is mathematically similar to the conservation of probability in classical stochastic dynamics. Further, interaction with a heat reservoir breaks the time-reversal symmetry in a qualitatively similar way in classical and quantum systems. Fundamental results in classical stochastic thermodynamics, such as fluctuation theorems, are based on these symmetries and therefore have been extended to the quantum realm during the early stages of stochastic thermodynamics [93]. On the other hand, quantum stochastic thermodynamics presents subtleties that are absent in the classical case. One of them concerns the nature of interactions with the heat reservoir. There are several different techniques to introduce these interactions, usually based on describing the degrees of freedom of the heat reservoir explicitly and then averaging them out to obtain an effective equation for the system alone. These different techniques are not equivalent, and each of them presents advantages and disadvantages. Another issue concerns with the definition of "work" in quantum mechanics. Operationally, measuring work requires measuring the energy of the system at two different times. However, in quantum mechanics, measurements can fundamentally alter the state of the system, and it has therefore been questioned whether work defined in this way is a legitimate observable. Nowadays quantum stochastic thermodynamics is a rather developed field, as reviewed in [49, 41].

Finally, stochastic thermodynamics provides a natural framework to study the performance of molecular machines operating out of equilibrium, such as molecular motors or information-processing enzymes, and indeed there exists a rich literature pursuing this direction. The application of stochastic thermodynamics in this context has been mostly carried on by the analysis of extremely simplified models, which helped to clarify the conceptual issues involved. At the same time, much progress has been made in quantitative experimental techniques in biophysics. This progress calls for the development and analysis of more realistic models that could be used to interpret experimental results and predict new, unexpected features of biological systems.

Appendixes

In these appendixes, we collect some introductory material and technical calculations whose exposition would have made heavy reading in the main chapters.

A.1 Convex functions and the Jensen inequality

Many useful inequalities are a consequence of a general relation called the Jensen inequality, which holds for **convex functions**. There is sometimes confusion on the definitions of concave and convex. The most used convention in mathematics is to define a ∪-shaped function as convex and a ∩-shaped function as concave. A way to memorize the definition is to think that a ∩-shape looks like a cave, hence it is concave. More formally, a real-valued function $f(x)$ is **convex** over an interval (a, b) if, for any $x_0, x_1 \in (a, b)$ and $0 \le \alpha \le 1$, one has

$$f\big((1 - \alpha)x_0 + \alpha x_1\big) \le (1 - \alpha)f(x_0) + \alpha f(x_1). \tag{A.1}$$

See the scheme in fig. A.1. A function f is **concave** if $(-f)$ is convex. If the inequality is strict for any $x_0 \ne x_1$ and $0 < \alpha < 1$, f is said to be **strictly convex**.

If a function $f(x)$ has a second derivative that is nonnegative (positive) in (a, b), then f is convex (strictly convex). Indeed, given x_0 and x, by the Taylor expansion and by the mean value theorem, we know that there is a point u between x_0 and x such that

$$f(x) = f(x_0) + f'(x_0)\,(x - x_0) + \frac{1}{2}f''(u)\,(x - x_0)^2, \tag{A.2}$$

where the last term is nonnegative. Let us now consider $x_\alpha = (1 - \alpha)x_0 + \alpha x_1$. We obtain

$$f(x_0) \ge f(x_\alpha) + f'(x_\alpha)\,(x_0 - x_\alpha) = f(x_\alpha) - f'(x_\alpha)\,\alpha\,(x_1 - x_0) \tag{A.3}$$

and

$$f(x_1) \ge f(x_\alpha) + f'(x_\alpha)\,(x_1 - x_\alpha) = f(x_\alpha) + f'(x_\alpha)\,(1 - \alpha)\,(x_1 - x_0). \tag{A.4}$$

Multiplying (A.3) by $(1 - \alpha)$ and (A.4) by α and summing, we obtain (A.1). The strict inequality is derived along the same lines when $f''(x) > 0$, $\forall x \in (x_0, x_1)$.

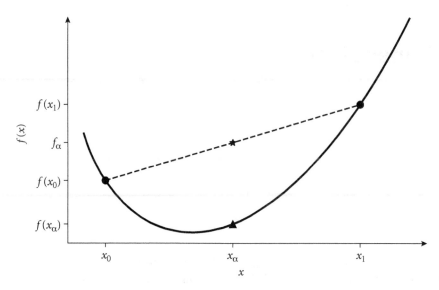

Figure A.1. Illustration of convexity. Given the point $x_\alpha = \alpha x_1 + (1 - \alpha)x_0$, with $0 \leq \alpha \leq 1$, we have $f(x_\alpha) \leq f_\alpha = (1 - \alpha)f(x_0) + \alpha f(x_1)$, $\forall \alpha$.

We also have the converse result: if the convex function $f(x)$ is twice derivable in the interval $[x_0, x_1]$, then $f''(x) \geq 0$ in the interval. Indeed, we have by definition, for any x for which $f''(x)$ exists,

$$f''(x) = \lim_{h \to 0} \frac{f(x+h) + f(x-h) - 2f(x)}{h^2}. \tag{A.5}$$

On the other hand, by convexity, we have

$$\frac{1}{2}\left(f(x+h) + f(x-h)\right) \geq f(x). \tag{A.6}$$

Thus the quantity on the right-hand side of the relation is nonnegative for all $h \neq 0$. In the limit, we obtain $f''(x) \geq 0$. The strict inequality does not necessarily hold, even if $f(x)$ is strictly convex. An example is the function $f(x) = x^4$.

We now take a convex function $f(x)$, which is twice differentiable in (a, b), and a point x_0 in the interior of the interval (a, b). Then $f(x) \geq f(x_0) + f'(x_0)(x - x_0)$ for $x \in (a, b)$. In other words, the graph of the convex function $f(x)$ lies above the tangent to it in x_0. The inequality is strict if $f(x)$ is strictly convex. Indeed, by the Taylor expansion and the mean value theorem, we have, for some $u \in (x_0, x)$,

$$f(x) = f(x_0) + f'(x_0)(x - x_0) + \frac{1}{2}f''(u)(x - x_0)^2, \tag{A.7}$$

where the last term is nonnegative. Thus we have proved the weak inequality. To prove the strict inequality, we assume that $f(x)$ is strictly convex and that $f(x) = f(x_0) + f'(x_0)(x - x_0)$ for some $x \neq x_0$. Convexity implies that, for any $u \in (x_0, x)$,

$$f(u) \leq f(x)\frac{u-x_0}{x-x_0} + f(x_0)\frac{x-u}{x-x_0} = f'(x_0)(u-x_0) + f(x_0). \qquad (A.8)$$

On the other hand, the result in eq. (A.7) implies

$$f(u) \geq f'(x_0)(u-x_0) + f(x_0) \qquad (A.9)$$

and therefore

$$f(u) = f(x)\frac{u-x_0}{x-x_0} + f(x_0)\frac{x-u}{x-x_0}, \qquad (A.10)$$

which violates the strict convexity.

The **Jensen inequality** states that if f is a convex function and x is a random variable, then

$$\langle f \rangle \geq f(\langle x \rangle), \qquad (A.11)$$

where $\langle x \rangle$ is the average of x. If f is strictly convex, the equality in (A.11) implies that x is a constant.

To prove this result, we define $g(x) = f(\langle x \rangle) + f'(\langle x \rangle)(x - \langle x \rangle)$. Since $f(x)$ is convex, we have $f(x) \geq g(x)$, $\forall x$, with equality for $x = \langle x \rangle$. Moreover, if $f(x)$ is strictly convex, we have $f(x) > g(x)$, $\forall x \neq \langle x \rangle$. Since $g(x)$ is linear, $\langle g(x) \rangle = g(\langle x \rangle)$. Thus we have

$$\langle f \rangle \geq \langle g \rangle = g(\langle x \rangle) = f(\langle x \rangle). \qquad (A.12)$$

The inequality is strict if $f(x)$ is strictly convex, unless x assumes only one value.

The Jensen inequality implies some useful corollaries. We often use it in cases in which $f(x)$ is an exponential function:

$$\langle e^{-x} \rangle \geq e^{-\langle x \rangle}, \qquad (A.13)$$

for any real-valued random variable x. Taking the logarithm, this relation implies

$$\langle x \rangle \geq -\ln \langle e^{-x} \rangle. \qquad (A.14)$$

We now consider two vectors (p_x) and (q_x) with nonnegative entries. The function $f(x) = x \ln x$ (for $x > 0$) is convex. We define

$$S = \sum_x p_x \ln \frac{p_x}{q_x} \qquad (A.15)$$

and the probability distribution

$$u_x = \frac{q_x}{q}, \qquad (A.16)$$

where $q = \sum_{x'} q_{x'}$. We then have

$$S = q \sum_x u_x \left(\frac{p_x}{q_x}\right) \ln \frac{p_x}{q_x}. \qquad (A.17)$$

By the Jensen inequality, we obtain

$$\mathcal{S} \ge qf\left(\sum_x u_x \frac{p_x}{q_x}\right) = \left(\sum_x p_x\right) \ln \frac{\sum_{x'} p_{x'}}{\sum_{x'} q_{x'}} = p \ln \frac{p}{q}, \qquad (A.18)$$

where $p = \sum_x p_x$. This relation is known as the **logsum inequality.** One of its immediate consequences holds when both p_x and q_x are probability distributions, satisfying $\sum_x p_x = 1$ and $\sum_x q_x = 1$. We then have

$$D_{KL}(p\|q) = \sum_x p_x \ln \frac{p_x}{q_x} \ge 0. \qquad (A.19)$$

The quantity on the left-hand side is known as the **Kullback-Leibler divergence** of the probability distributions p_x and q_x. It is a measure of how the distribution p_x is different from q_x.

We also mention a useful property of the Kullback-Leibler divergence. Consider a time-dependent solution $p_x(t)$ of a master equation with a unique stationary solution p_x^{st}, positive for all x. Then the Kullback-Leibler divergence $D_{KL}(p(t)\|p^{st})$ decreases monotonically:

$$\frac{d}{dt} D_{KL}(p(t)\|p^{st}) \le 0. \qquad (A.20)$$

This is a consequence of a more general result,

$$\frac{d}{dt}\Phi(p) = \frac{d}{dt}\sum_x p_x^{st} f\left(\frac{p_x(t)}{p_x^{st}}\right) \le 0, \qquad (A.21)$$

where $f(x)$ is an arbitrary convex function. We have indeed

$$\begin{aligned}
\frac{d}{dt}\Phi(p) &= \sum_{xx'} \left[k_{xx'}p_{x'} - k_{x'x}p_x\right]f'\left(\frac{p_x}{p_x^{st}}\right) \\
&= \sum_{xx'} k_{xx'}p_{x'}^{st}\left[u_{x'}f'(u_x) - u_{x'}f'(u_{x'})\right],
\end{aligned} \qquad (A.22)$$

where we introduce the shorthand $u_x = p_x/p_x^{st}$. We now have, for any arbitrary vector $w = (w_x)$,

$$\sum_{xx'} k_{xx'}p_{x'}^{st}(w_x - w_{x'}) = 0. \qquad (A.23)$$

Choosing $w_x = f(u_x) - u_x f'(u_x)$ and summing this relation to (A.22), we obtain

$$\frac{d}{dt}\Phi(p) = \sum_{xx'} k_{xx'}p_{x'}^{st}\left[(u_{x'} - u_x)f'(u_x) + f(u_x) - f(u_{x'})\right]. \qquad (A.24)$$

The quantity in brackets is nonpositive by (A.9), and the result (A.21) follows. Choosing $f(x) = x \ln x$, we obtain the result for the Kullback-Leibler divergence.

A.2   Legendre transformation

In this appendix, we review the Legendre transformation and its properties. We consider a function $f(x)$. For simplicity, we assume that it is twice differentiable and that its second derivative is positive. It is therefore possible to solve for x in the relation

$$\frac{df(x)}{dx} = q. \tag{A.25}$$

We would like to use the derivative q instead of x as an independent variable. If we solve for x in eq. (A.25) and substitute it in $f(x)$, we indeed obtain a function of q. However, we would obtain the same function if, instead of starting from $f(x)$, we started from a function $f_1(x) = f(x + C)$, where C is an arbitrary constant. Therefore, this procedure discards important information on f. For example, if f depends on other variables beyond x, C becomes an arbitrary function of the other variables. This ambiguity is removed by the Legendre transformation, which allows us to define a function $g(q)$ that is equivalent to $f(x)$, in the sense that its knowledge allows us to recover precisely the original $f(x)$.

With our hypotheses, the function $f(x)$ is convex. We can represent it by a curve in the (x, f) plane. The same curve is also identified as the envelope of its tangents. Given a tangent of slope q in the point x, its intercept with the y-axis is given by

$$\phi(q) = f(x) - q\,x. \tag{A.26}$$

Knowledge of $\phi(q)$ allows us to recover $f(x)$. Thus $\phi(q)$ is a good candidate for our task. It is more convenient to use $g(q) = -\phi(q)$, which can be obtained by a variational principle. Indeed, since $f(x)$ is convex, its graph lies always above any tangent to it. As a consequence, if x_0 is such that $f'(x_0) = q$, we have

$$f(x) \geq f(x_0) + q\,(x - x_0), \qquad \forall x. \tag{A.27}$$

Thus $g(q)$ satisfies

$$g(q) = q\,x_0 - f(x_0) \geq q\,x - f(x), \qquad \forall x. \tag{A.28}$$

Therefore, $g(q)$ is given by

$$g(q) = \max_x \left(q\,x - f(x) \right), \tag{A.29}$$

and the abscissa $x(q)$ of the maximum satisfies eq. (A.25). Equation (A.29) defines the **Legendre transform** $g(q)$ of $f(x)$. Figure A.2 shows a simple geometric construction representing eq. (A.29). We distinguish between the "Legendre transformation," which is the transformation applied to a function $f(x)$, and the "Legendre transform," which is the outcome of the transformation, i.e., the function $g(q)$.

Under our hypotheses, we have

$$\frac{dg(q)}{dq} = x(q). \tag{A.30}$$

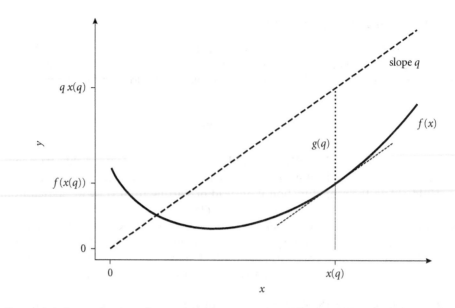

Figure A.2. Legendre transform of the function $f(x)$. To obtain the Legendre transform $g(q)$, draw a line through the origin with slope q, and take the point $x(q)$ at which the directed vertical distance from the graph of f to this line is maximized. Then $g(q)$ is given by the maximal value of this distance. If the tangent to the graph in $(x(q), f(x(q)))$ exists, it has slope q.

Moreover, the second derivative of g with respect to q is positive:

$$\frac{d^2 g}{dq^2} = \frac{dx}{dq} = \left(\frac{dq}{dx}\right)^{-1} = \frac{1}{f''(x(q))} > 0. \tag{A.31}$$

We can thus apply the same transformation to $g(q)$. By eq. (A.29), this procedure yields $f(x)$ again. Therefore, under our assumptions, the Legendre transformation is an involution.

The variational principle can be generalized to functions $f(x)$ that are nonconvex or not everywhere differentiable, by taking the supremum instead of the maximum:

$$g(q) = \sup_x \left(q x - f(x)\right). \tag{A.32}$$

This equation defines a transformation more general than the Legendre one, called the **Legendre-Fenchel transformation**. When $f(x)$ is convex and twice differentiable, we recover the Legendre transformation. However, the Legendre-Fenchel transform is defined in more general situations. In particular, if the function $f(x)$ is not convex, some tangents have more than one contact point with the graph. In these cases, eq. (A.25) has more than one solution, and, by eq. (A.30), there is a whole range of tangents to the $g(q)$ graphs for the corresponding values of q. Thus $g(q)$ exhibits angular points. When this happens, the Legendre-Fenchel transformation is no longer an involution: applying it twice, we recover the convex envelope of the graph of $f(x)$ rather than the graph itself.

A.3   Probabilities and probability distributions

In this appendix, we review basic properties of probability and probability distributions. Given a random variable x assuming discrete values, its **probability distribution** p_x represents the probability of observing the outcome x. For continuous variables, we denote with $p(x)$ the **probability density**, such that $p(x)\,\mathrm{d}x$ is the probability that the random variable x assumes a value between x and $x + \mathrm{d}x$.

This definition extends to multiple random variables. Given two discrete variables x and y, we define their **joint probability distribution** $p_{x,y}$ as the probability that the first one assumes the value x and the second one assumes a value y. Given the joint probability distribution $p_{x,y}$, the **marginal probability distribution**, or **marginal**, of the first variable is defined by

$$p_x = \sum_y p_{x,y}, \tag{A.33}$$

where the sum runs over all possible values of y. The **conditional probability** of x and y (or the probability of x, given y) is given by the **Bayes formula**

$$p_{x|y} = \frac{p_{x,y}}{p_y}. \tag{A.34}$$

An important consequence of eqs. (A.33) and (A.34) is the **law of total probability**:

$$p_x = \sum_y p_{x|y}p_y. \tag{A.35}$$

The law of total probability allows us to "decompose" the probability of an event into a weighted sum of conditioned probabilities.

Here are some common examples of discrete distributions:

The binomial distribution. We consider N independent, identically distributed events with two possible outcomes: success, with probability r, and failure, with probability $1 - r$. The distribution of the total number of successes is the binomial distribution:

$$p_x = \binom{N}{x} r^x (1 - r)^{N-x}, \qquad x \in \{0, 1, \dots, N\}, \tag{A.36}$$

where

$$\binom{N}{x} = \frac{N!}{(N - x)!\, x!} \tag{A.37}$$

is the binomial coefficient.

The Poisson distribution. The Poisson distribution is the limit of the binomial distribution for $N \to \infty$, where $\theta = Nr$ is kept constant. It is given by

$$p_x = \frac{\theta^x}{x!}\,\mathrm{e}^{-\theta}, \qquad x \in \{0, 1, 2, \dots\}. \tag{A.38}$$

We now consider the probability densities of continuous variables. The joint probability density $p(x, y)$ of two continuous random variables x and y is such that

$p(x, y)\, dx\, dy$ is the probability that the value of the first variable falls in the range between x and $x + dx$, and at the same time the value of the second variable falls between y and $y + dy$. The conditional probability density $p(x|y)$ is given by the Bayes formula

$$p(x|y) = \frac{p(x, y)}{p(y)}, \tag{A.39}$$

where $p(y)$ is the marginal probability density:

$$p(y) = \int dx\, p(x, y). \tag{A.40}$$

Here are some common examples of continuous probability distributions:

The uniform distribution. Given an interval $[a, b]$ of the x-axis, the uniform distribution expression reads

$$p(x) = \begin{cases} 1/(b - a), & \text{if } a < x < b; \\ 0, & \text{otherwise.} \end{cases} \tag{A.41}$$

The Gaussian distribution. The Gaussian distribution is defined on the whole real axis by

$$p(x) = \frac{1}{\sqrt{2\pi\sigma^2}} \exp\left[-\frac{(x - x_0)^2}{2\sigma^2}\right]. \tag{A.42}$$

The exponential distribution. The exponential distribution is defined by

$$p(x) = \begin{cases} x_0^{-1} e^{-x/x_0}, & \text{if } x > 0; \\ 0, & \text{otherwise.} \end{cases} \tag{A.43}$$

The delta distribution. The delta distribution describes continuous random variables that are bound to assume only one value x_0: $p(x) = \delta(x - x_0)$. It can be formally obtained as the limit of the Gaussian distribution for $\sigma \to 0$. The delta function satisfies $\int_a^b dx\, \delta(x - x_0) = 1$ if $a < x_0 < b$ and 0 otherwise. Moreover, the delta function possesses the following properties:

- $\int_{-\infty}^{\infty} dx\, \delta(x - x_0) f(x) = f(x_0)$, where $f(x)$ is an arbitrary function;
- $\delta(-x) = \delta(x)$;
- $\delta(ax) = |a|^{-1} \delta(x)$.

A.4 Generating functions and cumulant generating functions

The **generating function** of a random variable with probability density $p(x)$ is defined by

$$\phi(q) = \langle e^{qx} \rangle = \int_{-\infty}^{\infty} dx\, e^{qx} p(x). \tag{A.44}$$

For discrete probability distributions p_x, eq. (A.44) still holds with the integral replaced by a sum. Mathematically, the generating function can be seen as the Laplace transform of the probability distribution. The integral in eq. (A.44) converges when $-\infty \le q < a$, where the value of a depends on the behavior of the distribution as $x \to +\infty$. To obtain the probability distribution from its generating function, one can invert the Laplace transform by the Bromwich integral

$$p(x) = \int_{\gamma - i\infty}^{\gamma + i\infty} \frac{dq}{2\pi i} e^{-qx} \phi(q). \qquad (A.45)$$

In eq. (A.45), integration is performed on a path parallel to the imaginary axis. The location γ of the path has to be negative enough that all the singularities of the integrand lie on the right of the integration path. The generating function of a Gaussian distribution has the Gaussian form, i.e., it is the exponential of a quadratic form in q:

$$\int \frac{dx}{\sqrt{2\pi \sigma^2}} e^{qx} \exp\left[-\frac{(x-x_0)^2}{2\sigma^2}\right] = \exp\left(qx_0 + \frac{q^2}{2}\sigma^2\right). \qquad (A.46)$$

The generating function gets its name from the property that its derivatives "generate" moments of the distribution:

$$\left(\frac{d}{dq}\right)^n \phi(q)\bigg|_{q=0} = \langle x^n e^{qx} \rangle \big|_{q=0} = \langle x^n \rangle, \qquad n \in \{1, 2, \dots\}. \qquad (A.47)$$

Generating functions are also useful to compute sums of random variables. Let us consider two independent random variables x and y with probability densities $p_1(x)$ and $p_2(y)$, respectively. The probability density of their sum $z = x + y$ is given by

$$p(z) = \int_{-\infty}^{\infty} dx \int_{-\infty}^{\infty} dy\, p_1(x)\, p_2(y)\, \delta(z - x - y) = \int_{-\infty}^{\infty} dx\, p_1(x)\, p_2(z - x). \qquad (A.48)$$

Denoting by $\phi^{(x)}(q)$, $\phi^{(y)}(q)$, and $\phi^{(z)}(q)$ the generating functions of the variables x, y, and z, respectively, one has

$$\phi^{(z)}(q) = \int_{-\infty}^{\infty} dz\, e^{qz}\, p(z) = \int_{-\infty}^{\infty} dx\, p_1(x)\, e^{qx} \int_{-\infty}^{\infty} dz\, e^{q(z-x)}\, p_2(z - x)$$
$$= \phi^{(x)}(q)\, \phi^{(y)}(q), \qquad (A.49)$$

where in the second equality we use eq. (A.48). Therefore, the generating function of the sum is the product of the generating functions. This result extends to sums of more than two random variables. For example, consider the empirical mean

$$f_N = \frac{1}{N} \sum_{k=1}^{N} x_k, \qquad (A.50)$$

where the x_k are independent, identically distributed random variables with distribution $p(x)$. An important corollary of eq. (A.49) is that the generating function of the empirical mean is given by

Table A.1. The first four cumulants

Order	Moment	Cumulant
1	$\langle x \rangle$	$\langle\langle x \rangle\rangle = \langle x \rangle$
2	$\langle x^2 \rangle$	$\langle\langle x^2 \rangle\rangle = \langle x^2 \rangle - \langle x \rangle^2$
3	$\langle x^3 \rangle$	$\langle\langle x^3 \rangle\rangle = \langle x^3 \rangle - 3\langle x^2 \rangle \langle x \rangle + 2\langle x \rangle^3$
4	$\langle x^4 \rangle$	$\langle\langle x^4 \rangle\rangle = \langle x^4 \rangle - 4\langle x^3 \rangle \langle x \rangle - 3\langle x^2 \rangle^2 + 12\langle x^2 \rangle \langle x \rangle^2 - 6\langle x \rangle^4$

$$\phi^{(f_N)}(q) = \left[\phi^{(x)}\left(\frac{q}{N}\right) \right]^N. \tag{A.51}$$

Linear combinations of Gaussian random variables are also Gaussian, since the product of their Gaussian generating functions is a Gaussian.

The **cumulant generating function** is the logarithm of the generating function:

$$\Phi(q) = \ln \langle e^{qx} \rangle = \ln \phi(q). \tag{A.52}$$

The **cumulants** $\langle\langle x^n \rangle\rangle$ are useful combinations of the moments of a distribution. They are related to the cumulant generating function by

$$\langle\langle x^n \rangle\rangle = \left(\frac{\mathrm{d}}{\mathrm{d}q} \right)^n \Phi(q) \Bigg|_{q=0}. \tag{A.53}$$

The first cumulant is equal to the first moment, i.e., the average. The second cumulant is the variance, $\langle\langle x^2 \rangle\rangle = \langle x^2 \rangle - \langle x \rangle^2 = \sigma_x^2$. Roughly speaking, the expressions of each higher-order cumulant contain the moment of the same order minus its "best approximation" in terms of lower-order moments. A list of the expressions of the first four cumulants in terms of moments is presented in table A.1.

In table A.2, we report expressions of the generating functions and cumulant generating functions for the common distributions we introduced in appendix A.3. For a Gaussian distribution, the generating function has the same functional form as the distribution itself, and thus the cumulant generating function is a second-order polynomial. This latter fact, combined with the definition of the cumulants, eq. (A.53), implies that all cumulants of order three and higher vanish for a Gaussian distribution. This is a very peculiar property: **Marcinkiewicz's theorem** states that the cumulant generating function of a probability distribution, if it is a polynomial, cannot be of degree greater than two. This means that nearly all distributions have an infinite number of nonvanishing cumulants—the only exceptions being the Gaussian, with the first two nonvanishing cumulants, and the singular delta distribution, with only one nonvanishing cumulant.

This result might seem a mathematical curiosity, but it lies at the core of the **central limit theorem**. Let us return to eq. (A.51) and evaluate the cumulant generating function of the empirical mean

$$\Phi^{(f_N)}(q) = N\Phi^{(x)}\left(\frac{q}{N}\right). \tag{A.54}$$

Table A.2. Generating functions and cumulant generating functions for common distributions

Distribution	$p(x)$	$\phi(q)$	$\Phi(q)$
Exponential	$x_0^{-1}e^{-x/x_0}$	$1/(1-qx_0)$	$-\ln\left(1-qx_0\right)$
Gaussian	$e^{-(x-x_0)^2/(2\sigma^2)}/\sqrt{2\pi\sigma^2}$	$e^{qx_0+\sigma^2q^2/2}$	$qx_0+\sigma^2q^2/2$
Delta	$\delta(x-x_0)$	e^{qx_0}	qx_0
Binomial	$\binom{N}{x}r^x(1-r)^{N-x}$	$\left(1-r(1-e^q)\right)^N$	$N\ln\left(1-r(1-e^q)\right)$
Poisson	$(\theta^x/x!)\,e^{-\mu}$	$\exp\left(\theta\left(e^q-1\right)\right)$	$\theta\left(e^q-1\right)$

By taking derivatives with respect to q, we can use this expression to relate the cumulants of x to those of f_N:

$$\langle\!\langle f_N^n \rangle\!\rangle = \frac{1}{N^{n-1}}\langle\!\langle x^n \rangle\!\rangle. \tag{A.55}$$

According to eq. (A.55), the average of an empirical mean is equal to that of the original variables, its variance is N times smaller, and its higher moments are suppressed by higher powers of $1/N$ compared to those of the original variable. This means that, as N increases, the distribution of the empirical mean is better approximated by a distribution characterized by only the first two nonvanishing cumulants, which is a Gaussian as we have seen.

The average $\langle x \rangle$ and the variance $\sigma^2 = \langle\!\langle x^2 \rangle\!\rangle$ of a Poisson distribution are both equal to the parameter θ. This result can be obtained by expanding the cumulant generating function $\Phi(q)$.

A.5   Ergodic properties of Markov processes

In this appendix, we discuss the ergodic properties of master equations with a finite number of states N, i.e., the statistical behavior of their solutions in the long run. We prove in particular that, if the jump rates are time-independent and correspond to a connected jump network, the master equation has a unique stationary distribution $p^{\text{st}} = (p_x^{\text{st}})$, which is reached as $t\to\infty$ independent of the initial distribution. The proof is inspired by Meyer [112, ch. 8].

As usual, we call $k_{x'x}$ the jump rates from x to $x'\neq x$. We also define for convenience $k_{xx}=-k_x^{\text{out}}$, where $k_x^{\text{out}}=\sum_{x'}k_{x'x}$ is the escape rate from state x. With this convention, the generator is $L=(k_{xx'})$. The master equation reads

$$\frac{dp_x}{dt}=\sum_{x'}k_{xx'}p_{x'}=\left(Lp\right)_x. \tag{A.56}$$

Given $\Delta t>0$, the propagator $p_{x;t+\Delta t|x';t}$ is given by

$$P_{xx'}(\Delta t)=p_{x;t+\Delta t|x';t}=\left(\exp\left(\Delta t\,L\right)\right)_{xx'}. \tag{A.57}$$

Since the jump network is connected, there is a path with nonvanishing jump probabilities that connect any state x' with any different state x. This implies that, for $\Delta t > 0$, all the entries in the matrix $P_{xx'}(\Delta t)$ are positive. We choose one such value of Δt and denote by $P = (P_{xx'})$ the corresponding value of the propagator. The matrix P has all positive entries and satisfies the normalization condition

$$\sum_{x'} P_{x'x} = 1, \qquad \forall x. \tag{A.58}$$

The matrix P defines a **Markov chain**, i.e., a Markov process in discrete time. If this Markov chain has a stationary distribution p^{st}, then p^{st} is an eigenvector of P with eigenvalue $\lambda = 1$. Since L commutes with P, p^{st} is also an eigenvector of L with eigenvalue $(\ln \lambda)/\Delta t = 0$ and is therefore a stationary distribution for the Markov process in continuous time. Thus we only need to obtain the result for the Markov chain.

The matrix P has an eigenvalue equal to 1 since, by eq. (A.58), the vector $\mathcal{I}$ such that $\mathcal{I}_x = 1$, $\forall x$ is a left eigenvector belonging to this eigenvalue. We denote by u the corresponding right eigenvector. Since P is real, u can be chosen to be real. We now show that we can also choose it to have all positive entries. Let us define $\delta = \min_{xx'} P_{xx'} > 0$ and the norm $\|u\| = \max_x |u_x|$. Given any vector v, we denote by $(v)^+$ its positive part and by $-(v)^-$ its negative part. We then define $\alpha = \min\{\|(u)^+\|, \|(u)^-\|\}$. Then $\left(P(u)^+\right)_x \geq \delta\alpha$, $\forall x$ and $\left(P(u)^-\right)_x \geq \delta\alpha$, $\forall x$. Therefore,

$$\begin{aligned}\|Pu\| &= \|P(u)^+ - P(u)^-\| \leq \|P(u)^+ - \delta\alpha\mathcal{I}\| + \|P(u)^- - \delta\alpha\mathcal{I}\| \\ &\leq \|(u)^+\| + \|(u)^-\| - 2\delta\alpha = \|u\| - 2\delta\alpha.\end{aligned} \tag{A.59}$$

Since $Pu = u$ and $\delta > 0$, this implies that $\alpha = 0$, i.e., that u can be chosen to be either all positive or all negative. We therefore set it to be all positive. The stationary distribution is therefore equal to $p_x^{\text{st}} = u_x / \sum_{x'} u_{x'}$.

We now prove that the stationary distribution is unique. We suppose that this is not the case and that there are at least two stationary distributions, p and p'. Then $w = p - p'$ is also an eigenvector of P belonging to the eigenvalue 1. On the one hand, we should have $w_x \geq 0$, $\forall x$ by the argument above, and on the other, we should have $\sum_x w_x = \sum_x p_x - \sum_x p'_x = 0$. Thus $w = 0$, i.e., the two distributions p and p' must be equal. Therefore, the eigenvalue 1 of P is nondegenerate. We now consider an arbitrary distribution p and the stationary one p^{st}. Since $P_{xx'} \geq \delta$, $\forall x, x'$, we have $\|Pp\| > \delta \|p\|$ for every positive p. We then have

$$\begin{aligned}\|p^{\text{st}} - Pp\| &= \|P(p^{\text{st}} - p)\| = \|P(p^{\text{st}} - p)^+ - P(p^{\text{st}} - p)^-\| \\ &\leq \|P(p^{\text{st}} - p)^+ - \delta\,\mathcal{I}\,\|(p^{\text{st}} - p)^+\|\| \\ &\quad + \|P(p^{\text{st}} - p)^- - \delta\,\mathcal{I}\,\|(p^{\text{st}} - p)^-\|\| \\ &\leq (1 - \delta)\,\|p^{\text{st}} - p\|.\end{aligned} \tag{A.60}$$

This implies that the other eigenvalues of P satisfy the inequality

$$|\lambda| \leq 1 - \delta < 1. \tag{A.61}$$

Thus the iterates $P^k p$ converge exponentially to p^{st} for any initial probability distribution p. Equation (A.61) also implies that the nonvanishing eigenvalues λ of L satisfy the inequality $\text{Re}\lambda < -\varepsilon < 0$ for some $\varepsilon > 0$. Therefore, the exponential convergence also holds for the Markov process in continuous time.

We extend these results to positive generators that do not conserve the normalization by exploiting Brouwer's fixed point theorem [26]. We define the set Σ of normalized vectors with nonnegative entries: $u \in \Sigma$ iff $u_x \geq 0, \forall x$ and $\sum_x u_x = 1$. Then Σ is a compact and convex subset of the N-dimensional Euclidean space. Given the matrix P with positive entries, we define the continuous mapping $p \mapsto T(p)$ by

$$T(p) = \sum_{x'} P_{xx'} p_{x'} \Big/ \sum_{xx'} P_{xx'} p_{x'}. \qquad (A.62)$$

According to Brouwer's theorem, a continuous mapping from a compact and convex set of a Euclidean space to itself has a fixed point, i.e., there is a vector p^* such that $T(p^*) = p^*$. Let $\rho = \sum_x (Pp^*)_x$. Then p^* is an eigenvector of P with nonnegative (actually positive) entries belonging to the positive eigenvalue ρ. The other properties are obtained by adapting the reasoning leading to eqs. (A.59) and (A.60).

A.6 Gillespie algorithm

When the exact solution of a master equation is not available, one may wish to numerically simulate the corresponding stochastic process. An exact method to generate trajectories associated with a master equation was introduced by Doob in the 1940s and later developed and popularized by Daniel T. Gillespie in the 1970s [64, 65]. It is therefore known as the **Gillespie algorithm**. The Gillespie algorithm is nowadays widely used and has been generalized in several ways [67].

We consider a master equation

$$\frac{dp_x}{dt} = \sum_{x'\,(\neq x)} k_{xx'} p_{x'} - k_x^{out} p_x, \qquad (A.63)$$

where as usual

$$k_x^{out} = \sum_{x'\,(\neq x)} k_{x'x}. \qquad (A.64)$$

We focus on the case in which the rates $k_{xx'}$ are constant in time. Given the rates and the initial distribution $p_{x_0}(t_0)$, we generate a trajectory x of the system in the time interval $[t_0, t_f]$ in the following way:

1. Choose the initial state x_0 at random according to the probability distribution $p_{x_0}(t_0)$. Set the state x of the system to $x = x_0$ and the time t to $t = t_0$. Set $x = ((x_0, t_0))$.
2. Draw the waiting time Δt to the next jump from an exponential distribution with average $1/k_x^{out}$. In practice, extract u with uniform probability between 0 and 1, and set $\Delta t = -\ln u/k_x^{out}$. Add Δt to the time variable t.
3. Draw the new state x' of the system among the states different from x, with probability given by $k_{x'x}/k_x^{out}$. Set the new state of the system to x'.

4. Append (x, t) to x. Repeat from step 2 until the time $t + \Delta t$ of the next jump, obtained in step 2, exceeds the wished final time t_f.

This procedure is known as the **direct method**. This algorithm is exact, in the sense that it produces trajectories with the correct statistical weight in continuous time without a time discretization. However, it suffers from the drawback that, if the number of possible transitions from a given state x is large, the escape rate k_x^{out} can also be large, and therefore the expected value of Δt will be in general small, leading to a very slow evolution. Gillespie [64] also introduced an alternative method, called the **first-reaction method**. It is also exact, but it is less efficient than the direct method, because it requires to update, in step 2, the waiting times $\Delta t_{x'}$ for all possible jumps $x \longrightarrow x'$ and then to keep the smallest one and discard all the others. A generalization of this idea by Gibson and Bruck [63], known as the **next-reaction method**, exploits a clever way of storing waiting times to facilitate identifying the next-occurring transition and the updating of the waiting-time data. Due to the additional computational overhead, this method becomes advantageous only for systems with a large number of possible transitions, like reaction networks with many species and many reactions. For these systems, a very efficient (but approximate) method, the **tau-leaping method**, has been introduced by Gillespie [66]. The C++ implementation of the tau-leaping and next-reaction methods is discussed in Press et al. [132, § 17.7.1–2]. The next-reaction method has been generalized to tackle processes with time-dependent jump rates [1].

A.7   Derivation of the Fokker-Planck equation

The Fokker-Planck equation governs the evolution of the probability density $p(x; t)$ of the position x of a Brownian particle in a general case, where the particle is subject to applied forces and/or the embedding fluid is not uniform. At variance with the diffusion equation, in this case the distribution of the displacement Δx experienced by the particle during a time interval of duration Δt can depend on x and is not necessarily even. We consider the temporal evolution of the average of an arbitrary function $f(x)$:

$$\langle f \rangle_t = \int dx f(x) \, p(x; t). \tag{A.65}$$

On the one hand, we have

$$\frac{d \langle f \rangle_t}{dt} = \int dx f(x) \, \frac{\partial}{\partial t} p(x; t). \tag{A.66}$$

On the other hand, the Chapman-Kolmogorov equation (2.64) implies

$$\langle f \rangle_{t+\Delta t} = \int dx' \int dx f(x') \, p(x'; t + \Delta t | x; t) \, p(x; t)$$

$$= \int d\Delta x \int dx f(x + \Delta x) \, p(x + \Delta x; t + dt | x; t) \, p(x, t)$$

$$\approx \int d\Delta x \int dx \left[f(x) + \Delta x \frac{\partial}{\partial x} f(x) + \frac{\Delta x^2}{2} \frac{\partial^2}{\partial x^2} f(x) \right]$$

$$\times p(x + \Delta x; t + dt | x; t) \, p(x; t)$$

$$= \int dx \left[f(x) + \langle \Delta x \rangle_x \frac{\partial}{\partial x} f(x) + \frac{1}{2} \langle \Delta x^2 \rangle_x \frac{\partial^2}{\partial x^2} f(x) \right] p(x; t'), \qquad (A.67)$$

where we define

$$\langle \dots \rangle_x = \int d\Delta x \dots p(x + \Delta x; t + \Delta t | x; t). \qquad (A.68)$$

We now introduce the **drift** v and the **diffusion coefficient** D, respectively, by

$$v(x, t) = \lim_{\Delta t \to 0} \frac{\langle \Delta x \rangle_{x,t}}{\Delta t}; \qquad (A.69)$$

$$D(x, t) = \lim_{\Delta t \to 0} \frac{\langle \Delta x^2 \rangle_{x,t}}{2 \Delta t}. \qquad (A.70)$$

The averages are taken with the condition that the position of the particle is equal to x at time t. We integrate by parts the last expression in (A.67), obtaining

$$\lim_{\Delta t \to 0} \frac{\langle f \rangle_{t+\Delta t} - \langle f \rangle_t}{\Delta t} = \int dx f(x) \left\{ -\frac{\partial}{\partial x} \left[v(x, t) p(x; t) \right] + \frac{\partial^2}{\partial x^2} \left[D(x, t) p(x; t) \right] \right\}. \qquad (A.71)$$

Since $f(x)$ is arbitrary, by comparison with (A.66), we see that the term in braces must be equal to $\partial p(x; t)/\partial t$. We obtain therefore the **Fokker-Planck equation** for $p(x; t)$:

$$\frac{\partial}{\partial t} p(x; t) = -\frac{\partial}{\partial x} \left[v(x, t) p(x; t) \right] + \frac{\partial^2}{\partial x^2} \left[D(x, t) p(x; t) \right]. \qquad (A.72)$$

To obtain the Kolmogorov backward equation, we also start from the Chapman-Kolmogorov equation (2.64):

$$p(x; t | x_0, t_0 - \Delta t_0) = \int dx' \, p(x; t | x', t_0) \, p(x'; t_0 | x_0, t_0 - \Delta t_0)$$

$$= \int d\Delta x_0 \, p(x; t | x_0 + \Delta x_0, t_0) \, p(x_0 + \Delta x_0; t_0 | x_0, t_0 - \Delta t_0)$$

$$\approx \int d\Delta x_0 \left[p(x; t | x_0, t_0) + \Delta x_0 \frac{\partial}{\partial x_0} p + \frac{1}{2} \Delta x_0^2 \frac{\partial^2}{\partial x_0^2} p \right]$$

$$\times p(x_0 + \Delta x_0; t_0 | x_0, t_0 - \Delta t_0) \qquad (A.73)$$

$$= p(x; t | x_0, t_0) + \langle \Delta x_0 \rangle_{x_0, t_0 - \Delta t_0} \frac{\partial}{\partial x_0} p + \frac{1}{2} \langle \Delta x_0^2 \rangle_{x_0, t_0 - \Delta t_0} \frac{\partial^2}{\partial x_0^2} p$$

$$= p(x; t | x_0, t_0) + v(x_0, t_0 - \Delta t_0) \Delta t_0 \frac{\partial}{\partial x_0} p + D(x_0, t_0 - \Delta t_0) \Delta t_0 \frac{\partial^2}{\partial x_0^2} p.$$

In the limit $\Delta t_0 \to 0$, we recover eq. (2.106).

A.8   Ito formula and Stratonovich-Ito mapping

In this appendix, we sketch the derivation of eq. (2.127), which maps a Langevin equation interpreted in the Ito convention to the corresponding one in the Stratonovich convention, and backward, as well as the corresponding Fokker-Planck equation. We also sketch the proof of the Ito formula, eq. (2.128). Since the main difficulty lies in handling the x-dependence of the noise coefficient, for the sake of simplicity, we limit ourselves to Langevin equations without drift.

We first show that the rule for change of variables in the Stratonovich interpretation is the same as in ordinary calculus. We consider a continuous variable $x(t)$ that satisfies the Langevin equation

$$\frac{dx}{dt} = \sigma(x)\,\xi(t),\tag{A.74}$$

in the Stratonovich interpretation. The solution of eq. (A.74) is given by

$$x(t) = x(t_0) + \int_{t_0}^{t} dW(t') \circ \sigma(x(t')),\tag{A.75}$$

which in discretized form reads

$$x_{i+1} = x_i + \sigma\left(\frac{x_{i+1} + x_i}{2}\right)(W_{i+1} - W_i),\tag{A.76}$$

where $x_i = x(t_i)$, $W_i = W(t_i)$, etc. We wish to evaluate the Langevin equation satisfied by $f(x(t))$, where $f(x)$ is a given function. To do this, we expand $f(x)$ around the midpoint in each time interval. We now have

$$f(x+a) = f(x-a) + 2af'(x) + \frac{1}{3}a^3 f'''(x) + \cdots,\tag{A.77}$$

where the third and subsequent terms in the expansion are higher than second order in a and therefore negligible. We thus have

$$\begin{aligned}
f(x_{i+1}) &= f(x_i) + f'\left(\frac{x_{i+1} + x_i}{2}\right)(x_{i+1} - x_i)\\
&= f(x_i) + f'\left(\frac{x_{i+1} + x_i}{2}\right)\sigma\left(\frac{x_{i+1} + x_i}{2}\right)(W_{i+1} - W_i).
\end{aligned}\tag{A.78}$$

Therefore, $f(x(t))$ satisfies the Langevin equation in the Stratonovich sense,

$$\frac{df}{dt} = f'(x)\,\sigma(x)\,\xi,\tag{A.79}$$

which corresponds to the ordinary calculus rule. In the general case with nonvanishing drift and time-dependent coefficients, the equation reads

$$\frac{df}{dt} = f'(x)\left[v(x,t) + \sigma(x,t)\,\xi\right].\tag{A.80}$$

We now turn to the Ito formula, eq. (2.128). We assume that $x(t)$ satisfies eq. (A.74) in the Ito interpretation, and consider the same $f(x)$. We then have

$$
df(x(t+dt)) = f(x+dt) - f(x(t)) = f'(x(t))\,dx + \frac{1}{2}f''(x(t))\,dx^2 + \cdots
$$
$$
\approx f'(x(t))\,\sigma(x(t))\,dW + \frac{1}{2}f''(x(t))\,\sigma^2(x(t))\,dW^2. \tag{A.81}
$$

Now $dW^2 = dt$, up to higher-order terms. We obtain therefore the Ito formula

$$
df = f'(x)\,dx + \frac{1}{2}f''(x)\,\sigma^2(x)\,dt. \tag{A.82}
$$

Also this result can be extended to the general case with drift and time-dependent coefficients.

We now assume that $x(t)$ is a solution of the Langevin equation eq. (A.74) in the Stratonovich interpretation. To obtain the Langevin equation obeyed by $x(t)$ in the Ito interpretation, we expand $\sigma(x)$, evaluated at midpoint, around its initial point, obtaining

$$
x_{i+1} = x_i + \sigma\left(\frac{x_{i+1}+x_i}{2}\right)(W_{i+1}-W_i)
$$
$$
= x_i + \sigma(x_i)(W_{i+1}-W_i) + \frac{1}{2}\sigma'(x_i)(x_{i+1}-x_i)(W_{i+1}-W_i) \tag{A.83}
$$
$$
\approx x_i + \sigma(x_i)(W_{i+1}-W_i) + \frac{1}{2}\sigma'(x)\,\sigma(x_i)(W_{i+1}-W_i)^2.
$$

Using again $dW^2 = dt$, we obtain the Langevin equation

$$
\frac{dx}{dt} = \sigma(x)\frac{1}{2}\sigma'(x) + \sigma(x)\,\xi. \tag{A.84}
$$

Considering the general Langevin equation with drift, we obtain the mapping given in eq. (2.127).

We use this result to obtain the Fokker-Planck equation associated with the Langevin equation (A.74) in the Stratonovich interpretation. To this aim, we first transform the Langevin equation into the Ito convention, obtaining eq. (A.74). We then write its associated Fokker-Planck equation:

$$
\frac{\partial}{\partial t}p(x;t) = \frac{\partial}{\partial x}\left[-\sigma(x)\frac{1}{2}\sigma'(x)\,p + \frac{1}{2}\frac{\partial}{\partial x}\left(\sigma^2(x)\,p\right)\right]
$$
$$
= \frac{1}{2}\frac{\partial}{\partial x}\left[\sigma(x)\frac{\partial}{\partial x}\left(\sigma(x)\,p\right)\right]. \tag{A.85}
$$

In the general case, where a drift $v(x,t)$ is also present, we obtain eq. (2.124), which is known as the **Stratonovich form** of the Fokker-Planck equation.

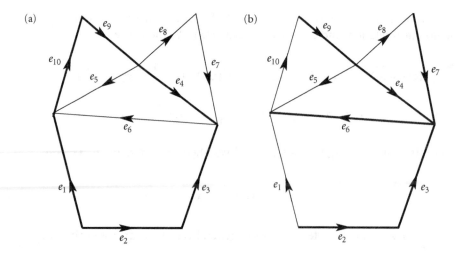

Figure A.3. Construction of the cycle space. (a) The cycle shown in bold, if oriented counterclockwise, is represented by the vector $v = -e_1 + e_2 + e_3 - e_4 - e_9 - e_{10}$. (b) Given the spanning tree shown in bold, the cycle defined by the chord e_{10} corresponds to the vector $v = e_{10} + e_9 + e_4 + e_6$. The cut corresponding to e_4 is given by $w = e_4 + e_5 + e_8 - e_{10}$. One has $v \cdot w = 0$.

A.9   Basis of the cycle space

In this appendix, we sketch the proof that the set of fundamental cycles constitutes a complete basis for the space of cycles associated with a jump network. Our proof loosely follows Bollobás [23, ch. 3]. We consider a connected jump network with N nodes and K edges, where each edge belongs to at least one cycle. This corresponds to the core network of a general connected jump network. We associate with it a set of fundamental cycles, using the construction described in section 3.9.

We now show that the set of fundamental cycles is a complete basis for the space of cycles. Since the spanning tree has $N - 1$ edges, the number v of cycles is equal to $K - N + 1$.

The proof is based on a geometrical construction. We define the K-dimensional linear space C_{edg}, where K is the number of edges, by giving an (arbitrary) orientation to each edge and associating it with a basis vector e_k. A generic vector x in this space is a linear combination $x = \sum_k x_k e_k$ of the basis vectors, with real coefficients x_k. We introduce in this space the scalar product $x \cdot y = \sum_k x_k y_k$.

We then introduce the space C_{cyc} associated with the directed cycles in the jump network. It is obtained by associating with each cycle C_α a vector c_α in C_{edg}, whose entries are $+1$ if a given edge is part of the cycle with the same orientation, -1 if the edge is part of the cycle with opposite orientation, and 0 if the edge is not part of the cycle. An example of this construction is shown in fig. A.3a. This representation allows us to show that the cycles in the fundamental set are linearly independent. If this were not the case, one could find a linear combination of them such that $\sum_{\alpha=1}^{v} \lambda_\alpha c_\alpha = 0$. But this is impossible since each cycle contains a unique chord not shared by the other cycles. Since the space spanned by the C_α contains at least $(K - N + 1)$ independent vectors, the dimension of the cycle space satisfies $\dim C_{\mathrm{cyc}} \geq v = K - N + 1$.

We now define the space C_{cut}. A **cut** is a partition of the nodes of the network in two disjoint and nonempty sets A and B, such that their union is the set of all nodes. In particular, each edge of the spanning tree defines, if removed, two disjoint sets of nodes and therefore a cut. We associate with each cut a cut vector in the edge space, whose entries are equal to 1 if the corresponding edge links a state in A to a state in B, -1 if it links a state in B to a state in A, and 0 otherwise; see fig. A.3b.

A useful result of graph theory states that, for a connected graph,

$$C_{edg} = C_{cyc} \oplus C_{cut}, \tag{A.86}$$

where $A \oplus B$ is the **direct sum** of the linear spaces A and B, whose only common element is the null vector. A generic element of $A \oplus B$ is a linear combination of an element of A and an element of B. From eq. (A.86), one obtains

$$\dim C_{cyc} = N - K + 1, \qquad \dim C_{cut} = N - 1. \tag{A.87}$$

To prove these results, we start from the observation that any cycle is orthogonal to any cut vector. Indeed, the scalar product between a cycle and a cut vector is the number of jumps in the cycle going from the set of states A to the set B minus the number of jumps in the cycle going from the set B to the set A. Since the cycle must return to the original state, this difference must vanish. This implies that C_{cyc} and C_{cut} are orthogonal spaces.

Given this fact, as well as the facts that $C_{cyc} \geq \nu = K - N + 1$ and $\dim C_{edg} = K$, to prove eqs. (A.86) and (A.87), it is sufficient to show that $\dim C_{cut} \geq N - 1$. This can be proved by another explicit construction. We consider the $N - 1$ cuts associated with each of the $N - 1$ edges of the spanning tree, as previously discussed. The cut vectors associated with these $N - 1$ cuts are linearly independent because of the same argument used for the cycle space: each of these $N - 1$ vectors has a nonzero component associated with the removed edge that is equal to zero for all other vectors. This completes the proof.

A.10 Actions and trajectory probabilities for Langevin equations

In this appendix, we derive in more detail the expression (4.96) of the difference of the action between the forward and backward trajectories of a Langevin equation. In particular, we better justify the use of the Stratonovich convention in the integral defining the action.

We rewrite the action $\mathcal{S}(x, \lambda)$ in eq. (4.90) with an explicit discretization,

$$\mathcal{S}(x, \lambda) = \frac{1}{4D} \int_{t_0}^{t_f} dt \left(\frac{dx}{dt} - \mu_P \mathcal{F}(x, t) \right)^2 =$$

$$= \lim_{\Delta t \to 0} \frac{1}{4D \, \Delta t} \sum_{\ell=1}^{\mathcal{N}} (x_\ell - x_{\ell-1} - \mu_P \mathcal{F}(x_{\ell-1}, t_{\ell-1}) \, \Delta t)^2, \tag{A.88}$$

and we rewrite the action of the backward trajectory in a similar way:

$$
S(\widehat{x}; \widehat{\lambda}) = \int_{t_0}^{t_f} dt\, \frac{1}{4D} \left(\frac{dx(t)}{dt} + \mu_P\, \mathcal{F}(x(t), \lambda(t)) \right)^2 =
$$

$$
= \lim_{\Delta t \to 0} \frac{1}{4D\, \Delta t} \sum_{\ell=1}^{\mathcal{N}} (x_\ell - x_{\ell-1} + \mu_P \mathcal{F}(x_\ell, t_\ell)\, \Delta t)^2 .
$$

(A.89)

In writing these expressions, we take into account that the forward and backward actions must be discretized in different ways: the former using the initial point of each time interval of duration Δt and the latter using its end point, which corresponds to its initial point in reverse time. In deriving the fluctuation relations, as in eq. (4.96), we evaluate the difference between $S(x, \lambda)$ and $S(\widehat{x}, \widehat{\lambda})$. We then have

$$
S(x, \lambda) - S(\widehat{x}, \widehat{\lambda})
$$

$$
= \lim_{\Delta t \to 0} \frac{1}{4D\, \Delta t} \sum_{\ell=1}^{n} \Big[(x_\ell - x_{\ell-1} - \mu_P \mathcal{F}(x_{\ell-1}, t_{\ell-1})\, \Delta t)^2
$$

$$
- (x_\ell - x_{\ell-1} + \mu_P \mathcal{F}(x_\ell, t_\ell)\, \Delta t)^2 \Big]
$$

(A.90)

$$
= \lim_{\Delta t \to 0} \frac{1}{4D} \sum_{\ell=1}^{n} \Big[-2(x_\ell - x_{\ell-1}) \times \mu_P \left(\mathcal{F}(x_{\ell-1}, t_{\ell-1} + \mathcal{F}(x_\ell, t_\ell)) \right)
$$

$$
+ \left(\mathcal{F}^2(x_{\ell-1}, t_{\ell-1}) - \mathcal{F}^2(x_\ell, t_\ell) \right) \Delta t \Big],
$$

where the terms $\left(\mathcal{F}^2(x_{\ell-1}, t_{\ell-1}) - \mathcal{F}^2(x_\ell, t_\ell) \right) \Delta t$ sum up to a boundary contribution proportional to Δt, which can be neglected in the $\Delta t \to 0$ limit. The factor $\left(\mathcal{F}(x_{\ell-1}, t_{\ell-1} + \mathcal{F}(x_\ell, t_\ell) \right)$ can be approximated by the midpoint expression $2\mathcal{F}((t_\ell + t_{\ell-1})/2, (x_\ell + x_{\ell-1})/2)$ up to higher-order terms. Taking into account the Einstein relation (3.84), we obtain

$$
S(x, \lambda) - S(\widehat{x}, \widehat{\lambda}) = -\frac{1}{k_B T} \int_{t_0}^{t_f} dx \circ \mathcal{F}(x, t),
$$

(A.91)

where $\circ$ reminds us that the integral has to be considered in the Stratonovich interpretation.

A.11   The Bennett-Crooks estimator for the free-energy difference

In this appendix, we sketch the derivation of the **Bennett-Crooks optimal estimator** of the free-energy difference, eq. (7.9), following Bennett [11] and Crooks [35]. We write the detailed fluctuation relation for the work w in the following way:

$$
p_F(w)\, e^{-w/k_B T} = p_B(-w)\, e^{-\Delta F/k_B T}.
$$

(A.92)

Given a function $f(w)$, it is convenient to define the averages:

$$
\phi_F = \left\langle f(w)e^{-w/k_BT} \right\rangle_F = \int dw\, p_F(w) f(w) e^{-w/k_BT},
$$

$$
\phi_B = \left\langle f(-w) \right\rangle_B = \int dw\, p_B(w) f(-w).
$$

(A.93)

The detailed fluctuation theorem permits us to express the free-energy difference in terms of these averages as

$$
\Delta F = -k_BT \left(\ln \phi_B - \ln \phi_F \right).
$$

(A.94)

Importantly, this expression is valid for any choice of the function $f(w)$. Our aim is to choose this function in such a way as to minimize the expected error on the free energy when the averages in eq. (A.93) are estimated empirically. In particular, we assume a large number of samples $\mathcal{N}_F$ for the forward and $\mathcal{N}_B$ for the backward protocols, such that the distributions of the data to be averaged are approximately Gaussian. Then the uncertainties on the empirical means of $\phi_{F,B}$ are respectively given by the standard errors

$$
\delta\phi_F^2 = \frac{\sigma_F^2}{\mathcal{N}_F}, \qquad \delta\phi_B^2 = \frac{\sigma_B^2}{\mathcal{N}_B},
$$

(A.95)

where we define the variances

$$
\sigma_F^2 = \int dw\, p_F(w)\, e^{-2w/k_BT} f^2(w) - \phi_F^2;
$$

$$
\sigma_B^2 = \int dw\, p_B(w) f^2(-w) - \phi_B^2.
$$

(A.96)

By propagating these uncertainties, we find that the expected squared error $\delta(\Delta F)^2$ in the estimate of ΔF is

$$
\frac{\delta(\Delta F)^2}{(k_BT)^2} = \frac{\sigma_F^2}{\mathcal{N}_F\phi_F^2} + \frac{\sigma_B^2}{\mathcal{N}_B\phi_B^2}.
$$

(A.97)

We want to minimize the expected squared error with respect to the function $f(w)$ for a given expectation of ΔF. We first express it as

$$
\frac{\delta(\Delta F)^2}{(k_BT)^2} = \frac{1}{\phi_F\phi_B} \left[e^{\Delta F/k_BT} \frac{\sigma_F^2}{\mathcal{N}_F} + e^{-\Delta F/k_BT} \frac{\sigma_B^2}{\mathcal{N}_B} \right] + \text{const.},
$$

(A.98)

where the constant does not depend on $f(w)$. Now

$$
e^{\Delta F/k_BT} \frac{\sigma_F^2}{\mathcal{N}_F} + e^{-\Delta F/k_BT} \frac{\sigma_B^2}{\mathcal{N}_B}
$$

$$
= \int dw\, p_B(-w) f^2(w) \left[\frac{e^{-w/k_BT}}{\mathcal{N}_F} + \frac{e^{-\Delta F/k_BT}}{\mathcal{N}_B} \right].
$$

(A.99)

Minimizing with respect to $f(w)$, we obtain

$$f(w) = \frac{\text{const.}}{(e^{-w/k_BT}/\mathcal{N}_F) + (e^{-\Delta F/k_BT}/\mathcal{N}_B)}. \tag{A.100}$$

Setting the constant to 1 and substituting in the definitions of ϕ_F, ϕ_B, we obtain

$$\phi_F = \int dw\, p_F(w) f(w) = \mathcal{N}_F \left\langle \frac{1}{1 + e^{(w-z)/k_BT}} \right\rangle,$$
$$\phi_B = \int dw\, p_B(w) f(-w) = \mathcal{N}_B \left\langle \frac{1}{1 + e^{(w+z)/k_BT}} \right\rangle, \tag{A.101}$$

where

$$z = \Delta F + k_BT \ln \frac{\mathcal{N}_B}{\mathcal{N}_F}. \tag{A.102}$$

Thus we have

$$\Delta F = -k_BT \left[\ln \phi_B - \ln \phi_F \right] \tag{A.103}$$
$$= -k_BT \left[\ln \left\langle \frac{1}{1 + e^{(w+z)/k_BT}} \right\rangle - \ln \left\langle \frac{1}{1 + e^{(w-z)/k_BT}} \right\rangle \right] - k_BT \ln \frac{\mathcal{N}_B}{\mathcal{N}_F}.$$

We then obtain the self-consistency equation

$$z = -k_BT \left[\ln \left\langle \frac{1}{1 + e^{(w+z)/k_BT}} \right\rangle - \ln \left\langle \frac{1}{1 + e^{(w-z)/k_BT}} \right\rangle \right], \tag{A.104}$$

from whose solution we evaluate

$$\Delta F = z - k_BT \ln \frac{\mathcal{N}_B}{\mathcal{N}_F}. \tag{A.105}$$

A.12   Cauchy-Schwarz inequality

The **Cauchy-Schwarz inequality** states that, given two vectors u and v for which a real scalar product $(u \cdot v)$ is defined, we have

$$(u \cdot v)^2 \le (u \cdot v)(v \cdot v). \tag{A.106}$$

The defining properties of a scalar product are

Commutativity: $(u \cdot v) = (v \cdot u)$.

Linearity: $(au \cdot v) = a(u \cdot v)$, $a \in \mathbb{R}$.

Additivity: $((u + v) \cdot w) = (u \cdot w) + (v \cdot w)$.

Positivity: $(u \cdot u) \ge 0$. If $(u \cdot u) = 0$, then $u = 0$.

Given u and v, we define $w(a)$ for $a \in \mathbb{R}$ by

$$w(a) = au + v. \tag{A.107}$$

We then have

$$(w(a) \cdot w(a)) = a^2(u \cdot u) + 2a(u \cdot v) + (v \cdot v) \geq 0, \qquad \forall a \in \mathbb{R}. \tag{A.108}$$

This is a quadratic form in a that can have at most one zero. Hence its discriminant Δ must be nonpositive. We therefore have

$$\Delta = 4\left[(u \cdot v)^2 - (u \cdot u)(v \cdot v)\right] \leq 0, \tag{A.109}$$

which is the Cauchy-Schwarz inequality. Note that the inequality is saturated when $v = \lambda u$, where λ is an arbitrary real scalar. The inequality readily generalizes to integrable functions, when the scalar product is defined by

$$(f, g) = \int dx\, f(x)\, g(x). \tag{A.110}$$

A.13   Bound for the current rate function

In this appendix, we sketch the derivation of the bound (8.33) for the current rate function, following Gingrich et al. [69].

As derived in eq. (6.116), the rate function $I(j)$ for the current $j = (j_{xx'})$ is bounded by

$$I(p^{\mathrm{st}}, j) = \frac{1}{4} \sum_{xx'} \lambda_{xx'} j_{xx'} - \frac{1}{2} \sum_{xx'} \left(t^*_{xx'} - t_{xx'}\right), \tag{A.111}$$

where p^{st} is the steady-state probability distribution, and

$$\lambda_{xx'} = 2\ln\left[\frac{1}{2k_{xx'} p^{\mathrm{st}}_{x'}}\left(j_{xx'} + \sqrt{j^2_{xx'} + \alpha^2_{xx'}}\right)\right] = -\lambda_{x'x}, \tag{A.112}$$

with $\alpha_{xx'} = \sqrt{k_{xx'} k_{x'x} f_{x'} f_x}$. The traffic variables t (cf. eq. (6.19)) are defined by

$$\begin{aligned}
t_{xx'} &= k_{xx'} p^{\mathrm{st}}_{x'} + k_{x'x} p^{\mathrm{st}}_x; \\
t^*_{xx'} &= \kappa^*_{xx'} p^{\mathrm{st}}_{x'} + \kappa^*_{x'x} p^{\mathrm{st}}_x.
\end{aligned} \tag{A.113}$$

The steady-state current is given by $J^{\mathrm{st}}_{xx'} = k_{xx'} p^{\mathrm{st}}_{x'} - k_{x'x} p^{\mathrm{st}}_x$. The traffic can also be expressed as functions of j and J^{st} by

$$t_{xx'} = \sqrt{(J^{\mathrm{st}}_{xx'})^2 + \alpha^2_{xx'}}; \qquad t^*_{xx'} = \sqrt{j^2_{xx'} + \alpha^2_{xx'}}. \tag{A.114}$$

In the first term on the right-hand side of eq. (A.111), we regroup the terms for each edge xx' and obtain

$$
\mathfrak{S} = \frac{1}{4} \sum_{xx'} j_{xx'} \lambda_{xx'} = \frac{1}{4} \sum_{x'<x} j_{xx'} (\lambda_{xx'} - \lambda_{x'x})
$$

$$
= \frac{1}{2} \sum_{x'<x} j_{xx'} \left[\ln \frac{j_{xx'} + \sqrt{j_{xx'}^2 + \alpha_{xx'}^2}}{j_{xx'} + \sqrt{-j_{xx'}^2 + \alpha_{xx'}^2}} + \ln \frac{k_{x'x} f_x}{k_{xx'} f_{x'}} \right]. \tag{A.115}
$$

We now have

$$
k_{xx'} p_{x'}^{\mathrm{st}} = J_{xx'}^{\mathrm{st}} + t_{xx'} = J_{xx'}^{\mathrm{st}} + \sqrt{(J_{xx'}^{\mathrm{st}})^2 + \alpha_{xx'}}, \tag{A.116}
$$

and we therefore obtain

$$
\mathfrak{S} = \frac{1}{2} \sum_{x'<x} j_{xx'} \left(\ln \frac{j_{xx'} + \sqrt{j_{xx'}^2 + \alpha_{xx'}^2}}{-j_{xx'} + \sqrt{j_{xx'}^2 + \alpha_{xx'}^2}} - \ln \frac{J_{xx'}^{\mathrm{st}} + \sqrt{(J_{xx'}^{\mathrm{st}})^2 + \alpha_{xx'}}}{-J_{xx'}^{\mathrm{st}} + \sqrt{(J_{xx'}^{\mathrm{st}})^2 + \alpha_{xx'}}} \right)
$$

$$
= \sum_{x<x'} j_{xx'} \left(\operatorname{arc\,sinh} \frac{j_{xx'}}{\alpha_{xx'}} - \operatorname{arc\,sinh} \frac{J_{xx'}^{\mathrm{st}}}{\alpha_{xx'}} \right), \tag{A.117}
$$

where we use the relation

$$
\operatorname{arc\,sinh} x = \ln \left(x + \sqrt{x^2 + 1} \right). \tag{A.118}
$$

We therefore obtain

$$
I(p^{\mathrm{st}}, j) = \sum_{x<x'} \Psi \left(j_{xx'}, J_{xx'}^{\mathrm{st}}, \alpha \right), \tag{A.119}
$$

where

$$
\Psi (j_{xx'}, J_{xx'}^{\mathrm{st}}, \alpha) = j_{xx'} \left(\operatorname{arc\,sinh} \frac{j_{xx'}}{\alpha} - \operatorname{arc\,sinh} \frac{J_{xx'}^{\mathrm{st}}}{\alpha} \right) + \sqrt{J_{xx'}^{\mathrm{st}2} + \alpha^2} - \sqrt{j_{xx'}^2 + \alpha^2}. \tag{A.120}
$$

We also introduce the quantity

$$
\Delta (j_{xx'}, J_{xx'}^{\mathrm{st}}, \alpha) = \Psi (j_{xx'}, J_{xx'}^{\mathrm{st}}, \alpha) - \frac{(j_{xx'} - J_{xx'}^{\mathrm{st}})^2}{2 J_{xx'}^{\mathrm{st}2}} \left(J_{xx'}^{\mathrm{st}} \operatorname{arc\,sinh} \frac{J_{xx'}^{\mathrm{st}}}{\alpha} \right) \le 0. \tag{A.121}
$$

To obtain this inequality, we observe that Δ is even in $j_{xx'}$, so that we only need to consider positive values of $j_{xx'}$. The derivative of Δ with respect to $j_{xx'}$ is

$$
\frac{\partial \Delta}{\partial j_{xx'}} = \operatorname{arc\,sinh} \frac{j_{xx'}}{\alpha} - \frac{j_{xx'}}{J_{xx'}^{\mathrm{st}}} \operatorname{arc\,sinh} \frac{J_{xx'}^{\mathrm{st}}}{\alpha}. \tag{A.122}
$$

The second term is a line that crosses the $\operatorname{arc\,sinh} (j_{xx'}/\alpha)$ curve at $j_{xx'} = 0$ and at $j_{xx'} = J_{xx'}^{\mathrm{st}}$. Now, $\sinh x$ is convex for $x \ge 0$, therefore $\operatorname{arc\,sinh} x$ is concave in the same region. Thus the derivative of Δ is positive for $0 < j_{xx'} < J^{\text{st}}_{xx'}$, vanishes for $j_{xx'} = J^{\text{st}}_{xx'}$ by inspection, and therefore, by concavity, is negative for $j_{xx'} > J^{\text{st}}_{xx'}$. Thus Δ, which vanishes at $j_{xx'} = J^{\text{st}}_{xx'}$, decreases as $\left| j_{xx'} - J^{\text{st}}_{xx'} \right|$ increases and is therefore nonpositive for $j_{xx'} \geq 0$. By symmetry, the same is true for $j_{xx'} < 0$.

The average entropy production rate in the steady state is given by

$$\dot{S}^{\text{tot}} = \frac{k_{\text{B}}}{2} \sum_{xx'} J^{\text{st}}_{xx'} \ln \frac{k_{xx'} p^{\text{st}}_{x'}}{k_{x'x} p^{\text{st}}_x}. \tag{A.123}$$

By means of eq. (A.116), we cast it in the form

$$\dot{S}^{\text{tot}} = \sum_{x' < x} \dot{S}^{\text{tot}}_{xx'} = \sum_{x' < x} 2k_{\text{B}} J^{\text{st}}_{xx'} \operatorname{arc\,sinh} \frac{J^{\text{st}}_{xx'}}{\alpha_{xx'}} = \sum_{x' < x} J^{\text{st}}_{xx'} \frac{A_{xx'}}{T}, \tag{A.124}$$

where

$$A_{xx'} = 2k_{\text{B}}T \operatorname{arc\,sinh} \frac{J^{\text{st}}_{xx'}}{\alpha_{xx'}}. \tag{A.125}$$

We therefore obtain the bound

$$I(j) \leq \sum_{x' < x} \frac{1}{4k_{\text{B}} (J^{\text{st}}_{xx'})^2} \left(j_{xx'} - J^{\text{st}}_{xx'} \right)^2 \dot{S}^{\text{tot}}_{xx'}. \tag{A.126}$$

We obtain eq. (8.36) from this result by the contraction principle. We define the scalar product in the space of currents by $(f, g) = \sum_{x' < x} f_{xx'} g_{xx'}$. Using the scalar product, the current j_d associated with a distance matrix d is expressed by $j_d = (d, j)$ (cf. eq. (6.17)). The empirical currents j must satisfy the conservation of probability. We impose this condition via the constraints $(h_y, j) = 0$, $\forall y$, where $h_{y,xx'} = \delta^K_{yx} - \delta^K_{yx'}$. By the contraction principle, we obtain.

$$I(j_d) = \inf_j I(j | (d, j) = j_d, (h_y, j) = 0, \forall y)$$

$$\leq I_{\text{LR}}(j | (d, j) = j_d, (h_y, j) = 0, \forall y). \tag{A.127}$$

We define j^* by

$$j^* = \frac{j_d}{J^{\text{st}}_d} J^{\text{st}}. \tag{A.128}$$

This current satisfies the conditions $(d, j) = j_d$, $(h_y, j) = 0$. Substituting it in $I_{\text{LR}}(j)$, we finally obtain the bound

$$I(j_d) \leq I_{\text{LR}}(j^*) = \frac{(j_d - J^{\text{st}}_d)}{4(J^{\text{st}}_d)^2} \dot{S}^{\text{tot}} = I_{\text{WLR}}(j_d). \tag{A.129}$$

Bibliography

[1] Anderson, D. F. (2007). A modified next reaction method for simulating chemical systems with time dependent propensities and delays. *Journal of Chemical Physics*, 127(21):214107.

[2] Andrieux, D., and Gaspard, P. (2004). Fluctuation theorem and Onsager reciprocity relations. *Journal of Chemical Physics*, 121:6167.

[3] Andrieux, D., and Gaspard, P. (2006). Erratum: "Fluctuation theorem and Onsager reciprocity relations" [*J. Chem. Phys.* 121:6167 (2004)]. *Journal of Chemical Physics*, 125:219902.

[4] Andrieux, D., and Gaspard, P. (2008). Nonequilibrium generation of information in copolymerization processes. *Proceedings of the National Academy of Sciences*, 105(28):9516–9521.

[5] Aurell, E., Gawedzki, K., Mejía-Monasterio, C., Mohayaee, R., and Muratore-Ginanneschi, P. (2012). Refined second law of thermodynamics for fast random processes. *Journal of Statistical Physics*, 147(3):487–505.

[6] Aurell, E., Mejía-Monasterio, C., and Muratore-Ginanneschi, P. (2011). Optimal protocols and optimal transport in stochastic thermodynamics. *Physical Review Letters*, 106(25):250601.

[7] Barato, A., and Seifert, U. (2014a). Unifying three perspectives on information processing in stochastic thermodynamics. *Physical Review Letters*, 112(9):090601.

[8] Barato, A. C., and Seifert, U. (2014b). Stochastic thermodynamics with information reservoirs. *Physical Review E*, 90(4):042150.

[9] Barato, A. C., and Seifert, U. (2015). Thermodynamic uncertainty relation for biomolecular processes. *Physical Review Letters*, 114(15):158101.

[10] Bennett, C. H. (1973). Logical reversibility of computation. *IBM Journal of Research and Development*, 17:525–532.

[11] Bennett, C. H. (1976). Efficient estimation of free energy differences from Monte Carlo data. *Journal of Computational Physics*, 22(2):245–268.

[12] Bennett, C. H. (1979). Dissipation-error tradeoff in proofreading. *Biosystems*, 11:85–91.

[13] Bennett, C. H. (1982). The thermodynamics of computation—a review. *International Journal of Theoretical Physics*, 21:905–940.

[14] Bennett, C. H. (2003). Notes on Landauer's principle, reversible computation and Maxwell's demon. *Studies in History and Philosophy of Modern Physics*, 34:501–510.

[15] Berg, H. C. (1993). *Random Walks in Biology*. Princeton University Press, Princeton, NJ.

[16] Bergmann, P. G., and Lebowitz, J. L. (1955). New approach to nonequilibrium processes. *Physical Review*, 99(2):578.

[17] Bertini, L., Faggionato, A., and Gabrielli, D. (2015). Flows, currents, and cycles for Markov chains: Large deviation asymptotics. *Stochastic Processes and Their Applications*, 125(7):2786–2819.

[18] Bo, S., and Celani, A. (2017). Multiple-scale stochastic processes: Decimation, averaging and beyond. *Physics Reports*, 670:1–59.

[19] Bochkov, G. N., and Kuzovlev, Y. E. (1977). General theory of thermal fluctuations in nonlinear systems. *Soviet Physics JETP*, 45:125–129.

[20] Bochkov, G. N., and Kuzovlev, Y. E. (1979). Fluctuation-dissipation relations for nonequilibrium processes in open systems. *Soviet Physics JETP*, 49:543–551.

[21] Bochkov, G. N., and Kuzovlev, Y. E. (1981a). Nonlinear fluctuation-dissipation relations and stochastic models in nonequilibrium thermodynamics: I. Generalized fluctuation-dissipation theorem. *Physica A: Statistical Mechanics and Its Applications*, 106(3):443–479.

[22] Bochkov, G. N., and Kuzovlev, Y. E. (1981b). Nonlinear fluctuation-dissipation relations and stochastic models in nonequilibrium thermodynamics: II. Kinetic potential and variational principles for nonlinear irreversible processes. *Physica A: Statistical Mechanics and Its Applications*, 106(3):480–520.

[23] Bollobás, B. (2013). *Modern Graph Theory*. Volume 184. Springer Science & Business Media, New York.

[24] Boltzmann, L. (1877). Uber die Beziehung zwischen dem zweiten Hauptsatze der mechanischen Wärmetheorie und der Wahrscheinlichkeitsrechnung respecktive den Sätzen über das Wärmegleichgewicht. *Wiener Berichte*, 2:373–435.

[25] Brillouin, L. (1949). Life, thermodynamics and cybernetics. *American Scientist*, 37:554–568.

[26] Brouwer, L. E. J. (1912). Über Abbildung von Mannigfaltigkeiten. *Mathematische Annalen*, 7:97–115.

[27] Bérut, A., Arakelyan, A., Petrosyan, A., Ciliberto, S., Dillenschneider, R., and Lutz, E. (2012). Experimental verification of Landauer's principle linking information and thermodynamics. *Nature*, 483:187–190.

[28] Callen, H. B. (1985). *Thermodynamics and an Introduction to Thermostatistics*. Wiley, New York.

[29] Chandler, D. (1987). *Introduction to Modern Statistical Mechanics*. Oxford University Press, New York.

[30] Chetrite, R., and Touchette, H. (2015). Nonequilibrium Markov processes conditioned on large deviations. *Annales Henri Poincaré*, 16(9):2005–2057.

[31] Collin, D., Ritort, F., Jarzynski, C., Smith, S. B., Tinoco Jr., I., and Bustamante, C. (2005). Verification of the Crooks fluctuation theorem and recovery of RNA folding free energies. *Nature*, 437(7056):231.

[32] Cover, T. M., and Thomas, J. A. (2006). *Elements of Information Theory*. 2nd edition. Wiley, Hoboken, NJ.

[33] Crooks, G. E. (1998). Nonequilibrium measurements of free energy differences for microscopically reversible Markovian systems. *Journal of Statistical Physics*, 90:1481–1487.

[34] Crooks, G. E. (1999). The entropy production fluctuation theorem and the nonequilibrium work relation for free energy differences. *Physical Review E*, 60:2721–2728.

[35] Crooks, G. E. (2000). Path-ensemble averages in systems driven far from equilibrium. *Physical Review E*, 61:2361–2366.

[36] Cuetara, G. B., Esposito, M., and Imparato, A. (2014). Exact fluctuation theorem without ensemble quantities. *Physical Review E*, 89(5):052119.

[37] De Groot, S. R., and Mazur, P. (1984). *Non-equilibrium Thermodynamics*. Dover, New York. (Reprinted by Courier (2013).)

[38] Dechant, A. (2018). Multidimensional thermodynamic uncertainty relations. *Journal of Physics A: Mathematical and Theoretical*, 52(3):035001.

[39] Dechant, A., and Sasa, S. (2018). Current fluctuations and transport efficiency for general Langevin systems. *Journal of Statistical Mechanics: Theory and Experiment*, 2018(6):063209.

[40] Dechant, A., and Sasa, S. (2020). Fluctuation-response inequality out of equilibrium. *Proceedings of the National Academy of Sciences*, 117(12):6430–6436.

[41] Deffner, S., and Campbell, S. (2019). *Quantum Thermodynamics: An Introduction to the Thermodynamics of Quantum Information*. Morgan & Claypool Publishers, San Rafael, CA.

[42] Dembo, A., and Zeitouni, O. (2010). *Large Deviations Techniques and Applications*. Volume 38 of *Stochastic Modelling and Applied Probability*. Springer, Berlin.

[43] Den Hollander, F. (2008). *Large Deviations*. Volume 14 of *Fields Institute Monographs*. American Mathematical Society, Providence, RI.

[44] Einstein, A. (1905). Über die von der molekularkinetischen Theorie der Wärme geforderte Bewegung von in ruhenden Flüssigkeiten suspendierten Teilchen. *Ann. Physik*, 17(8):549–560.

[45] Einstein, A. (1910). Theorie der Opaleszenz von homogenen Flüssigkeiten und Flüssigkeits-gemischen in der Nähe des kritischen Zustandes. *Ann. Physik*, 33:1275–1298.

[46] Einstein, A. (1956). *Investigations on the Theory of the Brownian Movement*. Edited by R. Fürth. Dover, New York.

[47] Ellis, R. S. (1985). *Entropy, Large Deviations and Statistical Mechanics*. Springer, Berlin.

[48] Esposito, M. (2012). Stochastic thermodynamics under coarse graining. *Physical Review E*, 85:041125.

[49] Esposito, M., Harbola, U., and Mukamel, S. (2009). Nonequilibrium fluctuations, fluctuation theorems, and counting statistics in quantum systems. *Reviews of Modern Physics*, 81(4): 1665.

[50] Esposito, M., and Van den Broeck, C. (2010). Three detailed fluctuation theorems. *Physical Review Letters*, 104:090601.

[51] Evans, D. J., Cohen, E. G. D., and Morriss, G. P. (1993). Probability of second law violations in shearing steady states. *Physical Review Letters*, 71:2401–2404.

[52] Falcioni, M., Palatella, L., Pigolotti, S., Rondoni, L., and Vulpiani, A. (2007). Initial growth of Boltzmann entropy and chaos in a large assembly of weakly interacting systems. *Physica A: Statistical Mechanics and Its Applications*, 385(1):170–184.

[53] Feynman, R. P. (1996). *Feynman Lectures on Computation*. Edited by A. J. G. Hey, and R. W. Allen. Addison-Wesley, Reading, MA.

[54] Feynman, R. P., Leighton, R. B., and Sands, M. (2011). *The Feynman Lectures on Physics: The New Millennium Edition*. Volume 1, *Mainly Mechanics, Radiation, and Heat*. Basic Books, New York.

[55] Fisher, R. A. (1930). *The Genetical Theory of Natural Selection*. Clarendon Press, New York. (Reprinted by Dover (1958).)

[56] Gallavotti, G., and Cohen, E. G. D. (1995a). Dynamical ensembles in nonequilibrium statistical mechanics. *Physical Review Letters*, 74:2694–2697.

[57] Gallavotti, G., and Cohen, E. G. D. (1995b). Dynamical ensembles in stationary states. *Journal of Statistical Physics*, 80:931–970.

[58] García-García, R., Genthon, A., and Lacoste, D. (2019). Linking lineage and population observables in biological branching processes. *Physical Review E*, 99(4):042413.

[59] Gardiner, C. (2009). *Stochastic Methods*. Volume 4 of Springer Series in Synergetics. 4th edition. Springer, Berlin.

[60] Garrahan, J. P. (2017). Simple bounds on fluctuations and uncertainty relations for first-passage times of counting observables. *Physical Review E*, 95(3):032134.

[61] Gawedzki, K. (2013). Fluctuation relations in stochastic thermodynamics. *arXiv*, 1308.1518.

[62] Giardinà, C., Kurchan, J., and Peliti, L. (2006). Direct evaluation of large-deviation functions. *Physical Review Letters*, 96:120603.

[63] Gibson, M. A., and Bruck, J. (2000). Exact stochastic simulation of chemical systems with many species and many channels. *Journal of Physical Chemistry*, 105:1876–1879.

[64] Gillespie, D. T. (1976). A general method for numerically simulating the stochastic time evolution of coupled chemical reactions. *Journal of Computational Physics*, 22(4):403–434.

[65] Gillespie, D. T. (1977). Exact stochastic simulation of coupled chemical reactions. *Journal of Physical Chemistry*, 81(25):2340–2361.

[66] Gillespie, D. T. (2001). Approximate accelerated stochastic simulation of chemically reacting systems. *Journal of Chemical Physics*, 115:1716–1733.

[67] Gillespie, D. T. (2007). Stochastic simulation of chemical kinetics. *Annual Reviews of Physical Chemistry*, 58:35–55.

[68] Gingrich, T. R., and Horowitz, J. M. (2017). Fundamental bounds on first passage time fluctuations for currents. *Physical Review Letters*, 119(17):170601.

[69] Gingrich, T. R., Horowitz, J. M., Perunov, N., and England, J. L. (2016). Dissipation bounds all steady-state current fluctuations. *Physical Review Letters*, 116(12):120601.

[70] Gingrich, T. R., Rotskoff, G. M., and Horowitz, J. M. (2017). Inferring dissipation from current fluctuations. *Journal of Physics A: Mathematical and Theoretical*, 50:184004.

[71] Gingrich, T. R., Rotskoff, G. M., Vaikuntanathan, S., and Geissler, P. L. (2014). Efficiency and large deviations in time-asymmetric stochastic heat engines. *New Journal of Physics*, 16(10): 102003.

[72] Goldstein, S., Huse, D. A., Lebowitz, J. L., and Sartori, P. (2017). On the nonequilibrium entropy of large and small systems. In Giacomin, G., Olla, S., Saada, E., Spohn, H., and Stoltz, G., editors, *Stochastic Dynamics Out of Equilibrium*. Volume 282 of *Springer Proceedings in Mathematics and Statistics*, pages 581–596, Springer, New York.

[73] Harris, R. J., and Schütz, G. M. (2007). Fluctuation theorems for stochastic dynamics. *Journal of Statistical Mechanics: Theory and Experiment*, 2007(07):P07020.

[74] Hasegawa, Y., and Van Vu, T. (2019). Fluctuation theorem uncertainty relation. *Physical Review Letters*, 123(11):110602.

[75] Hatano, T., and Sasa, S. (2001). Steady state thermodynamics of Langevin systems. *Physical Review Letters*, 86:3463–3466.

[76] Hong, J., Lambson, B., Dhuey, S., and Bokor, J. (2016). Experimental test of Landauer's principle in single-bit operations on nanomagnetic memory bits. *Science Advances*, 2(3):e1501492.

[77] Horowitz, J. M., and Esposito, M. (2014). Thermodynamics with continuous information flow. *Physical Review X*, 4(3):031015.

[78] Horowitz, J. M., and Gingrich, T. R. (2017). Proof of the finite-time thermodynamic uncertainty relation for steady-state currents. *Physical Review E*, 96(2):020103.

[79] Horowitz, J. M., and Gingrich, T. R. (2019). Thermodynamic uncertainty relations constrain non-equilibrium fluctuations. *Nature Physics*, 16(1):15–20.

[80] Howard, J. (2001). *Mechanics of Motor Proteins and the Cytoskeleton*. Sinauer Associates, Sunderland, MA.

[81] Jarzynski, C. (1997). Nonequilibrium equality for free energy differences. *Physical Review Letters*, 78(14):2690.

[82] Jarzynski, C. (2007). Comparison of far-from-equilibrium work relations. *Comptes Rendus Physique*, 8:495–506.

[83] Jarzynski, C. (2017). Stochastic and macroscopic thermodynamics of strongly coupled systems. *Physical Review X*, 7(1):011008.

[84] Jun, Y., Gavrilov, M., and Bechhoefer, J. (2014). High-precision test of Landauer's principle in a feedback trap. *Physical Review Letters*, 113(19):190601.

[85] Kawaguchi, K., and Nakayama, Y. (2013). Fluctuation theorem for hidden entropy production. *Physical Review E*, 88(2):022147.

[86] Kawai, R., Parrondo, J., and Van den Broeck, C. (2007). Dissipation: The phase-space perspective. *Physical Review Letters*, 98:080602.

[87] Khinchin, A. I. (1957). *Mathematical Foundations of Information Theory*. Translated by R. A. Silverman and M. D. Friedman. Dover, New York.

[88] Kobayashi, T. J., and Sughiyama, Y. (2015). Fluctuation relations of fitness and information in population dynamics. *Physical Review Letters*, 115(23):238102.

[89] Koski, J. V., Maisi, V. F., Pekola, J. P., and Averin, D. V. (2014). Experimental realization of a Szilard engine with a single electron. *Proceedings of the National Academy of Sciences*, 111(38):13786–13789.

[90] Koski, J. V., Maisi, V. F., Sagawa, T., and Pekola, J. P. (2014). Experimental observation of the role of mutual information in the nonequilibrium dynamics of a Maxwell demon. *Physical Review Letters*, 113(3):030601.

[91] Koza, Z. (1999). General technique of calculating the drift velocity and diffusion coefficient in arbitrary periodic systems. *Journal of Physics A: Mathematical and General*, 32(44): 7637.

[92] Kurchan, J. (1998). Fluctuation theorem for stochastic dynamics. *Journal of Physics A: Mathematical and General*, 31(16):3719.

[93] Kurchan, J. (2000). A quantum fluctuation theorem. *arXiv*, preprint cond-mat/0007360.

[94] Lacoste, D., Lau, A. W., and Mallick, K. (2008). Fluctuation theorem and large deviation function for a solvable model of a molecular motor. *Physical Review E*, 78:011915.

[95] Landau, L. D., Lifshitz, E. M., and Pitaevskii, L. P. (1980). *Statistical Physics, Part I*. Volume 5 of *Course of Theoretical Physics*. Pergamon Press, Oxford, UK.

[96] Landauer, R. (1961). Irreversibility and heat generation in the computing process. *IBM Journal of Research and Development*, 3:183–191.

[97] Landauer, R. (1991). Information is physical. *Physics Today*, 44(5):23–29.

[98] Lebowitz, J. L., and Spohn, H. (1999). A Gallavotti-Cohen-type symmetry in the large deviation functional for stochastic dynamics. *Journal of Statistical Physics*, 95(1):333–365.

[99] Lecomte, V., and Tailleur, J. (2007). A numerical approach to large deviations in continuous time. *Journal of Statistical Mechanics: Theory and Experiment*, 2007(03):P03004.

[100] Lee, H. K., Kwon, C., and Park, H. (2013). Fluctuation theorems and entropy production with odd-parity variables. *Physical Review Letters*, 110(5):050602.

[101] Leff, H., and Rex, A. F., editors (2003). *Maxwell's Demon 2: Entropy, Classical and Quantum Information, Computing*. IOP Publishing, Bristol, UK.

[102] Leibler, S., and Kussell, E. (2010). Individual histories and selection in heterogeneous populations. *Proceedings of the National Academy of Sciences*, 107(29):13183–13188.

[103] Liphardt, J., Dumont, S., Smith, S. B., Tinoco, I., and Bustamante, C. (2002). Equilibrium information from nonequilibrium measurements in an experimental test of Jarzynski's equality. *Science*, 296(5574):1832–1835.

[104] Liphardt, J., Onoa, B., Smith, S. B., Tinoco, I., and Bustamante, C. (2001). Reversible unfolding of single RNA molecules by mechanical force. *Science*, 292(5517):733–737.

[105] MacKay, D. J. C. (2003). *Information Theory, Inference and Learning Algorithms*. Cambridge University Press, Cambridge, UK.

[106] Maes, C. (1999). The fluctuation theorem as a Gibbs property. *Journal of Statistical Physics*, 95(1–2):367–392.

[107] Maes, C., and Netočný, K. (2008). Canonical structure of dynamical fluctuations in mesoscopic nonequilibrium steady states. *EPL (Europhysics Letters)*, 82(3):30003.

[108] Mandal, D., and Jarzynski, C. (2012). Work and information processing in a solvable model of Maxwell's demon. *Proceedings of the National Academy of Sciences*, 109(29):11641–11645.

[109] Marchetti, M. C., Joanny, J.-F., Ramaswamy, S., Liverpool, T. B., Prost, J., Rao, M., and Simha, R. A. (2013). Hydrodynamics of soft active matter. *Reviews of Modern Physics*, 85(3):1143.

[110] Marini Bettolo Marconi, U., Puglisi, A., Rondoni, L., and Vulpiani, A. (2008). Fluctuation-dissipation: Response theory in statistical physics. *Physics Reports*, 461(4-6):111–195.

[111] Maxwell, J. C. (1871). *Theory of Heat*. Longmans, Green and Co., London.

[112] Meyer, C. D. (2001). *Matrix Analysis and Applied Linear Algebra*. SIAM, Society for Industrial and Applied Mathematics, Philadelphia, PA.

[113] Mossa, A., Manosas, M., Forns, N., Huguet, J. M., and Ritort, F. (2009). Dynamic force spectroscopy of DNA hairpins: I. Force kinetics and free energy landscapes. *Journal of Statistical Mechanics: Theory and Experiment*, 2009(02):P02060.

[114] Murashita, Y., Funo, K., and Ueda, M. (2014). Nonequilibrium equalities in absolutely irreversible processes. *Physical Review E*, 90(4):042110.

[115] Murashita, Y., and Ueda, M. (2017). Gibbs paradox revisited from the fluctuation theorem with absolute irreversibility. *Physical Review Letters*, 118(6):060601.

[116] Mustonen, V., and Lässig, M. (2010). Fitness flux and ubiquity of adaptive evolution. *Proceedings of the National Academy of Sciences*, 107(9):4248–4253.

[117] Neri, I., Roldán, É., and Jülicher, F. (2017). Statistics of infima and stopping times of entropy production and applications to active molecular processes. *Physical Review X*, 7(1):011019.

[118] Neri, I., Roldán, É., Pigolotti, S., and Jülicher, F. (2019). Integral fluctuation relations for entropy production at stopping times. *Journal of Statistical Mechanics: Theory and Experiment*, 2019(10):104006.

[119] Nozoe, T., Kussell, E., and Wakamoto, Y. (2017). Inferring fitness landscapes and selection on phenotypic states from single-cell genealogical data. *PLoS Genetics*, 13(3):e1006653.

[120] Øksendal, B. (2003). *Stochastic Differential Equations*. Springer, Berlin.

[121] Onsager, L. (1931a). Reciprocal relations in irreversible processes. I. *Physical Review*, 37(4):405.

[122] Onsager, L. (1931b). Reciprocal relations in irreversible processes. II. *Physical Review*, 38(12):2265.

[123] Oono, Y., and Paniconi, M. (1998). Steady state thermodynamics. *Progress of Theoretical Physics Supplement*, 130:29–44.

[124] Parrondo, J. M., Horowitz, J. M., and Sagawa, T. (2015). Thermodynamics of information. *Nature Physics*, 11(2):131.

[125] Peliti, L. (2011). *Statistical Mechanics in a Nutshell*. Princeton University Press, Princeton, NJ.

[126] Penrose, O. (1970). *Foundations of Statistical Mechanics: A Deductive Treatment*. Dover, New York.

[127] Pietzonka, P., Barato, A. C., and Seifert, U. (2016). Universal bound on the efficiency of molecular motors. *Journal of Statistical Mechanics: Theory and Experiment*, 2016(12):124004.

[128] Pigolotti, S., Neri, I., Roldán, E., and Jülicher, F. (2017). Generic properties of stochastic entropy production. *Physical Review Letters*, 119(14):140604.

[129] Pigolotti, S., and Vulpiani, A. (2008). Coarse graining of master equations with fast and slow states. *Journal of Chemical Physics*, 128(15):154114.

[130] Pippard, A. B. (1957). *Elements of Classical Thermodynamics*. Cambridge University Press, Cambridge, UK.

[131] Polettini, M., Verley, G., and Esposito, M. (2015). Efficiency statistics at all times: Carnot limit at finite power. *Physical Review Letters*, 114(5):050601.

[132] Press, W. H., Teukolsky, S. A., Vetterling, W. T., and Flannery, B. P. (2007). *Numerical Recipes: The Art of Scientific Computing*. 3rd edition. Cambridge University Press, New York.

[133] Puglisi, A., Pigolotti, S., Rondoni, L., and Vulpiani, A. (2010). Entropy production and coarse graining in Markov processes. *Journal of Statistical Mechanics: Theory and Experiment*, 2010(05):P05015.

[134] Qian, H. (2002). Mesoscopic nonequilibrium thermodynamics of single macromolecules and dynamic entropy-energy compensation. *Physical Review E*, 65:01602.

[135] Rao, R., and Esposito, M. (2016). Nonequilibrium thermodynamics of chemical reaction networks: Wisdom from stochastic thermodynamics. *Physical Review X*, 6(4):041064.

[136] Rao, R., and Esposito, M. (2018a). Conservation laws shape dissipation. *New Journal of Physics*, 20:023007.

[137] Rao, R., and Esposito, M. (2018b). Detailed fluctuation theorems: A unifying perspective. *Entropy*, 20:635.

[138] Redner, S. (2001). *A Guide to First-Passage Processes*. Cambridge University Press, Cambridge, UK.

[139] Risken, H. (1996). *The Fokker-Planck Equation*. Springer Series in Synergetics. 2nd edition. Springer, Berlin.

[140] Sagawa, T., and Ueda, M. (2010). Generalized Jarzynski equality under nonequilibrium feedback control. *Physical Review Letters*, 104:090602.

[141] Saito, K., and Dhar, A. (2016). Waiting for rare entropic fluctuations. *EPL (Europhysics Letters)*, 114(5):50004.

[142] Sartori, P., Granger, L., Lee, C. F., and Horowitz, J. M. (2014). Thermodynamic costs of information processing in sensory adaptation. *PLoS Computational Biology*, 10(12):e1003974.

[143] Sartori, P., and Pigolotti, S. (2013). Kinetic versus energetic discrimination in biological copying. *Physical Review Letters*, 110:188101.

[144] Sartori, P., and Pigolotti, S. (2015). Thermodynamics of error correction. *Physical Review X*, 5(4):041039.

[145] Schmiedl, T., and Seifert, U. (2007). Optimal finite-time processes in stochastic thermodynamics. *Physical Review Letters*, 98(10):108301.

[146] Schnakenberg, J. (1976). Network theory of microscopic and macroscopic behavior of master equation systems. *Reviews of Modern Physics*, 48:571–585.

[147] Schuler, S., Speck, T., Tietz, C., Wrachtrup, J., and Seifert, U. (2005). Experimental test of the fluctuation theorem for a driven two-level system with time-dependent rates. *Physical Review Letters*, 94(18):180602.

[148] Seifert, U. (2005). Entropy production along a stochastic trajectory and an integral fluctuation theorem. *Physical Review Letters*, 95:040602.

[149] Seifert, U. (2012). Stochastic thermodynamics, fluctuation theorems, and molecular machines. *Reports on Progress in Physics*, 75:126001.

[150] Seifert, U. (2018). Stochastic thermodynamics: From principles to the cost of precision. *Physica A: Statistical Mechanics and Its Applications*, 504:176–191.

[151] Sekimoto, K. (1997). Kinetic characterization of heat bath and the energetics of thermal ratchet models. *Journal of the Physical Society of Japan*, 66:1234–1237.

[152] Sekimoto, K. (1998). Langevin equation and thermodynamics. *Progress of Theoretical Physics Supplement*, 130:17–27.

[153] Sekimoto, K. (2010). *Stochastic Energetics*. Number 799 in Lecture Notes in Physics. Springer, Heidelberg, Germany.

[154] Shannon, C. E. (1948). A mathematical theory of communication. *Bell System Technical Journal*, 27:379–423, 623–656.

[155] Shiraishi, N., Matsumoto, T., and Sagawa, T. (2016). Measurement-feedback formalism meets information reservoirs. *New Journal of Physics*, 18(1):013044.

[156] Singh, S., Menczel, P., Golubev, D. S., Khaymovich, I. M., Peltonen, J. T., Flindt, C., Saito, K., Roldán, E., and Pekola, J. P. (2019). Universal first-passage-time distribution of non-Gaussian currents. *Physical Review Letters*, 122(23):230602.

[157] Sivak, D. A., and Crooks, G. E. (2012). Thermodynamic metrics and optimal paths. *Physical Review Letters*, 108(19):190602.

[158] Speck, T. (2011). Work distribution for the driven harmonic oscillator with time-dependent strength: Exact solution and slow driving. *Journal of Physics A: Mathematical and Theoretical*, 44(30):305001.

[159] Spinney, R. E., and Ford, I. J. (2012). Nonequilibrium thermodynamics of stochastic systems with odd and even variables. *Physical Review Letters*, 108(17):170603.

[160] Strasberg, P., and Esposito, M. (2020). Measurability of nonequilibrium thermodynamics in terms of the Hamiltonian of mean force. *Physical Review E*, 101(5):050101.

[161] Sughiyama, Y., Kobayashi, T. J., Tsumura, K., and Aihara, K. (2015). Pathwise thermodynamic structure in population dynamics. *Physical Review E*, 91(3):032120.

[162] Szilard, L. (1929). Über die Entropieverminderung in einem thermodynamischen System bei Eingreffen intelligenter Wesen. *Zeitschrift für Physik*, 53:840–856.

[163] Szilard, L. (2006). On the decrease of entropy in a thermodynamic system by the intervention of intelligent beings. *Behavioral Science*, 9:301–310.

[164] Tafoya, S., Large, S. J., Liu, S., Bustamante, C., and Sivak, D. A. (2019). Using a system's equilibrium behavior to reduce its energy dissipation in nonequilibrium processes. *Proceedings of the National Academy of Sciences*, 116(13):5920–5924.

[165] Talkner, P., and Hänggi, P. (2016). Open system trajectories specify fluctuating work but not heat. *Physical Review E*, 94(2):022143.

[166] Tietz, C., Schuler, S., Speck, T., Seifert, U., and Wrachtrup, J. (2006). Measurement of stochastic entropy production. *Physical Review Letters*, 97(5):050602.

[167] Tizón-Escamilla, N., Lecomte, V., and Bertin, E. (2019). Effective driven dynamics for one-dimensional conditioned Langevin processes in the weak-noise limit. *Journal of Statistical Mechanics: Theory and Experiment*, 2019(1):013201.

[168] Touchette, H. (2009). The large deviation approach to statistical mechanics. *Physics Reports*, 478(1–3):1–69.

[169] Toyabe, S., Sagawa, T., Ueda, M., Muneyuki, E., and Sano, M. (2010). Experimental demonstration of information-to-energy conversion and validation of the generalized Jarzynski equality. *Nature Physics*, 6:988–992.

[170] Van den Broeck, C., and Esposito, M. (2015). Ensemble and trajectory thermodynamics: A brief introduction. *Physica A: Statistical Mechanics and Its Applications*, 418:6–16.

[171] van Kampen, N. G. (2007). *Stochastic Processes in Physics and Chemistry*. 3rd edition. North Holland, Amsterdam.

[172] Verley, G., Esposito, M., Willaert, T., and Van den Broeck, C. (2014). The unlikely Carnot efficiency. *Nature Communications*, 5:4721.

[173] Verley, G., and Lacoste, D. (2012a). Fluctuation theorems and inequalities generalizing the second law of thermodynamics out of equilibrium. *Physical Review E*, 86(5):051127.

[174] Verley, G., and Lacoste, D. (2012b). Fluctuations and response from a Hatano and Sasa approach. *Physica Scripta*, 86:058505.

[175] Verley, G., Willaert, T., Van den Broeck, C., and Esposito, M. (2014). Universal theory of efficiency fluctuations. *Physical Review E*, 90(5):052145.

[176] Wolpert, D. H. (2019). The stochastic thermodynamics of computation. *Journal of Physics A: Mathematical and Theoretical*, 52(19):193001.

Author Index

Index